INTRODUCTION TO STRING FIELD THEORY

ADVANCED SERIES IN MATHEMATICAL PHYSICS

Also in the Series

Volume 1: Mathematical Aspects of String Theory
edited by S-T Yau (*Harvard*)
Volume 2: Bombay Lectures on Highest Weight Representations of Infinite Dimensional Lie Algebras
V G Kac (*MIT*) & A K Raina (*Tata*)
Volume 3: Kac-Moody and Virasoro Algebras: A Reprint Volume for Physicists
edited by P Goddard (*Cambridge*) & D Olive (*Imperial College*)

INTRODUCTION TO STRING FIELD THEORY

Advanced Series in Mathematical Physics
Vol. 8

INTRODUCTION TO STRING FIELD THEORY

WARREN SIEGEL
State University of New York
Stony Brook

World Scientific
Singapore • New Jersey • London • Hong Kong

Published by

World Scientific Publishing Co. Pte. Ltd.
P O Box 128, Farrer Road, Singapore 9128

USA office: World Scientific Publishing Co., Inc.
687 Hartwell Street, Teaneck, NJ 07666, USA

UK office: World Scientific Publishing Co. Pte. Ltd.
P O Box 379, London N12 7JS, England

INTRODUCTION TO STRING FIELD THEORY

ISBN 9971-50-731-5
9971-50-732-3 (pbk)

Printed in Singapore by JBW Printers & Binders Pte. Ltd.

PREFACE

First, I'd like to explain the title of this book. I always hated books whose titles began "Introduction to..." In particular, when I was a grad student, books titled "Introduction to Quantum Field Theory" were the most difficult and advanced textbooks available, and I always feared what a quantum field theory book which was not introductory would look like. There is now a standard reference on relativistic string theory by Green, Schwarz, and Witten, *Superstring Theory* [0.1], which consists of two volumes, is over 1,000 pages long, and yet admits to having some major omissions. Now that I see, from an author's point of view, how much effort is necessary to produce a non-introductory text, the words "Introduction to" take a more tranquilizing character. (I have worked on a one-volume, non-introductory text on another topic, but that was in association with three coauthors.) Furthermore, these words leave me the option of omitting topics which I don't understand, or at least being more heuristic in the areas which I haven't studied in detail yet.

The rest of the title is "String Field Theory." This is the newest approach to string theory, although the older approaches are continuously developing new twists and improvements. The main alternative approach is the quantum mechanical (/analog-model/path-integral/interacting-string-picture/Polyakov/conformal- "field-theory") one, which necessarily treats a fixed number of fields, corresponding to homogeneous equations in the field theory. (For example, there is no analog in the mechanics approach of even the nonabelian gauge transformation of the field theory, which includes such fundamental concepts as general coordinate invariance.) It is also an S-matrix approach, and can thus calculate only quantities which are gauge-fixed (although limited background-field techniques allow the calculation of 1-loop effective actions with only some coefficients gauge-dependent). In the old S-matrix approach to field theory, the basic idea was to start with the S-matrix, and then analytically continue to obtain quantities which are off-shell (and perhaps in more general gauges). However, in the long run, it turned out to be more practical to work directly with field theory Lagrangians, even for semiclassical results such as spontaneous symmetry breaking and instantons, which change the meaning of "on-shell" by redefining the vacuum to be a state which is not as obvious from looking at the unphysical-vacuum S-matrix. Of course, S-matrix methods are always valuable for perturbation theory, but even in perturbation theory it is far more convenient to start with the field theory in order to determine which vacuum to perturb about, which gauges to use, and what power-counting rules can be used to determine divergence structure without specific S-matrix calculations. (More details on this comparison are in the Introduction.)

Unfortunately, string field theory is in a rather primitive state right now, and not even close to being as well understood as ordinary (particle) field theory. Of course, this is exactly the reason why the present is the best time to do research in this area. (Anyone who can honestly say, "I'll learn it when it's better understood," should mark a date on his calendar for returning to graduate school.) It is therefore simultaneously the best time for someone to read a book on the topic and the worst time for someone to write one. I have tried to compensate for this problem somewhat by expanding on the more introductory parts of the topic. Several of the early chapters are actually on the topic of general (particle/string) field theory, but explained from a new point of view resulting from insights gained from string field theory. (A more standard course on quantum field theory is assumed as a prerequisite.) This includes the use of a universal method for treating free field theories, which allows the derivation of a single, simple, free, local, Poincaré-invariant, gauge-invariant action that can be applied directly to any field. (Previously, only some special cases had been treated, and each in a different way.) As a result, even though the fact that I have tried to make this book self-contained with regard to string theory in general means that there is significant overlap with other treatments, within this overlap the approaches are sometimes quite different, and perhaps in some ways complementary. (The treatments of ref. [0.2] are also quite different, but for quite different reasons.)

Exercises are given at the end of each chapter (except the introduction) to guide the reader to examples which illustrate the ideas in the chapter, and to encourage him to perform calculations which have been omitted to avoid making the length of this book diverge.

This work was done at the University of Maryland, with partial support from the National Science Foundation. It is partly based on courses I gave in the falls of 1985 and 1986. I received valuable comments from Aleksandar Miković, Christian Preitschopf, Anton van de Ven, and Harold Mark Weiser. I especially thank Barton Zwiebach, who collaborated with me on most of the work on which this book was based.

June 16, 1988 Warren Siegel

CONTENTS

1. INTRODUCTION

1.1. Motivation

The experiments which gave us quantum theory and general relativity are now quite old, but a satisfactory theory which is consistent with both of them has yet to be found. Although the importance of such a theory is undeniable, the urgency of finding it may not be so obvious, since the quantum effects of gravity are not yet accessible to experiment. However, recent progress in the problem has indicated that the restrictions imposed by quantum mechanics on a field theory of gravitation are so stringent as to *require* that it also be a unified theory of all interactions, and thus quantum gravity would lead to predictions for other interactions which can be subjected to present-day experiment. Such indications were given by supergravity theories [1.1], where finiteness was found at some higher-order loops as a consequence of supersymmetry, which requires the presence of matter fields whose quantum effects cancel the ultraviolet divergences of the graviton field. Thus, quantum consistency led to higher symmetry which in turn led to unification. However, even this symmetry was found insufficient to guarantee finiteness at all loops [1.2] (unless perhaps the graviton were found to be a bound-state of a truly finite theory). Interest then returned to theories which had already presented the possibility of consistent quantum gravity theories as a consequence of even larger (hidden) symmetries: theories of relativistic strings [1.3-5]. Strings thus offer a possibility of consistently describing all of nature. However, even if strings eventually turn out to disagree with nature, or to be too intractable to be useful for phenomenological applications, they are still the only consistent toy models of quantum gravity (especially for the theory of the graviton as a bound state), so their study will still be useful for discovering new properties of quantum gravity.

The fundamental difference between a particle and a string is that a particle is a 0-dimensional object in space, with a 1-dimensional world-line describing its trajectory in spacetime, while a string is a (finite, open or closed) 1-dimensional object in space, which sweeps out a 2-dimensional world-sheet as it propagates through spacetime:

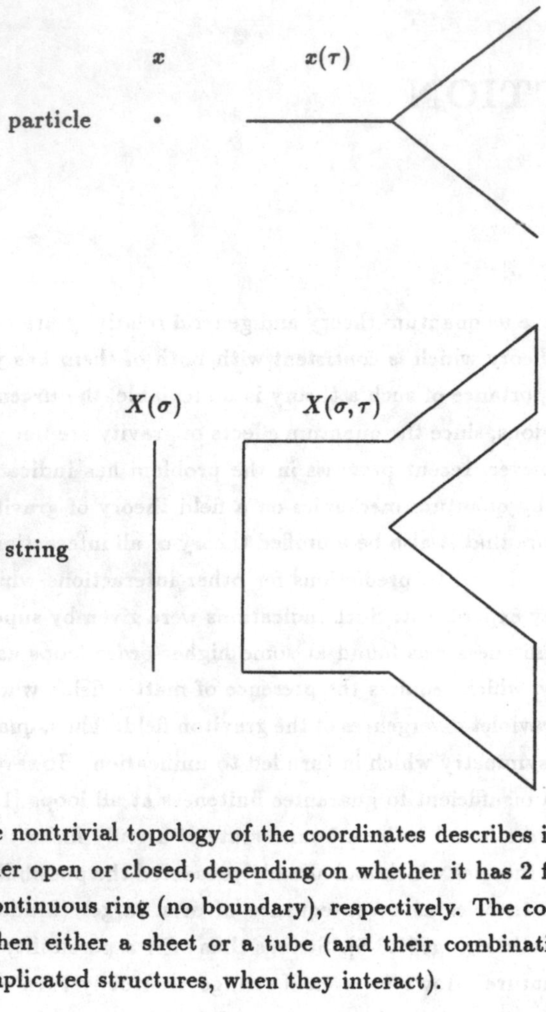

The nontrivial topology of the coordinates describes interactions. A string can be either open or closed, depending on whether it has 2 free ends (its boundary) or is a continuous ring (no boundary), respectively. The corresponding spacetime figure is then either a sheet or a tube (and their combinations, and topologically more complicated structures, when they interact).

Strings were originally intended to describe hadrons directly, since the observed spectrum and high-energy behavior of hadrons (linearly rising Regge trajectories, which in a perturbative framework implies the property of hadronic duality) seems realizable only in a string framework. After a quark structure for hadrons became generally accepted, it was shown that confinement would naturally lead to a string formulation of hadrons, since the topological expansion which follows from using $1/N_{color}$ as a perturbation parameter (the only dimensionless one in massless QCD, besides $1/N_{flavor}$), after summation in the other parameter (the gluon coupling, which becomes the hadronic mass scale after dimensional transmutation), is the same perturbation expansion as occurs in theories of fundamental strings [1.6]. Certain string theories can thus be considered alternative and equivalent formulations of QCD, just as general field theories can be equivalently formulated either in terms of "fundamental" particles or in terms of the particles which arise as bound states. However, in practice certain criteria, in particular renormalizability, can be simply formulated only in one formalism: For example, QCD is easier to use than a theory where gluons are treated as bound states of self-interacting quarks, the latter being a nonrenormalizable theory which needs an unwieldy criterion ("asymptotic safety" [1.7]) to restrict the available infinite number of couplings to a finite subset. On the other hand, atomic physics is easier to use as a theory of electrons, nuclei, and photons than a formulation in terms of fields describing self-interacting atoms whose excitations lie on Regge trajectories (particularly since QED is not confining). Thus, the choice of formulation is dependent on the dynamics of the particular theory, and perhaps even on the region in momentum space for that particular application: perhaps quarks for large transverse momenta and strings for small. In particular, the running of the gluon coupling may lead to nonrenormalizability problems for small transverse momenta [1.8] (where an infinite number of arbitrary couplings may show up as nonperturbative vacuum values of operators of arbitrarily high dimension), and thus QCD may be best considered as an effective theory at large transverse momenta (in the same way as a perturbatively nonrenormalizable theory at low energies, like the Fermi theory of weak interactions, unless asymptotic safety is applied). Hence, a string formulation, where mesons are the fundamental fields (and baryons appear as skyrmeon-type solitons [1.9]) may be unavoidable. Thus, strings may be important for hadronic physics as well as for gravity and unified theories; however, the presently known string models seem to apply only to the latter, since they contain massless particles and have (maximum) spacetime dimension $D = 10$ (whereas confinement in QCD occurs for $D \leq 4$).

1.2. Known models (interacting)

Although many string theories have been invented which are consistent at the tree level, most have problems at the one-loop level. (There are also theories which are already so complicated at the free level that the interacting theories have been too difficult to formulate to test at the one-loop level, and these will not be discussed here.) These one-loop problems generally show up as anomalies. It turns out that the anomaly-free theories are exactly the ones which are finite. Generally, topological arguments based on reparametrization invariance (the "stretchiness" of the string world sheet) show that any multiloop string graph can be represented as a tree graph with many one-loop insertions [1.10], so all divergences should be representable as just one-loop divergences. The fact that one-loop divergences should generate overlapping divergences then implies that one-loop divergences cause anomalies in reparametrization invariance, since the resultant multiloop divergences are in conflict with the one-loop-insertion structure implied by the invariance. Therefore, finiteness should be a necessary requirement for string theories (even purely bosonic ones) in order to avoid anomalies in reparametrization invariance. Furthermore, the absence of anomalies in such global transformations determines the dimension of spacetime, which in all known nonanomalous theories is $D = 10$. (This is also known as the "critical," or maximum, dimension, since some of the dimensions can be compactified or otherwise made unobservable, although the number of degrees of freedom is unchanged.)

In fact, there are only four such theories:

I:	N=1 supersymmetry, SO(32) gauge group, open [1.11]
IIA,B:	N=2 nonchiral or chiral supersymmetry [1.12]
heterotic:	N=1 supersymmetry, SO(32) or $E_8 \otimes E_8$ [1.13]
	or broken N=1 supersymmetry, SO(16)⊗SO(16) [1.14]

All except the first describe only closed strings; the first describes open strings, which produce closed strings as bound states. (There are also many cases of each of these theories due to the various possibilities for compactification of the extra dimensions onto tori or other manifolds, including some which have tachyons.) However, for simplicity we will first consider certain inconsistent theories: the bosonic string, which has global reparametrization anomalies unless $D = 26$ (and for which the local anomalies described above even for $D = 26$ have not yet been explicitly derived), and the spinning string, which is nonanomalous only when it is truncated to the above strings. The heterotic strings are actually closed strings for which modes propagating in the clockwise direction are nonsupersymmetric and 26-dimensional, while the counterclockwise ones are $N = 1$ (perhaps-broken) supersymmetric and 10-dimensional, or vice versa.

1.3. Aspects

There are several aspects of, or approaches to, string theory which can best be classified by the spacetime dimension in which they work: $D = 2,4,6,10$. The 2D approach is the method of first-quantization in the two-dimensional world sheet swept out by the string as it propagates, and is applicable solely to (second-quantized) perturbation theory, for which it is the only tractable method of calculation. Since it discusses only the properties of individual graphs, it can't discuss properties which involve an unfixed number of string fields: gauge transformations, spontaneous symmetry breaking, semiclassical solutions to the string field equations, etc. Also, it can describe only the gauge-fixed theory, and only in a limited set of gauges. (However, by introducing external particle fields, a limited amount of information on the gauge-invariant theory can be obtained.) Recently most of the effort in this area has been concentrated on applying this approach to higher loops. However, in particle field theory, particularly for Yang-Mills, gravity, and supersymmetric theories (all of which are contained in various string theories), significant (and sometimes indispensable) improvements in higher-loop calculations have required techniques using the gauge-invariant field theory action. Since such techniques, whose string versions have not yet been derived, could drastically affect the S-matrix techniques of the 2D approach, we do not give the most recent details of the 2D approach here, but some of the basic ideas, and the ones we suspect most likely to survive future reformulations, will be described in chapters 6-9.

The 4D approach is concerned with the phenomenological applications of the low-energy effective theories obtained from the string theory. Since these theories are still very tentative (and still too ambiguous for many applications), they will not be discussed here. (See [1.15,0.1].)

The 6D approach describes the compactifications (or equivalent eliminations) of the 6 additional dimensions which must shrink from sight in order to obtain the observed dimensionality of the macroscopic world. Unfortunately, this approach has several problems which inhibit a useful treatment in a book: (1) So far, no justification has been given as to why the compactification occurs to the desired models, or to 4 dimensions, or at all; (2) the style of compactification (Kałuża-Klein, Calabi-Yau, toroidal, orbifold, fermionization, etc.) deemed most promising changes from year to year; and (3) the string model chosen to compactify (see previous section) also changes every few years. Therefore, the 6D approach won't be discussed here, either (see [1.16,0.1]).

What is discussed here is primarily the 10D approach, or second quantization, which seeks to obtain a more systematic understanding of string theory that would allow treatment of nonperturbative as well as perturbative aspects, and describe the enlarged hidden gauge symmetries which give string theories their finiteness and other unusual properties. In particular, it would be desirable to have a formalism in which all the symmetries (gauge, Lorentz, spacetime supersymmetry) are manifest, finiteness follows from simple power-counting rules, and all possible models (including possible 4D models whose existence is implied by the $1/N$ expansion of QCD and hadronic duality) can be straightforwardly classified. In ordinary (particle) supersymmetric field theories [1.17], such a formalism (*superfields* or *superspace*) has resulted in much simpler rules for constructing general actions, calculating quantum corrections (*supergraphs*), and explaining all finiteness properties (independent from, but verified by, explicit supergraph calculations). The finiteness results make use of the background field gauge, which can be defined only in a field theory formulation where all symmetries are manifest, and in this gauge divergence cancellations are automatic, requiring no explicit evaluation of integrals.

1.4. Outline

String theory can be considered a particular kind of particle theory, in that its modes of excitation correspond to different particles. All these particles, which differ in spin and other quantum numbers, are related by a symmetry which reflects the properties of the string. As discussed above, quantum field theory is the most complete framework within which to study the properties of particles. Not only is this framework not yet well understood for strings, but the study of string field theory has brought attention to aspects which are not well understood even for general types of particles. (This is another respect in which the study of strings resembles the study of supersymmetry.) We therefore devote chapts. 2-4 to a general study of field theory. Rather than trying to describe strings in the language of old quantum field theory, we recast the formalism of field theory in a mold prescribed by techniques learned from the study of strings. This language clarifies the relationship between physical states and gauge degrees of freedom, as well as giving a general and straightforward method for writing free actions for arbitrary theories.

In chapts. 5-6 we discuss the mechanics of the particle and string. As mentioned above, this approach is a useful calculational tool for evaluating graphs in perturbation theory, including the interaction vertices themselves. The quantum mechanics of the string is developed in chapts. 7-8, but it is primarily discussed directly as an operator algebra for the field theory, although it follows from quantization of the classical mechanics of the previous chapter, and vice versa. In general,

the procedure of first-quantization of a relativistic system serves only to identify its constraint algebra, which directly corresponds to both the field equations and gauge transformations of the free field theory. However, as described in chapts. 2-4, such a first-quantization procedure does not exist for general particle theories, but the constraint system can be derived by other means. The free gauge-covariant theory then follows in a straightforward way. String perturbation theory is discussed in chapt. 9.

Finally, the methods of chapts. 2-4 are applied to strings in chapts. 10-12, where string field theory is discussed. These chapters are still rather introductory, since many problems still remain in formulating interacting string field theory, even in the light-cone formalism. However, a more complete understanding of the extension of the methods of chapts. 2-4 to just particle field theory should help in the understanding of strings.

Chapts. 2-5 can be considered almost as an independent book, an attempt at a general approach to all of field theory. For those few high energy physicists who are not intensely interested in strings (or do not have high enough energy to study them), it can be read as a new introduction to ordinary field theory, although familiarity with quantum field theory as it is usually taught is assumed. Strings can then be left for later as an example. On the other hand, for those who want just a brief introduction to strings, a straightforward, though less elegant, treatment can be found via the light cone in chapts. 6,7,9,10 (with perhaps some help from sects. 2.1 and 2.5). These chapters overlap with most other treatments of string theory. The remainder of the book (chapts. 8,11,12) is basically the synthesis of these two topics.

2. GENERAL LIGHT CONE

2.1. Actions

Before discussing the string we first consider some general properties of gauge theories and field theories, starting with the light-cone formalism.

In general, light-cone field theory [2.1] looks like *non*relativistic field theory. Using light-cone notation, for vector indices a and the Minkowski inner product $A \cdot B = \eta^{ab} A_b B_a = A^a B_a$,

$$a = (+, -, i) \quad , \quad A \cdot B = A_+ B_- + A_- B_+ + A_i B_i \quad , \tag{2.1.1}$$

we interpret x_+ as being the "time" coordinate (even though it points in a lightlike direction), in terms of which the evolution of the system is described. The metric can be diagonalized by $A_\pm \equiv 2^{-1/2}(A_1 \mp A_0)$. For positive energy $E(= p^0 = -p_0)$, we have on shell $p_+ \geq 0$ and $p_- \leq 0$ (corresponding to paths with $\Delta x_+ \geq 0$ and $\Delta x_- \leq 0$), with the opposite signs for negative energy (antiparticles). For example, for a real scalar field the lagrangian is rewritten as

$$-\tfrac{1}{2}\phi(p^2 + m^2)\phi = -\phi p_+ \left(p_- + \frac{{p_i}^2 + m^2}{2p_+} \right) \phi = -\phi p_+ (p_- + H)\phi \quad , \tag{2.1.2}$$

where the momentum $p_a \equiv i\partial_a$, $p_- = i\partial/\partial x_+$ with respect to the "time" x_+, and p_+ appears like a mass in the "hamiltonian" H. (In the light-cone formalism, p_+ is assumed to be invertible.) Thus, the field equations are first-order in these time derivatives, and the field satisfies a nonrelativistic-style Schrödinger equation. The field equation can then be solved explicitly: In the free theory,

$$\phi(x_+) = e^{ix_+ H}\phi(0) \quad . \tag{2.1.3}$$

p_- can then be effectively replaced with $-H$. Note that, unlike the nonrelativistic case, the hamiltonian H, although hermitian, is imaginary (in coordinate space), due to the i in $p_+ = i\partial_+$. Thus, (2.1.3) is consistent with a (coordinate-space) reality condition on the field.

For a spinor, half the components are auxiliary (nonpropagating, since the field equation is only first-order in momenta), and all auxiliary components are eliminated in the light-cone formalism by their equations of motion (which, by definition, don't involve inverting time derivatives p_-):

$$\begin{aligned} -\tfrac{1}{2}\bar{\psi}(\not{p}+im)\psi &= -\tfrac{1}{2}2^{1/4}(\psi_+{}^\dagger \quad \psi_-{}^\dagger)\begin{pmatrix} \sqrt{2}p_- & \sigma_i p_i + im \\ \sigma_i p_i - im & -\sqrt{2}p_+ \end{pmatrix} 2^{1/4}\begin{pmatrix}\psi_+ \\ \psi_-\end{pmatrix} \\ &= -\psi_+{}^\dagger p_- \psi_+ + \psi_-{}^\dagger p_+ \psi_- \\ &\quad - \frac{1}{\sqrt{2}}\psi_-{}^\dagger(\sigma_i p_i - im)\psi_+ - \frac{1}{\sqrt{2}}\psi_+{}^\dagger(\sigma_i p_i + im)\psi_- \\ &\to \quad -\psi_+{}^\dagger(p_- + H)\psi_+ \quad , \end{aligned} \tag{2.1.4}$$

where H is the *same* hamiltonian as in (2.1.2). (There is an extra overall factor of 2 in (2.1.4) for complex spinors. We have assumed real (Majorana) spinors.)

For the case of Yang-Mills, the covariant action is

$$S = \frac{1}{g^2}\int d^D x \; tr\,\mathcal{L} \quad , \quad \mathcal{L} = \tfrac{1}{4}F_{ab}{}^2 \quad , \tag{2.1.5a}$$

$$F_{ab} \equiv [\nabla_a, \nabla_b] \quad , \quad \nabla_a \equiv p_a + A_a \quad , \quad \nabla_a{}' = e^{i\lambda}\nabla_a e^{-i\lambda} \quad . \tag{2.1.5b}$$

(Contraction with a matrix representation of the group generators is implicit.) The light-cone gauge is then defined as

$$A_+ = 0 \quad . \tag{2.1.6}$$

Since the gauge transformation of the gauge condition doesn't involve the time derivative ∂_-, the Faddeev-Popov ghosts are nonpropagating, and can be ignored. The field equation of A_- contains no time derivatives, so A_- is an auxiliary field. We therefore eliminate it by its equation of motion:

$$0 = [\nabla^a, F_{+a}] = p_+{}^2 A_- + [\nabla^i, p_+ A_i] \quad \to \quad A_- = -\frac{1}{p_+{}^2}[\nabla^i, p_+ A_i] \quad . \tag{2.1.7}$$

The only remaining fields are A_i, corresponding to the physical transverse polarizations. The lagrangian is then

$$\begin{aligned} \mathcal{L} &= \tfrac{1}{2}A_i \Box A_i + [A_i, A_j]p_i A_j + \tfrac{1}{4}[A_i, A_j]^2 \\ &\quad + (p_j A_j)\frac{1}{p_+}[A_i, p_+ A_i] + \tfrac{1}{2}\left(\frac{1}{p_+}[A_i, p_+ A_i]\right)^2 \quad . \end{aligned} \tag{2.1.8}$$

In fact, for *arbitrary* spin, after gauge-fixing ($A_{+\cdots} = 0$) and eliminating auxiliary fields ($A_{-\cdots} = \cdots$), we get for the free theory

$$\mathcal{L} = -\psi^\dagger (p_+)^k (p_- + H)\psi \quad , \tag{2.1.9}$$

where $k = 1$ for bosons and 0 for fermions.

The choice of light-cone gauges in particle mechanics will be discussed in chapt. 5, and for string mechanics in sect. 6.3 and chapt. 7. Light-cone field theory for strings will be discussed in chapt. 10.

2.2. Conformal algebra

Since the free kinetic operator of any light-cone field is just $\square$ (up to factors of ∂_+), the only nontrivial part of any free light-cone field theory is the representation of the Poincaré group ISO(D−1,1) (see, e.g., [2.2]). In the next section we will derive this representation for arbitrary massless theories (and will later extend it to the massive case) [2.3]. These representations are nonlinear in the coordinates, and are constructed from all the irreducible (matrix) representations of the light-cone's SO(D−2) rotation subgroup of the spin part of the SO(D−1,1) Lorentz group. One simple method of derivation involves the use of the conformal group, which is SO(D,2) for D-dimensional spacetime (for $D > 2$). We therefore use SO(D,2) notation by writing (D+2)-dimensional vector indices which take the values $\pm$ as well as the usual D a's: $\mathcal{A} = (\pm, a)$. The metric is as in (2.1.1) for the $\pm$ indices. (These $\pm$'s should not be confused with the light-cone indices $\pm$, which are related but are a subset of the a's.) We then write the conformal group generators as

$$J_{\mathcal{AB}} = (J_{+a} = -ip_a, \quad J_{-a} = -iK_a, \quad J_{-+} = \Delta, \quad J_{ab}) \quad , \tag{2.2.1}$$

where J_{ab} are the Lorentz generators, Δ is the dilatation generator, and K_a are the conformal boosts. An obvious linear coordinate representation in terms of D+2 coordinates is

$$J_{\mathcal{AB}} = x_{[\mathcal{A}}\partial_{\mathcal{B}]} + M_{\mathcal{AB}} \quad , \tag{2.2.2}$$

where [ ] means antisymmetrization and $M_{\mathcal{AB}}$ is the intrinsic (matrix, or coordinate-independent) part (with the same commutation relations that follow directly for the orbital part). The usual representation in terms of D coordinates is obtained by imposing the SO(D,2)-covariant constraints

$$x^{\mathcal{A}}x_{\mathcal{A}} = x^{\mathcal{A}}\partial_{\mathcal{A}} = M_{\mathcal{A}}{}^{\mathcal{B}}x_{\mathcal{B}} + d\, x_{\mathcal{A}} = 0 \tag{2.2.3a}$$

for some constant d (the canonical dimension, or scale weight). Corresponding to these constraints, which can be solved for everything with a "−" index, are the "gauge conditions" which determine everything with a "+" index but no "−" index:

$$\partial_+ = x_+ - 1 = M_{+a} = 0 \quad . \tag{2.2.3b}$$

This gauge can be obtained by a unitary transformation. The solution to (2.2.3) is then

$$J_{+a} = \partial_a \quad , \quad J_{-a} = -\tfrac{1}{2}x_b{}^2\partial_a + x_a x^b \partial_b + M_a{}^b x_b + d\!\!\!/ x_a \quad ,$$
$$J_{-+} = x^a\partial_a + d\!\!\!/ \quad , \quad J_{ab} = x_{[a}\partial_{b]} + M_{ab} \quad . \qquad (2.2.4)$$

This realization can also be obtained by the usual coset space methods (see, e.g., [2.4]), for the space SO(D,2)/ISO(D-1,1)⊗GL(1). The subgroup corresponds to all the generators *except* J_{+a}. One way to perform this construction is: First assign the coset space generators J_{+a} to be partial derivatives ∂_a (since they all commute, according to the commutation relations which follow from (2.2.2)). We next equate this first-quantized coordinate representation with a second-quantized field representation: In general,

$$0 = \delta\left\langle x \middle| \Phi \right\rangle = \left\langle Jx \middle| \Phi \right\rangle + \left\langle x \middle| \hat{J}\Phi \right\rangle$$

$$\rightarrow \quad J\left\langle x \middle| \Phi \right\rangle = \left\langle Jx \middle| \Phi \right\rangle = -\hat{J}\left\langle x \middle| \Phi \right\rangle = -\left\langle x \middle| \hat{J}\Phi \right\rangle \quad , \qquad (2.2.5)$$

where J (which acts directly on $\langle x|$) is expressed in terms of the coordinates and their derivatives (plus "spin" pieces), while $\hat{J}$ (which acts directly on $|\Phi\rangle$) is expressed in terms of the *fields* Φ and their *functional* derivatives. The minus sign expresses the usual relation between active and passive transformations. The structure constants of the second-quantized algebra have the same sign as the first-quantized ones. We can then solve the "constraint" $J_{+a} = -\hat{J}_{+a}$ on $\langle x|\Phi\rangle$ as

$$\left\langle x \middle| \Phi \right\rangle \equiv \Phi(x) = U\Phi(0) = e^{-x^a J_{+a}}\Phi(0) \quad . \qquad (2.2.6)$$

The other generators can then be determined by evaluating

$$J\Phi(x) = -\hat{J}\Phi(x) \quad \rightarrow \quad U^{-1}JU\Phi(0) = -U^{-1}\hat{J}U\Phi(0) \quad . \qquad (2.2.7)$$

On the left-hand side, the unitary transformation replaces any ∂_a with a $-\hat{J}_{+a}$ (the ∂_a itself getting killed by the $\Phi(0)$). On the right-hand side, the transformation gives terms with x dependence and other $\hat{J}$'s (as determined by the commutator algebra). (The calculations are performed by expressing the transformation as a sum of multiple commutators, which in this case has a finite number of terms.) The net result is (2.2.4), where $d\!\!\!/$ is $-\hat{J}_{-+}$ on $\Phi(0)$, M_{ab} is $-\hat{J}_{ab}$, and J_{-a} can have the additional term $-\hat{J}_{-a}$. However, $\hat{J}_{-a}$ on $\Phi(0)$ can be set to zero consistently in (2.2.4), and does vanish for physically interesting representations.

From now on, we use ± as in the light-cone notation, not SO(D,2) notation.

The conformal equations of motion are all those which can be obtained from $p_a{}^2 = 0$ by conformal transformations (or, equivalently, the irreducible tensor operator quadratic in conformal generators which includes p^2 as a component). Since conformal theories are a subset of massless ones, the massless equations of motion are a subset of the conformal ones (i.e., the massless theories satisfy fewer constraints). In particular, since massless theories are scale invariant but not always invariant under conformal boosts, the equations which contain the generators of conformal boosts must be dropped.

The complete set of equations of motion for an arbitrary massless representation of the Poincaré group are thus obtained simply by performing a conformal boost on the defining equation, $p^2 = 0$ [2.5,6]:

$$0 = \tfrac{1}{2}[K_a, p^2] = \tfrac{1}{2}\{J_a{}^b, p_b\} + \tfrac{1}{2}\{\Delta, p_a\} = M_a{}^b p_b + \left(d - \frac{D-2}{2}\right) p_a \quad . \tag{2.2.8}$$

d is determined by the requirement that the representation be nontrivial (for other values of d this equation implies $p = 0$). For nonzero spin ($M_{ab} \neq 0$) this equation implies $p^2 = 0$ by itself. For example, for scalars the equation implies only $d = (D-2)/2$. For a Dirac spinor, $M_{ab} = \frac{1}{4}[\gamma_a, \gamma_b]$ implies $d = (D-1)/2$ and the Dirac equation (in the form $\gamma_a \gamma \cdot p\psi = 0$). For a second-rank antisymmetric tensor, we find $d = D/2$ and Maxwell's equations. In this covariant approach to solving these equations, all the solutions are in terms of field strengths, not gauge fields (since the latter are not unitary representations). We can solve these equations in light-cone *notation*: Choosing a reference frame where the only nonvanishing component of the momentum is p_+, (2.2.8) reduces to the equations $M_{-i} = 0$ and $M_{-+} = d - (D-2)/2$. The equation $M_{-i} = 0$ says that the only nonvanishing components are the ones with as many (lower) "+" indices as possible (and for spinors, project with γ_+), and no "−" indices. In terms of Young tableaux, this means 1 "+" for each column. M_{-+} then just counts the number of "+" 's (plus 1/2 for a γ_+-projected spinor index), so we find that $d - (D-2)/2 =$ the number of columns (+ 1/2 for a spinor). We also find that the *on-shell* gauge field is the representation found by subtracting one box from each column of the Young tableau, and in the field strength those subtracted indices are associated with factors of momentum.

These results for massless representations can be extended to massive representations by the standard trick of adding one spatial dimension and constraining the extra momentum component to be the mass (operator): Writing

$$a \to (a, m) \quad , \quad p_m = M \quad , \tag{2.2.9}$$

where the index m takes one value, $p^2 = 0$ becomes $p^2 + M^2 = 0$, and (2.2.8) becomes

$$M_a{}^b p_b + M_{am} M + \left(d - \frac{D-2}{2} \right) p_a = 0 \quad . \tag{2.2.10}$$

The fields (or states) are now representations of an SO(D,1) spin group generated by M_{ab} and M_{am} (instead of the usual SO(D-1,1) of just M_{ab} for the massless case). The fields additional to those obtained in the massless case (on-shell field strengths) correspond to the on-shell gauge fields in the massless limit, resulting in a first-order formalism. For example, for spin 1 the additional field is the usual vector. For spin 2, the extra fields correspond to the on-shell, and thus traceless, parts of the Lorentz connection and metric tensor.

For field theory, we'll be interested in real representations. For the massive case, since (2.2.9) forces us to work in momentum space with respect to p_m, the reality condition should include an extra factor of the reflection operator which reverses the "m" direction. For example, for tensor fields, those components with an odd number of m indices should be imaginary (and those with an even number real).

In chapt. 4 we'll show how to obtain the off-shell fields, and thus the trace parts, by working directly in terms of the gauge fields. The method is based on the light-cone representation of the Poincaré algebra discussed in the next section.

2.3. Poincaré algebra

In contrast to the above covariant approach to solving (2.2.8,10), we now consider solving them in unitary gauges (such as the light-cone gauge), since in such gauges the gauge fields are essentially field strengths anyway because the gauge has been fixed: e.g., for Yang-Mills $A_a = \nabla_+{}^{-1} F_{+a}$, since $A_+ = 0$. In such gauges we work in terms of only the physical degrees of freedom (as in the case of the on-shell field strengths), which satisfy $p^2 = 0$ (unlike the auxiliary degrees of freedom, which satisfy algebraic equations, and the gauge degrees of freedom, which don't appear in any field equations).

In the light-cone formalism, the object is to construct all the Poincaré generators from just the manifest ones of the $(D-2)$-dimensional Poincaré subgroup, p_+, and the coordinates conjugate to these momenta. The light-cone gauge is imposed by the condition

$$M_{+i} = 0 \quad , \tag{2.3.1}$$

which, when acting on the independent fields (those with only i indices), says that all fields with $+$ indices have been set to vanish. The fields with $-$ indices (auxiliary fields) are then determined as usual by the field equations: by solving (2.2.8) for M_{-i}. The solution to the i, $+$, and $-$ parts of (2.2.8) gives

$$M_{-i} = \frac{1}{p_+}(M_i{}^j p_j + kp_i) \quad ,$$

$$M_{-+} = d\!\!\!/ - \frac{D-2}{2} \equiv k \quad ,$$

$$kp^2 = 0 \quad . \tag{2.3.2}$$

If (2.2.8) is solved without the condition (2.3.1), then M_{+i} can still be removed (and (2.3.2) regained) by a unitary transformation. (In a first-quantized formalism, this corresponds to a gauge choice: see sect. 5.3 for spin 1/2.) The appearance of k is related to ordering ambiguities, and we can also choose $M_{-+} = 0$ by a *non*unitary transformation (a rescaling of the field by a power of p_+). Of course, we also solve $p^2 = 0$ as

$$p_- = -\frac{p_i{}^2}{2p_+} \quad . \tag{2.3.3}$$

These equations, together with the gauge condition for M_{+i}, determine all the Poincaré generators in terms of M_{ij}, p_i, p_+, x_i, and x_-. In the orbital pieces of J_{ab}, x_+ can be set to vanish, since p_- is no longer conjugate: i.e., we work at "time" $x_+ = 0$ for the "hamiltonian" p_-, or equivalently in the Schrödinger picture. (Of course, this also corresponds to removing x_+ by a unitary transformation, i.e., a time translation via p_-. This is also a gauge choice in a first-quantized formalism: see sect. 5.1.) The final result is

$$p_i = i\partial_i \quad , \quad p_+ = i\partial_+ \quad , \quad p_- = -\frac{p_i{}^2}{2p_+} \quad ,$$

$$J_{ij} = -ix_{[i}p_{j]} + M_{ij} \quad , \quad J_{+i} = ix_i p_+ \quad , \quad J_{-+} = -ix_- p_+ + k \quad ,$$

$$J_{-i} = -ix_- p_i - ix_i \frac{p_j{}^2}{2p_+} + \frac{1}{p_+}(M_i{}^j p_j + kp_i) \quad . \tag{2.3.4}$$

The generators are (anti)hermitian for the choice $k = \frac{1}{2}$; otherwise, the Hilbert space metric must include a factor of ${p_+}^{1-2k}$, with respect to which all the generators are pseudo(anti)hermitian. In this light-cone approach to Poincaré representations, where we work with the fundamental fields rather than field strengths, $k = 0$ for bosons and $\frac{1}{2}$ for fermions (giving the usual dimensions $d = \frac{1}{2}(D-2)$ for bosons and $\frac{1}{2}(D-1)$ for fermions), and thus the metric is p_+ for bosons and 1 for fermions, so the light-cone kinetic operator (metric)$\cdot 2(i\partial_- - p_-) \sim \Box$ for bosons and $\Box/p_+$ for fermions.

This construction of the D-dimensional Poincaré algebra in terms of D−1 coordinates is analogous to the construction in the previous section of the D-dimensional conformal algebra SO(D,2) in terms of D coordinates, except that in the conformal case (1) we start with D+2 coordinates instead of D, (2) x's and p's are switched, and (3) the further constraint $x \cdot p = 0$ and gauge condition $x_+ = 1$ are used. Thus, J_{ab} of (2.3.4) becomes J_{AB} of (2.2.4) if x_- is replaced with $-(1/p_+)x^j p_j$, p_+ is set to 1, and we then switch $p \to x, x \to -p$. Just as the conformal representation (2.2.4) can be obtained from the Poincaré representation (in 2 extra dimensions, by $i \to a$) (2.3.4) by eliminating one coordinate (x_-), (2.3.4) can be reobtained from (2.2.4) by reintroducing this coordinate: First choose $d = -ix_-p_+ + k$. Then switch $x_i \to p_i$, $p_i \to -x_i$. Finally, make the (almost unitary) transformation generated by $exp[-ip^i x_i(ln\ p_+)]$, which takes $x_i \to p_+ x_i$, $p_i \to p_i/p_+$, $x_- \to x_- + p^i x_i/p_+$.

To extend these results to arbitrary representations, we use the trick (2.2.9), or directly solve (2.2.10), giving the light-cone form of the Poincaré algebra for arbitrary representations: (2.3.4) becomes

$$p_i = i\partial_i \quad , \quad p_+ = i\partial_+ \quad , \quad p_- = -\frac{{p_i}^2 + M^2}{2p_+} \quad ,$$

$$J_{ij} = -ix_{[i}p_{j]} + M_{ij} \quad , \quad J_{+i} = ix_i p_+ \quad , \quad J_{-+} = -ix_- p_+ + k \quad ,$$

$$J_{-i} = -ix_- p_i - ix_i \frac{{p_j}^2 + M^2}{2p_+} + \frac{1}{p_+}({M_i}^j p_j + M_{im}M + kp_i) \quad . \tag{2.3.5}$$

Thus, massless irreducible representations of the Poincaré group ISO(D−1,1) are irreducible representations of the spin subgroup SO(D−2) (generated by M_{ij}) which also depend on the coordinates (x_i, x_-), and irreducible massive ones are irreducible representations of the spin subgroup SO(D−1) (generated by (M_{ij}, M_{im})) for some nonvanishing constant M. Notice that the introduction of masses has modified only p_- and J_{-i}. These are also the only generators modified when interactions are introduced, where they become nonlinear in the fields.

The light-cone representation of the Poincaré algebra will be used in sect. 3.4 to derive BRST algebras, used for enforcing unitarity in covariant formalisms, which in turn will be used extensively to derive gauge-invariant actions for particles and strings in the following chapters. The general light-cone analysis of this section will be applied to the special case of the free string in chapt. 7.

2.4. Interactions

For interacting theories, the derivation of the Poincaré algebra is not so general, but depends on the details of the particular type of interactions in the theory. We again consider the case of Yang-Mills. Since only p_- and J_{-i} obtain interacting contributions, we consider the derivation of only those operators. The expression for $p_- A_i$ is then given directly by the field equation of A_i

$$0 = [\nabla^a, F_{ai}] = [\nabla^j, F_{ji}] + [\nabla_+, F_{-i}] + [\nabla_-, F_{+i}] = [\nabla^j, F_{ji}] + 2[\nabla_+, F_{-i}] + [\nabla_i, F_{+-}]$$

$$\rightarrow \quad p_- A_i = [\nabla_i, A_-] - \frac{1}{2p_+}\left([\nabla^j, F_{ji}] + [\nabla_i, p_+ A_-]\right) \quad , \tag{2.4.1}$$

where we have used the Bianchi identity $[\nabla_{[+}, F_{-i]}] = 0$. This expression for p_- is also used in the orbital piece of $J_{-i}A_j$. In the spin piece M_{-i} we start with the covariant-formalism equation $M_{-i}A_j = -\delta_{ij}A_-$, substitute the solution to A_-'s field equation, and then add a gauge transformation to cancel the change of gauge induced by the covariant-formalism transformation $M_{-i}A_+ = A_i$. The net result is that in the light-cone formalism

$$J_{-i}A_j = -i(x_- p_i - x_i p_-)A_j - \left(\delta_{ij}A_- + [\nabla_j, \frac{1}{p_+}A_i]\right) \quad , \tag{2.4.2}$$

with A_- given by (2.1.7) and $p_- A_j$ by (2.4.1). In the abelian case, these expressions agree with those obtained by a different method in (2.3.4). All transformations can then be written in functional second-quantized form as

$$\delta = -\int d^{D-2}x_i dx_-\ tr\ (\delta A_i)\frac{\delta}{\delta A_i} \quad \rightarrow \quad [\delta, A_i] = -(\delta A_i) \quad . \tag{2.4.3}$$

The minus sign is as in (2.2.5) for relating first- and second-quantized operators.

As an alternative, we can consider canonical second-quantization, which has certain advantages in the light cone, and has an interesting generalization in the covariant case (see sect. 3.4). From the light-cone lagrangian

$$L = -i\int \Phi^\dagger p_+ \dot{\Phi} - H(\Phi) \quad , \tag{2.4.4}$$

where $\dot{}$ is the "time"-derivative $i\partial/\partial x_+$, we find that the fields have equal-time commutators similar to those in nonrelativistic field theory:

$$[\Phi^\dagger(1), \Phi(2)] = -\frac{1}{2p_{+2}}\delta(2-1) \quad , \tag{2.4.5}$$

where the δ-function is over the transverse coordinates and x_- (and may include a Kronecker δ in indices, if Φ has components). Unlike nonrelativistic field theory, the fields satisfy a reality condition, in coordinate space:

$$\Phi^* = \Omega\Phi \quad , \tag{2.4.6}$$

where Ω is the identity or some symmetric, unitary matrix (the "charge conjugation" matrix; * here is the hermitian conjugate, or adjoint, in the operator sense, i.e., unlike †, it excludes matrix transposition). As in quantum mechanics (or the Poisson bracket approach to classical mechanics), the generators can then be written as functions of the dynamical variables:

$$V = \sum_n \frac{1}{n!} \int dz_1 \cdots dz_n \, \mathcal{V}^{(n)}(z_1, \ldots, z_n)\Phi(z_1)\cdots\Phi(z_n) \quad , \tag{2.4.7}$$

where the arguments z stand for either coordinates or momenta and the $\mathcal{V}$'s are the vertex functions, which are just functions of the coordinates (not operators). Without loss of generality they can be chosen to be cyclically symmetric in the fields (or totally symmetric, if group-theory indices are also permuted). (Any asymmetric piece can be seen to contribute to a lower-point function by the use of (2.4.5,6).) In light-cone theories the coordinate-space integrals are over all coordinates except x_+. The action of the second-quantized operator V on fields is calculated using (2.4.5):

$$[V, \Phi(z_1)^\dagger] = -\frac{1}{2p_{+1}} \sum_n \frac{1}{(n-1)!} \int dz_2 \cdots dz_n \, \mathcal{V}^{(n)}(z_1, \ldots, z_n)\Phi(z_2)\cdots\Phi(z_n) \quad . \tag{2.4.8}$$

A particular case of the above equations is the free case, where the operator V is quadratic in Φ. We will then generally write the second-quantized operator V in terms of a first-quantized operator $\mathcal{V}$ with a single integration:

$$V = \int dz \, \Phi^\dagger p_+ \mathcal{V}\Phi \quad \rightarrow \quad [V, \Phi] = -\mathcal{V}\Phi \quad . \tag{2.4.9}$$

This can be checked to relate to (2.4.7) as $\mathcal{V}^{(2)}(z_1, z_2) = 2\Omega_1 p_{+1} \mathcal{V}_1 \delta(2-1)$ (with the symmetry of $\mathcal{V}^{(2)}$ imposing corresponding conditions on the operator $\mathcal{V}$). In the interacting case, the generalization of (2.4.9) is

$$V = \frac{1}{N} \int dz \, \Phi^\dagger 2p_+(\mathcal{V}\Phi) \quad , \tag{2.4.10}$$

where N is just the number of fields in any particular term. (In the free case $N = 2$, giving (2.4.9).)

For example, for Yang-Mills, we find

$$p_- = \int \tfrac{1}{4}(F_{ij})^2 + \tfrac{1}{2}(p_+ A_-)^2 \quad , \qquad (2.4.11a)$$

$$J_{-i} = \int ix_-(p_+ A_j)(p_i A_j) + ix_i \left[\tfrac{1}{4}(F_{jk})^2 + \tfrac{1}{2}(p_+ A_-)^2\right] - A_i p_+ A_- \quad . \qquad (2.4.11b)$$

(The other generators follow trivially from (2.4.9).) p_- is minus the hamiltonian H (as in the free case (2.1.2,4,9)), as also follows from performing the usual Legendre transformation on the lagrangian.

In general, all the explicit x_i-dependence of all the Poincaré generators can be determined from the commutation relations with the momenta (translation generators) p_i. Furthermore, since only p_- and J_{-i} get contributions from interactions, we need consider only those. Let's first consider the "hamiltonian" p_-. Since it commutes with p_i, it is translation invariant. In terms of the vertex functions, this translates into the condition:

$$(p_1 + \cdots + p_n)\tilde{\mathcal{V}}^{(n)}(p_1, \ldots, p_n) = 0 \quad , \qquad (2.4.12)$$

where the $\widetilde{\ }$ indicates Fourier transformation with respect to the coordinate-space expression, implying that most generally

$$\tilde{\mathcal{V}}^{(n)}(p_1, \ldots, p_n) = \tilde{f}(p_1, \ldots, p_{n-1})\delta(p_1 + \cdots + p_n) \quad , \qquad (2.4.13)$$

or in coordinate space

$$\mathcal{V}^{(n)}(x_1, \ldots, x_n) = \tilde{f}\left(i\frac{\partial}{\partial x_1}, \ldots, i\frac{\partial}{\partial x_{n-1}}\right)\delta(x_1 - x_n)\cdots\delta(x_{n-1} - x_n)$$
$$= f(x_1 - x_n, \ldots, x_{n-1} - x_n) \quad . \qquad (2.4.14)$$

In this coordinate representation one can see that when $\mathcal{V}$ is inserted back in (2.4.7) we have the usual expression for a translation-invariant vertex used in field theory. Namely, fields at the same point in coordinate space, with derivatives acting on them, are multiplied and integrated over coordinate space. In this form it is clear that there is no explicit coordinate dependence in the vertex. As can be seen in (2.4.14), the most general translationally invariant vertex involves an arbitrary function of coordinate differences, denoted as f above. For the case of bosonic coordinates, the function $\tilde{f}$ may contain inverse derivatives (that is, translational invariance does not imply locality.) For the case of anticommuting coordinates (see sect. 2.6) the situation is simpler: There is no locality issue, since the most general function f can always be obtained from a function $\tilde{f}$ polynomial in derivatives, acting on δ-functions.

We now consider J_{-i}. From the commutation relations we find:

$$[p_i, J_{-j}] = -\eta_{ij}p_- \quad \rightarrow \quad [J_{-i}, \Phi] = ix_i[p_-, \Phi] + [\Delta J_{-i}, \Phi] \quad , \tag{2.4.15}$$

where ΔJ_{-i} is translationally invariant (commutes with p_i), and can therefore be represented without explicit x^i's. For the Yang-Mills case, this can be seen to agree with (2.4.2) or (2.4.11).

This light-cone analysis will be applied to interacting strings in chapt. 10.

2.5. Graphs

Feynman graphs for any interacting light-cone field theory can be derived as in covariant field theory, but an alternative not available there is to use a nonrelativistic style of perturbation (i.e., just expanding e^{iHt} in H_{INT}), since the field equations are now linear in the time derivative $p_- = i\partial/\partial x_+ = i\partial/\partial\tau$. (As in sect. 2.1, but unlike sects. 2.3 and 2.4, we now use p_- to refer to this partial derivative, as in covariant formalisms, while $-H$ refers to the corresponding light-cone Poincaré generator, the two being equal on shell.) This formalism can be derived straightforwardly from the usual Feynman rules (after choosing the light-cone gauge and eliminating auxiliary fields) by simply Fourier transforming from p_- to $x_+ = \tau$ (but keeping all other momenta):

$$\int_{-\infty}^{\infty} \frac{dp_-}{2\pi} e^{-ip_-\tau} \frac{1}{2p_+p_- + p_i{}^2 + m^2 + i\epsilon} = -i\Theta(p_+\tau)\frac{1}{2p_+} e^{i\tau(p_i{}^2+m^2)/2p_+} \quad . \tag{2.5.1}$$

($\Theta(u) = 1$ for $u > 1$, 0 for $u < 1$.) We now draw all graphs to represent the τ coordinate, so that graphs with different τ-orderings of the vertices must be considered as separate contributions. Then we direct all the propagators toward increasing τ, so the change in τ between the ends of the propagator (as appears in (2.5.1)) is always positive (i.e., the orientation of the momenta is defined to be toward increasing τ). We next Wick rotate $\tau \rightarrow i\tau$. We also introduce external line factors which transform H back to $-p_-$ on external lines. The resulting rules are:

(a) Assign a τ to each vertex, and order them with respect to τ.

(b) Assign (p_-, p_+, p_i) to each external line, but only (p_+, p_i) to each internal line, all directed toward increasing τ. Enforce conservation of (p_+, p_i) at each vertex, and total conservation of p_-.

(c) Give each internal line a propagator

$$\Theta(p_+)\frac{1}{2p_+}e^{-\tau(p_i{}^2+m^2)/2p_+}$$

for the (p_+, p_i) of that line and the positive difference τ in the proper time between the ends.

(d) Give each external line a factor

$$e^{\tau p_-}$$

for the p_- of that line and the τ of the vertex to which it connects.

(e) Read off the vertices from the action as usual.

(f) Integrate

$$\int_0^\infty d\tau$$

for each τ difference between consecutive (though not necessarily connected) vertices. (Performing just this integration gives the usual old-fashioned perturbation theory in terms of energy denominators [2.1], except that our external-line factors differ off shell in order to reproduce the usual Feynman rules.)

(g) Integrate

$$\int_{-\infty}^{\infty} \frac{dp_+ \, d^{D-2}p_i}{(2\pi)^{D-1}}$$

for each loop.

The use of such methods for strings will be discussed in chapt. 10.

2.6. Covariantized light cone

There is a covariant formalism for any field theory that has the interesting property that it can be obtained directly and easily from the light-cone formalism, without any additional gauge-fixing procedure [2.7]. Although this covariant gauge is not as general or convenient as the usual covariant gauges (in particular, it sometimes has additional off-shell infrared divergences), it bears strong relationship to both the light-cone and BRST formalisms, and can be used as a conceptual bridge. The basic idea of the formalism is: Consider a covariant theory in D dimensions. This is equivalent to a covariant theory in $(D+2)-2$ dimensions, where the notation indicates the addition of 2 extra commuting coordinates (1 space, 1 time) and 2 (real) anticommuting coordinates, with a similar extension of Lorentz indices [2.8]. (A similar use of OSp groups in gauge-fixed theories, but applied to only the Lorentz indices and not the coordinates, appears in [2.9].) This extends the Poincaré group ISO(D−1,1) to a graded analog IOSp(D,2|2). In practice, this means we just take the light-cone transverse indices to be graded, watching out for signs introduced by the corresponding change in statistics, and replace the Euclidean SO(D-2) metric with the corresponding graded OSp(D-1,1|2) metric:

$$i = (a, \alpha) \quad , \quad \delta_{ij} \rightarrow \eta_{ij} = (\eta_{ab}, C_{\alpha\beta}) \quad , \tag{2.6.1}$$

where η_{ab} is the usual Lorentz metric and

$$C_{\alpha\beta} = C^{\beta\alpha} = \sigma_2 \tag{2.6.2}$$

is the Sp(2) metric, which satisfies the useful identity

$$C_{\alpha\beta}C^{\gamma\delta} = \delta_{[\alpha}{}^{\gamma}\delta_{\beta]}{}^{\delta} \quad \rightarrow \quad A_{[\alpha}B_{\beta]} = C_{\alpha\beta}C^{\gamma\delta}A_{\gamma}B_{\delta} \quad . \tag{2.6.3}$$

The OSp metric is used to raise and lower graded indices as:

$$x^i = \eta^{ij}x_j \quad , \quad x_i = x^j\eta_{ji} \quad ; \quad \eta^{ik}\eta_{jk} = \delta_j{}^i \quad . \tag{2.6.4}$$

The sign conventions are that adjacent indices are contracted with the contravariant (up) index first. The equivalence follows from the fact that, for momentum-space Feynman graphs, the trees will be the same if we constrain the 2 − 2 extra "ghost" momenta to vanish on external lines (since they'll then vanish on internal lines by momentum conservation); and the loops are then the same because, when the momentum integrands are written as gaussians, the determinant factors coming from the 2 extra anticommuting dimensions exactly cancel those from the 2 extra commuting ones. For example, using the proper-time form ("Schwinger parametrization") of the propagators (cf. (2.5.1)),

$$\frac{1}{p^2 + m^2} = \int_0^\infty d\tau \; e^{-\tau(p^2+m^2)} \quad , \tag{2.6.5}$$

all momentum integrations take the form

$$\begin{aligned} \frac{1}{\pi}\int d^{D+2}p \; d^2p_\alpha \; e^{-f(2p_+p_- + p^a p_a + p^\alpha p_\alpha + m^2)} &= \int d^D p \; e^{-f(p^a p_a + m^2)} \\ &= \left(\frac{\pi}{f}\right)^{D/2} e^{-fm^2} \end{aligned} \quad , \tag{2.6.6}$$

where f is a function of the proper-time parameters.

The covariant theory is thus obtained from the light-cone one by the substitution

$$(p_-, p_+; p_i) \rightarrow (p_-, p_+; p_a, p_\alpha) \quad , \tag{2.6.7a}$$

where

$$p_- = p_\alpha = 0 \tag{2.6.7b}$$

on physical states. It's not necessary to set $p_+ = 0$, since it only appears in the combination $p_- p_+$ in OSp(D,2|2)-invariant products. Thus, p_+ can be chosen arbitrarily on external lines (but should be nonvanishing due to the appearance of factors of $1/p_+$). We now interpret $x^\pm$ and x^α as the unphysical coordinates. Vector indices on fields are treated similarly: Having been reduced to transverse ones by the light-cone formalism, they now become covariant vector indices with 2 additional anticommuting values ((2.6.1)). For example, in Yang-Mills the vector field becomes the usual vector field plus two anticommuting scalars A_α, corresponding to Faddeev-Popov ghosts.

The graphical rules become:

(a) Assign a τ to each vertex, and order them with respect to τ.

(b) Assign (p_+, p_a) to each external line, but (p_+, p_a, p_α) to each internal line, all directed toward increasing τ. Enforce conservation of (p_+, p_a, p_α) at each vertex (with $p_\alpha = 0$ on external lines).

(c) Give each internal line a propagator

$$\Theta(p_+)\frac{1}{2p_+}e^{-\tau(p_a{}^2+p^\alpha p_\alpha+m^2)/2p_+}$$

for the (p_+, p_a, p_α) of that line and the positive difference τ in the proper time between the ends.

(d) Give each external line a factor

$$1 \quad .$$

(e) Read off the vertices from the action as usual.

(f) Integrate

$$\int_0^\infty d\tau$$

for each τ difference between consecutive (though not necessarily connected) vertices.

(g) Integrate

$$\int d^2 p_\alpha$$

for each loop (remembering that for any anticommuting variable θ, $\int d\theta\, 1 = 0$, $\int d\theta\, \theta = 1$, $\theta^2 = 0$).

(h) Integrate

$$2\int_{-\infty}^{\infty} dp_+$$

for each loop.

(i) Integrate

$$\int \frac{d^D p}{(2\pi)^D}$$

for each loop.

For theories with only scalars, integrating just (f-h) gives the usual Feynman graphs (although it may be necessary to add several graphs due to the τ-ordering of non-adjacent vertices). Besides the correspondence of the τ parameters to the usual Schwinger parameters, after integrating out just the anticommuting parameters the p_+ parameters resemble Feynman parameters.

These methods can also be applied to strings (chapt. 10).

Exercises

(1) Find the light-cone formulation of QED. Compare with the Coulomb gauge formulation.

(2) Derive the commutation relations of the conformal group from (2.2.2). Check that (2.2.4) satisfies them. Evaluate the commutators implicit in (2.2.7) for each generator.

(3) Find the Lorentz transformation M_{ab} of a vector (consistent with the conventions of (2.2.2)). (Hint: Look at the transformations of x and p.) Find the explicit form of (2.2.8) for that case. Solve these equations of motion. To what simpler representation is this equivalent? Study this equivalence with the light-cone analysis given below (2.2.8). Generalize the analysis to totally antisymmetric tensors of arbitrary rank.

(4) Repeat problem (3) for the massive case. Looking at the separate SO(D-1,1) representations contained in the SO(D,1) representations, show that first-order formalisms in terms of the usual fields have been obtained, and find the corresponding second-order formulations.

(5) Check that the explicit forms of the Poincaré generators given in (2.3.5) satisfy the correct algebra (see problem (2)). Find the explicit transformations acting on the vector representation of the spin group SO(D-1). Compare with (2.4.1-2).

(6) Derive (2.4.11). Compare that p_- with the light-cone hamiltonian which follows from (2.1.5).

(7) Calculate the 4-point amplitude in ϕ^3 theory with light-cone graphs, and compare with the usual covariant Feynman graph calculation. Calculate the 1-loop propagator correction in the same theory using the *covariantized* light-cone rules, and again compare with ordinary Feynman graphs, paying special attention to Feynman parameters.

3. GENERAL BRST

3.1. Gauge invariance and constraints

In the previous chapter we saw that a gauge theory can be described either in a manifestly covariant way by using gauge degrees of freedom, or in a manifestly unitary way (with only physical degrees of freedom) with Poincaré transformations which are nonlinear (in both coordinates and fields). In the gauge-covariant formalism there is a D-dimensional manifest Lorentz covariance, and in the light-cone formalism a $D-2$-dimensional one, and in each case a corresponding number of degrees of freedom. There is also an intermediate formalism, more familiar from nonrelativistic theory: The hamiltonian formalism has a $D-1$-dimensional manifest Lorentz covariance (rotations). As in the light-cone formalism, the notational separation of coordinates into time and space suggests a particular type of gauge condition: temporal (timelike) gauges, where time-components of gauge fields are set to vanish. In chapt. 5, this formalism will be seen to have a particular advantage for first-quantization of relativistic theories: In the classical mechanics of relativistic theories, the coordinates are treated as functions of a "proper time" so that the usual time coordinate can be treated on an equal footing with the space coordinates. Thus, canonical quantization with respect to this unobservable (proper) "time" coordinate doesn't destroy manifest Poincaré covariance, so use of a hamiltonian formalism can be advantageous, particularly in deriving BRST transformations, and the corresponding second-quantized theory, where the proper-time doesn't appear anyway.

We'll first consider Yang-Mills, and then generalize to arbitrary gauge theories. In order to study the temporal gauge, instead of the decomposition (2.1.1) we simply separate into time and spatial components

$$a = (0, i) \quad , \quad A \cdot B = -A_0 B_0 + A_i B_i \quad . \tag{3.1.1}$$

The lagrangian (2.1.5) is then

$$\mathcal{L} = \tfrac{1}{4} {F_{ij}}^2 - \tfrac{1}{2}(p_0 A_i - [\nabla_i, A_0])^2 \quad . \tag{3.1.2}$$

The gauge condition

$$A_0 = 0 \tag{3.1.3}$$

transforms under a gauge transformation with a time derivative: Under an infinitesimal transformation about $A_0 = 0$,

$$\delta A_0 \approx \partial_0 \lambda \quad , \tag{3.1.4}$$

so the Faddeev-Popov ghosts are propagating. Furthermore, the gauge transformation (3.1.4) does not allow the gauge choice (3.1.3) everywhere: For example, if we choose periodic boundary conditions in time (to simplify the argument), then

$$\delta \int_{-\infty}^{\infty} dx^0\, A_0 \approx 0 \quad . \tag{3.1.5}$$

A_0 can then be fixed by an appropriate initial condition, e.g., $A_0|_{x^0=0} = 0$, but then the corresponding field equation is lost. Therefore, we must impose

$$0 = \frac{\delta S}{\delta A_0} = -[\nabla_i, F_{0i}] = -[\nabla_i, p_0 A_i] \quad \mathit{at}\ x^0 = 0 \tag{3.1.6}$$

as an initial condition. Another way to understand this is to note that gauge fixing eliminates only degrees of freedom which don't occur in the lagrangian, and thus can eliminate only redundant equations of motion: Since $[\nabla_i, F_{0i}] = 0$ followed from the gauge-invariant action, the fact that it doesn't follow after setting $A_0 = 0$ means some piece of A_0 can't truly be gauged away, and so we must compensate by imposing the equation of motion for that piece. Due to the original gauge invariance, (3.1.6) then holds for all time from the remaining field equations: In the gauge (3.1.3), the lagrangian (3.1.2) becomes

$$\mathcal{L} = \tfrac{1}{2} A_i \Box A_i - \tfrac{1}{2}(p_i A_i)^2 + [A_i, A_j] p_i A_j + \tfrac{1}{4}[A_i, A_j]^2 \quad , \tag{3.1.7}$$

and the covariant divergence of the implied field equations yields the *time derivative* of (3.1.6). (This follows from the identity $[\nabla^b, [\nabla^a, F_{ab}]] = 0$ upon applying the field equations $[\nabla^a, F_{ia}] = 0$. In unitary gauges, the corresponding constraint can be derived without time derivatives, and hence is implied by the remaining field equations under suitable boundary conditions.) Equivalently, if we notice that (3.1.4) does not fix the gauge completely, but leaves time-independent gauge transformations, we need to impose a constraint on the initial states to make them gauge invariant. But the generator of the residual gauge transformations on the remaining fields A_i is

$$\mathcal{G}(x_i) = \left[\nabla_i, i\frac{\delta}{\delta A_i}\right] \quad , \tag{3.1.8}$$

which is the same as the constraint (3.1.6) under canonical quantization of (3.1.7). Thus, the same operator (1) gives the constraint which must be imposed in addition to the field equations because too much of A_0 was dropped, and (2) (its transpose) gives the gauge transformations remaining because they left the gauge-fixing function A_0 invariant. The fact that these are identical is not surprising, since in Faddeev-Popov quantization the latter corresponds to the Faddeev-Popov ghost while the former corresponds to the antighost.

These properties appear very naturally in a hamiltonian formulation: We start again with the gauge-invariant lagrangian (3.1.2). Since A_0 has no time-derivative terms, we Legendre transform with respect to just $\dot{A}_i$. The result is

$$S_H = \frac{1}{g^2}\int d^Dx\ tr\ \mathcal{L}_H \quad , \quad \mathcal{L}_H = \dot{A}^i\Pi_i - \mathcal{H} \quad , \quad \mathcal{H} = \mathcal{H}_0 + A_0 i\mathcal{G} \quad ,$$

$$\mathcal{H}_0 = \tfrac{1}{2}\Pi_i{}^2 - \tfrac{1}{4}F_{ij}{}^2 \quad , \quad \mathcal{G} = [\nabla_i, \Pi_i] \quad , \tag{3.1.9}$$

where $\dot{} = \partial_0$. As in ordinary nonrelativistic classical mechanics, eliminating the momentum Π_i from the hamiltonian form of the action (first order in time derivatives) by its equation of motion gives back the lagrangian form (second order in time derivatives). Note that A_0 appears linearly, as a Lagrange multiplier.

The gauge-invariant hamiltonian formalism of (3.1.9) can be generalized [3.1]: Consider a lagrangian of the form

$$\mathcal{L}_H = \dot{z}^M e_M{}^A(z)\pi_A - \mathcal{H} \quad , \quad \mathcal{H} = \mathcal{H}_0(z,\pi) + \lambda^i i\mathcal{G}_i(z,\pi) \quad , \tag{3.1.10}$$

where z, π, and λ are the variables, representing "coordinates," covariant "momenta," and Lagrange multipliers, respectively. They depend on the time, and also have indices (which may include continuous indices, such as spatial coordinates). e, which is a function of z, has been introduced to allow for cases with a symmetry (such as supersymmetry) under which $dz^M e_M{}^A$ (but not dz itself) is covariant, so that π will be covariant, and thus a more convenient variable in terms of which to express the constraints $\mathcal{G}$. When $\mathcal{H}_0$ commutes with $\mathcal{G}$ (quantum mechanically, or in terms of Poisson brackets for a classical treatment), this action has a gauge invariance generated by $\mathcal{G}$, for which λ is the gauge field:

$$\delta(z,\pi) = [\zeta^i\mathcal{G}_i, (z,\pi)] \quad ,$$

$$\delta\left(\frac{\partial}{\partial t} - \lambda^i\mathcal{G}_i\right) = 0 \quad \rightarrow \quad (\delta\lambda^i)\mathcal{G}_i = \dot{\zeta}^i\mathcal{G}_i + [\lambda^j\mathcal{G}_j, \zeta^i\mathcal{G}_i] \quad , \tag{3.1.11}$$

where the gauge transformation of λ has been determined by the invariance of the "total" time-derivative $d/dt = \partial/\partial t + i\mathcal{H}$. (More generally, if $[\zeta^i\mathcal{G}_i, \mathcal{H}_0] = f^i i\mathcal{G}_i$, then $\delta\lambda^i$ has an extra term $-f^i$.) Using the chain rule ((d/dt) on $f(t,q_k(t))$ equals $\partial/\partial t + \dot{q}_k(\partial/\partial q_k)$) to evaluate the time derivative of $\mathcal{G}$, we find the lagrangian transforms as a total derivative

$$\delta\mathcal{L}_H = \frac{d}{dt}\left[(\delta z^M)e_M{}^A\pi_A - \zeta^i i\mathcal{G}_i\right] \quad , \tag{3.1.12}$$

which is the usual transformation law for an action with local symmetry generated by the current $\mathcal{G}$. When $\mathcal{H}_0$ vanishes (as in relativistic mechanics), the special case $\zeta^i = \zeta\lambda^i$ of the transformations of (3.1.11) are τ reparametrizations, generated by the hamiltonian $\lambda^i\mathcal{G}_i$. In general, after canonical quantization, the wave function satisfies the Schrödinger equation $\partial/\partial t + i\mathcal{H}_0 = 0$, as well as the constraints $\mathcal{G} = 0$ (and thus $\partial/\partial t + i\mathcal{H} = 0$ in any gauge choice for λ). Since $[\mathcal{H}_0, \mathcal{G}] = 0$, $\mathcal{G} = 0$ at $t = 0$ implies $\mathcal{G} = 0$ for all t.

In some cases (such as Yang-Mills), the Lorentz covariant form of the action can be obtained by eliminating all the π's. A covariant first-order form can generally be obtained by introducing additional auxiliary degrees of freedom which enlarge π to make it Lorentz covariant. For example, for Yang-Mills we can rewrite (3.1.9) as

$$\mathcal{L}_H = \tfrac{1}{2}G_{0i}{}^2 - G_0{}^iF_{0i} + \tfrac{1}{4}F_{ij}{}^2$$

$$\rightarrow \quad \mathcal{L}_1 = -\tfrac{1}{4}G_{ab}{}^2 + G^{ab}F_{ab} \quad , \tag{3.1.13}$$

where $G_{0i} = i\Pi_i$, and the independent (auxiliary) fields G_{ab} also include G_{ij}, which have been introduced to put $\frac{1}{4}F_{ij}{}^2$ into first-order form and thus make the lagrangian manifestly Lorentz covariant. Eliminating G_{ij} by their field equations gives back the hamiltonian form.

Many examples will be given in chapts. 5-6 for relativistic first-quantization, where $\mathcal{H}_0$ vanishes, and thus the Schrödinger equation implies the wave function is proper-time-independent (i.e., we require $\mathcal{H}_0 = 0$ because the proper time is not physically observable). Here we give an interesting example in D=2 which will also be useful for strings. Consider a single field A with canonical momentum P and choose

$$i\mathcal{G} = \tfrac{1}{4}(P + A')^2 \quad , \quad \mathcal{H}_0 = \tfrac{1}{4}(P - A')^2 \quad , \tag{3.1.14}$$

where $'$ is the derivative with respect to the 1 space coordinate (which acts as the index M or i from above). From the algebra of $P \pm A'$, it's easy to check, at least at the Poisson bracket level, that the $\mathcal{G}$ algebra closes and $\mathcal{H}_0$ is invariant. (This algebra, with particular boundary conditions, will be important in string theory:

See chapt. 8. Note that $P + A'$ does not form an algebra, so its square must be used.) The transformation laws (3.1.11) are found to be

$$\delta A = \zeta \tfrac{1}{2}(P + A') \quad , \quad \delta\lambda = \dot{\zeta} - \lambda \overleftrightarrow{\partial_1} \zeta \quad . \tag{3.1.15}$$

In the gauge $\lambda = 1$ the action becomes the usual hamiltonian one for a massless scalar, but the constraint implies $P + A' = 0$, which means that modes propagate only to the right and not the left. The lagrangian form again results from eliminating P, and after the redefinitions

$$\hat{\lambda} = 2\frac{1-\lambda}{1+\lambda} \quad , \quad \hat{\zeta} = \sqrt{2}\frac{1}{1+\lambda}\zeta \quad , \tag{3.1.16}$$

we find [3.2]

$$\mathcal{L} = -(\partial_+ A)(\partial_- A) + \tfrac{1}{2}\hat{\lambda}(\partial_- A)^2 \quad ;$$

$$\delta A = \hat{\zeta}\partial_- A \quad , \quad \delta\hat{\lambda} = 2\partial_+\hat{\zeta} + \hat{\zeta}\overleftrightarrow{\partial}_-\hat{\lambda} \quad ; \tag{3.1.17}$$

where $\partial_\pm$ are defined as in sect. 2.1.

The gauge fixing (including Faddeev-Popov ghosts) and initial condition can be described in a very concise way by the BRST method. The basic idea is to construct a symmetry relating the Faddeev-Popov ghosts to the unphysical modes of the gauge field. For example, in Yang-Mills only $D-2$ Lorentz components of the gauge field are physical, so the Lorentz-gauge D-component gauge field requires 2 Faddeev-Popov ghosts while the temporal-gauge $D-1$-component field requires only 1. The BRST symmetry rotates the additional gauge-field components into the FP ghosts, and vice versa. Since the FP ghosts are anticommuting, the generator of this symmetry must be, also.

3.2. IGL(1)

We will find that the methods of Becchi, Rouet, Stora, and Tyutin [3.3] are the most useful way not only to perform quantization in Lorentz-covariant and general nonunitary gauges, but also to derive gauge-invariant theories. BRST quantization is a more general way of quantizing gauge theories than either canonical or path-integral (Faddeev-Popov), because it (1) allows more general gauges, (2) gives the Slavnov-Taylor identities (conditions for unitarity) directly (they're just the Ward identities for BRST invariance), and (3) can separate the gauge-invariant part of a gauge-fixed action. It is defined by the conditions: (1) BRST transformations form a global group with a single (abelian) anticommuting generator Q. The group property then implies $Q^2 = 0$ for closure. (2) Q acts on physical fields as a gauge transformation with the gauge parameter replaced by the (real) ghost. (3) Q on the (real) antighost gives a BRST auxiliary field (necessary for closure of the algebra off shell). Nilpotence of Q then implies that the auxiliary field is BRST invariant. Physical states are defined to be those which are BRST invariant (modulo null states, which can be expressed as Q on something) and have vanishing ghost number (the number of ghosts minus antighosts).

There are two types of BRST formalisms: (1) first-quantized-style BRST, originally found in string theory [3.4] but also applicable to ordinary field theory, which contains all the field equations as well as the gauge transformations; and (2) second-quantized-style BRST, the original form of BRST, which contains only the gauge transformations, corresponding in a hamiltonian formalism to those field equations (constraints) found from varying the time components of the gauge fields. However, we'll find (in sect. 4.4) that, after restriction to a certain subset of the fields, BRST1 is equivalent to BRST2. (It's the BRST variation of the additional fields of BRST1 that leads to the field equations for the physical fields.) The BRST2 transformations were originally found from Yang-Mills theory. We will first derive the YM BRST2 transformations, and by a simple generalization find BRST operators for arbitrary theories, applicable to BRST1 or BRST2 and to lagrangian or hamiltonian formalisms.

In the general case, there are two forms for the BRST operators, corresponding to different classes of gauges. The gauges commonly used in field theory fall into three classes: (1) unitary (Coulomb, Arnowitt-Fickler/axial, light-cone) gauges, where the ghosts are nonpropagating, and the constraints are solved explicitly (since they contain no time derivatives); (2) temporal/timelike gauges, where the ghosts have equations of motion first-order in time derivatives (making them canonically conjugate to the antighosts); and (3) Lorentz (Landau, Fermi-Feynman) gauges, where the ghost equations are second-order (so ghosts are independent of antighosts), and the Nakanishi-Lautrup auxiliary fields [3.5] (Lagrange multipliers for the gauge conditions) are canonically conjugate to the auxiliary time-components of the gauge fields. Unitary gauges have only physical polarizations; temporal gauges have an additional pair of unphysical polarizations of opposite statistics for each gauge generator; Lorentz gauges have two pairs. In unitary gauges the BRST operator vanishes identically; in temporal gauges it is constructed from group generators, or constraints, multiplied by the corresponding ghosts, plus terms for nilpotence; in Lorentz gauges it has an extra "abelian" term consisting of the products of the second set of unphysical fields. Temporal-gauge BRST is defined in terms of a ghost number operator in addition to the BRST operator, which itself has ghost number 1. We therefore refer to this formalism by the corresponding symmetry group with two generators, IGL(1). Lorentz-gauge BRST has also an antiBRST operator [3.6], and this and BRST transform as an "isospin" doublet, giving the larger group ISp(2), which can be extended further to OSp(1,1|2) [2.3,3.7]. Although the BRST2 OSp operators are generally of little value (only the IGL is required for quantization), the BRST1 OSp gives a powerful method for obtaining free gauge-invariant formalisms for arbitrary (particle or string) field theories. In particular, for arbitrary representations of the Poincaré group a certain OSp(1,1|2) can be extended to IOSp(D,2|2) [2.3], which is derived from (but does not directly correspond to quantization in) the light-cone gauge.

One simple way to formulate anticommuting symmetries (such as supersymmetry) is through the use of anticommuting coordinates [3.8]. We therefore extend spacetime to include one extra, anticommuting coordinate, corresponding to the one anticommuting symmetry:

$$a \rightarrow (a, \alpha) \tag{3.2.1}$$

for all vector indices, including those on coordinates, with Fermi statistics for all quantities with an odd number of anticommuting indices. (α takes only one value.) Covariant derivatives and gauge transformations are then defined by the corresponding generalization of (2.1.5b), and field strengths with graded commutators (commutators or anticommutators, according to the statistics). However, unlike supersymmetry, the extra coordinate does not represent extra physical degrees of freedom, and so we constrain all field strengths with anticommuting indices to vanish [3.9]: For Yang-Mills,

$$F_{\alpha a} = F_{\alpha \beta} = 0 \quad , \tag{3.2.2a}$$

so that gauge-invariant quantities can be constructed only from the usual F_{ab}. When Yang-Mills is coupled to matter fields ϕ, we similarly have the constraints

$$\nabla_\alpha \phi = \nabla_\alpha \nabla_a \phi = 0 \quad , \tag{3.2.2b}$$

and these in fact imply (3.2.2a) (consider $\{\nabla_\alpha, \nabla_\beta\}$ and $[\nabla_\alpha, \nabla_a]$ acting on ϕ). These constraints can be solved easily:

$$F_{\alpha a} = 0 \quad \rightarrow \quad p_\alpha A_a = [\nabla_a, A_\alpha] \quad ,$$
$$F_{\alpha \beta} = 0 \quad \rightarrow \quad p_\alpha A_\beta = -\tfrac{1}{2}\{A_\alpha, A_\beta\} = -A_\alpha A_\beta \quad ;$$
$$\nabla_\alpha \phi = 0 \quad \rightarrow \quad p_\alpha \phi = -A_\alpha \phi \quad . \tag{3.2.3}$$

(In the second line we have used the fact that α takes only one value.) Defining " $|$ " to mean $|_{x^\alpha=0}$, we now interpret $A_\alpha|$ as the usual gauge field, $iA_\alpha|$ as the FP ghost, and the BRST operator Q as $Q(\psi|) = (p_\alpha\psi)|$. (Similarly, $\phi|$ is the usual matter field.) Then $\partial_\alpha\partial_\beta = 0$ (since α takes only one value and ∂_α is anticommuting) implies nilpotence

$$Q^2 = 0 \quad . \tag{3.2.4}$$

In a hamiltonian approach [3.10] these transformations are sufficient to perform quantization in a temporal gauge, but for the lagrangian approach or Lorentz gauges we also need the FP antighost and Nakanishi-Lautrup auxiliary field, which we define in terms of an unconstrained scalar field $\widetilde{A}$: $\widetilde{A}|$ is the antighost, and

$$B = (p_\alpha i\widetilde{A})| \tag{3.2.5}$$

is the auxiliary field.

The BRST transformations (3.2.3) can be represented in operator form as

$$Q = C^i\mathcal{G}_i + \tfrac{1}{2}C^jC^if_{ij}{}^k\frac{\partial}{\partial C^k} - iB^i\frac{\partial}{\partial \widetilde{C}^i} \quad , \tag{3.2.6a}$$

where i is a combined space(time)/internal-symmetry index, C is the FP ghost, $\widetilde{C}$ is the FP antighost, B is the NL auxiliary field, and the action on the physical fields is given by the constraint/gauge-transformation $\mathcal{G}$ satisfying the algebra

$$[\mathcal{G}_i, \mathcal{G}_j\} = f_{ij}{}^k\mathcal{G}_k \quad , \tag{3.2.6b}$$

where we have generalized to graded algebras with graded commutator $[\ ,\ \}$ (commutator or anticommutator, as appropriate). In this case,

$$\mathcal{G} = \left[\nabla, \cdot i\frac{\delta}{\delta A}\right] \quad , \tag{3.2.7}$$

where the structure constants in (3.2.6b) are the usual group structure constants times δ-functions in the coordinates. Q of (3.2.6a) is antihermitian when C, $\widetilde{C}$, and B are hermitian and $\mathcal{G}$ is antihermitian, and is nilpotent (3.2.4) as a consequence of (3.2.6b). Since $\widetilde{C}$ and B appear only in the last term in (3.2.6a), these properties also hold if that term is dropped. (In the notation of (3.2.1-5), the fields A and $\widetilde{A}$ are independent.)

When $[\mathcal{G}_i, f_{jk}{}^l\} \neq 0$, (3.2.6a) still gives $Q^2 = 0$. However, when the gauge invariance has a gauge invariance of its own, i.e., $\Lambda^i\mathcal{G}_i = 0$ for some nontrivial Λ depending on the physical variables implicit in $\mathcal{G}$, then, although (3.2.6a) is still nilpotent, it requires extra terms in order to allow gauge fixing this invariance of the ghosts. In some cases (see sect. 5.4) this requires an infinite number of new terms (and ghosts). In general, the procedure of adding in the additional ghosts and invariances can be tedious, but in sect. 3.4 we'll find a method which automatically gives them all at once.

The gauge-fixed action is required to be BRST-invariant. The gauge-invariant part already is, since Q on physical fields is a special case of a gauge transformation. The gauge-invariant lagrangian is quantized by adding terms which are Q on something (corresponding to integration over x^α), and thus BRST-invariant (since $Q^2 = 0$): For example, rewriting (3.2.3,5) in the present notation,

$$\begin{aligned} QA_a &= -i[\nabla_a, C] \quad , \\ QC &= iC^2 \quad , \\ Q\widetilde{C} &= -iB \quad , \\ QB &= 0 \quad , \end{aligned} \qquad (3.2.8)$$

we can choose

$$\mathcal{L}_{GF} = iQ\left\{\widetilde{C}\,[f(A) + g(B)]\right\} = B\,[f(A) + g(B)] - \widetilde{C}\frac{\partial f}{\partial A_a}[\nabla_a, C] \quad , \qquad (3.2.9)$$

which gives the usual FP term for gauge condition $f(A) = 0$ with gauge-averaging function $Bg(B)$. However, gauges more general than FP can be obtained by putting more complicated ghost-dependence into the function on which Q acts, giving terms more than quadratic in ghosts. In the temporal gauge

$$f(A) = A_0 \qquad (3.2.10)$$

and g contains no time derivatives in (3.2.9), so upon quantization B is eliminated (it's nonpropagating) and $\widetilde{C}$ is canonically conjugate to C. Thus, in the hamiltonian formalism (3.2.6a) gives the correct BRST transformations without the last term, where the fields are now functions of just space and not time, the sum in (3.2.7) runs over just the spatial values of the spacetime index as in (3.1.8), and the derivatives correspond to functional derivatives which give δ functions in just spatial coordinates. On the other hand, in Lorentz gauges the ghost and antighost are independent even after quantization, and the last term in Q is needed in both lagrangian and hamiltonian formalisms; but the product in (3.2.7) and the arguments of the fields and δ functions are as in the temporal gauge. Therefore, in the lagrangian approach Q is gauge independent, while in the hamiltonian approach the only gauge dependence is the set of unphysical fields, and thus the last term in Q. Specifically, for Lorentz gauges we choose

$$f(A) = \partial \cdot A \quad , \quad g(B) = \tfrac{1}{2}\zeta B \quad \rightarrow$$

$$\begin{aligned} \mathcal{L}_{GF} &= \zeta \tfrac{1}{2} B^2 + B\partial \cdot A - \widetilde{C}\partial \cdot [\nabla, C] \\ &= -\frac{1}{\zeta}\tfrac{1}{2}(\partial \cdot A)^2 + \zeta \tfrac{1}{2}\widetilde{B}^2 - \widetilde{C}\partial \cdot [\nabla, C] \quad , \end{aligned}$$

$$\widetilde{B} = B + \frac{1}{\zeta}\partial \cdot A \quad , \tag{3.2.11}$$

using (3.2.9).

The main result is that (3.2.6a) gives a general BRST operator for arbitrary algebras (3.2.6b), for hamiltonian or lagrangian formalisms, for arbitrary gauges (including temporal and Lorentz), where the last term contains arbitrary numbers (perhaps 0) of sets of $(\widetilde{C}, B)$ fields. Since $\mathcal{G} = 0$ is the field equation (3.1.6), physical states must satisfy $Q\psi = 0$. Actually, $\mathcal{G} = 0$ is satisfied only as a Gupta-Bleuler condition, but still $Q\psi = 0$ because in the $C^i\mathcal{G}_i$ term in (3.2.6a) positive-energy parts of C^i multiply negative-energy parts of $\mathcal{G}_i$, and vice versa. Thus, for any value of an appropriate index i, either $C^i|\psi\rangle = \langle\psi|\mathcal{G}_i = 0$ or $\mathcal{G}_i|\psi\rangle = \langle\psi|C^i = 0$, modulo contributions from the $C^2\partial/\partial C$ term. However, since $\mathcal{G}$ is also the generator of gauge transformations (3.1.8), any state of the form $\psi + Q\lambda$ is equivalent to ψ. The physical states are therefore said to belong to the "cohomology" of Q: those satisfying $Q\psi = 0$ modulo gauge transformations $\delta\psi = Q\lambda$. ("Physical" has a more restrictive meaning in BRST1 than BRST2: In BRST2 the physical states are just the gauge-invariant ones, while in BRST1 they must also be on shell.) In addition, physical states must have a specified value of the ghost number, defined by the ghost number operator

$$J^3 = C^i \frac{\partial}{\partial C^i} - \widetilde{C}^i \frac{\partial}{\partial \widetilde{C}^i} \quad , \tag{3.2.12a}$$

where

$$[J^3, Q] = Q \quad , \tag{3.2.12b}$$

and the latter term in (3.2.12a) is dropped if the last term in (3.2.6a) is. The two operators Q and J^3 form the algebra IGL(1), which can be interpreted as a translation and scale transformation, respectively, with respect to the coordinate x^α (i.e., the conformal group in 1 anticommuting dimension).

From the gauge generators $\mathcal{G}_i$, which act on only the physical variables, we can define IGL(1)-invariant generalizations which transform also C, as the adjoint representation:

$$\widehat{\mathcal{G}}_i = \left\{ Q, \frac{\partial}{\partial C^i} \right\} = \mathcal{G}_i + C^j f_{ji}{}^k \frac{\partial}{\partial C^k} \quad . \tag{3.2.13}$$

The $\widehat{\mathcal{G}}$'s are gauge-fixed versions of the gauge generators $\mathcal{G}$.

Types of gauges for first-quantized theories will be discussed in chapt. 5 for particles and chapt. 6 and sect. 8.3 for strings. Gauge fixing for general field theories using BRST will be described in sect. 4.4, and for closed string field theory in sect. 11.1. IGL(1) algebras will be used for deriving general gauge-invariant free actions in sect. 4.2. The algebra will be derived from first-quantization for the particle in sect. 5.2 and for the string in sect. 8.1. However, in the next section we'll find that IGL(1) can always be derived as a subgroup of OSp(1,1|2), which can be derived in a more general way than by first-quantization.

3.3. OSp(1,1|2)

Although the IGL(1) algebra is sufficient for quantization in arbitrary gauges, in the following section we will find the larger OSp(1,1|2) algebra useful for the BRST1 formalism, so we give a derivation here for BRST2 and again generalize to arbitrary BRST. The basic idea is to introduce a second BRST, "antiBRST," corresponding to the antighost. We therefore repeat the procedure of (3.2.1-7) with 2 anticommuting coordinates [3.11] by simply letting the index α run over 2 values (cf. sect. 2.6). The solution to (3.2.2) is now

$$\begin{aligned} F_{\alpha a} = 0 \quad &\rightarrow \quad p_\alpha A_a = [\nabla_a, A_\alpha] \quad , \\ F_{\alpha\beta} = 0 \quad &\rightarrow \quad p_\alpha A_\beta = -\tfrac{1}{2}\{A_\alpha, A_\beta\} - iC_{\alpha\beta}B \quad ; \\ \nabla_\alpha \phi = 0 \quad &\rightarrow \quad p_\alpha \phi = -A_\alpha \phi \quad ; \end{aligned} \tag{3.3.1a}$$

where A_α now includes both ghost and antighost. The appearance of the NL field is due to the ambiguity in the constraint $F_{\alpha\beta} = p_{(\alpha}A_{\beta)} + \cdots$. The remaining (anti)BRST transformation then follows from further differentiation:

$$\{p_\alpha, p_\beta\}A_\gamma = 0 \quad \rightarrow \quad p_\alpha B = -\tfrac{1}{2}[A_\alpha, B] + i\tfrac{1}{12}\left[A^\beta, \{A_\alpha, A_\beta\}\right] \quad . \tag{3.3.1b}$$

The generalization of (3.2.6a) is then [3.12], defining $Q^\alpha(\psi|) = (\partial^\alpha\psi)|$ (and renaming $C^\alpha = A^\alpha$),

$$Q^\alpha = C^{i\alpha}\mathcal{G}_i + \tfrac{1}{2}C^{j\alpha}C^{i\beta}f_{ij}{}^k\frac{\partial}{\partial C^{k\beta}} - B^i\frac{\partial}{\partial C^i{}_\alpha} + \tfrac{1}{2}C^{j\alpha}B^i f_{ij}{}^k\frac{\partial}{\partial B^k}$$

$$-\tfrac{1}{12}C^{k\beta}C^{j\alpha}C^{i}{}_{\beta}f_{ij}{}^{l}f_{lk}{}^{m}\frac{\partial}{\partial B^m} \quad , \tag{3.3.2}$$

and of (3.2.12a) is

$$J_{\alpha\beta} = C^{i}{}_{(\alpha}\frac{\partial}{\partial C^{i\beta)}} \, , \tag{3.3.3}$$

where ( ) means index symmetrization. These operators form an ISp(2) algebra consisting of the translations Q_α and rotations $J_{\alpha\beta}$ on the coordinates x^α:

$$\{Q_\alpha, Q_\beta\} = 0 \quad ,$$

$$[J_{\alpha\beta}, Q_\gamma] = -C_{\gamma(\alpha}Q_{\beta)} \quad , \quad [J_{\alpha\beta}, J_{\gamma\delta}] = -C_{(\gamma(\alpha}J_{\beta)\delta)} \quad . \tag{3.3.4}$$

In order to relate to the IGL(1) formalism, we write

$$Q^\alpha = (Q, \tilde{Q}) \quad , \quad C^\alpha = (C, \tilde{C}) \quad , \quad J^{\alpha\beta} = \begin{pmatrix} J^+ & -iJ^3 \\ -iJ^3 & J^- \end{pmatrix} \quad , \tag{3.3.5}$$

and make the unitary transformation

$$ln\ U = \tfrac{1}{2}C^j\tilde{C}^i f_{ij}{}^k i\frac{\partial}{\partial B^k} \quad . \tag{3.3.6}$$

Then UQU^{-1} is Q of (3.2.6a) and $UJ^3U^{-1} = J^3$ is J^3 of (3.2.12a). However, whereas there is an arbitrariness in the IGL(1) algebra in redefining J^3 by a constant, there is no such ambiguity in the OSp(1,1|2) algebra (since it is "simple").

Unlike the IGL case, the NL fields now are an essential part of the algebra. Consequently, the algebra can be enlarged to OSp(1,1|2) [3.7]:

$$J_{-\alpha} = Q_\alpha \quad , \quad J_{+\alpha} = 2C^{i}{}_{\alpha}\frac{\partial}{\partial B^i} \quad ,$$

$$J_{\alpha\beta} = C^{i}{}_{(\alpha}\frac{\partial}{\partial C^{i\beta)}} \quad , \quad J_{-+} = 2B^i\frac{\partial}{\partial B^i} + C^{i\alpha}\frac{\partial}{\partial C^{i\alpha}} \quad , \tag{3.3.7}$$

with Q_α as in (3.3.2), satisfy

$$[J_{\alpha\beta}, J_{\gamma\delta}] = -C_{(\gamma(\alpha}J_{\beta)\delta)} \quad ,$$

$$[J_{\alpha\beta}, J_{\pm\gamma}] = -C_{\gamma(\alpha}J_{\pm\beta)} \, ,$$

$$\{J_{-\alpha}, J_{+\beta}\} = -C_{\alpha\beta}J_{-+} - J_{\alpha\beta} \quad ,$$

$$[J_{-+}, J_{\pm\alpha}] = \mp J_{\pm\alpha} \quad ,$$

$$rest = 0 \quad . \tag{3.3.8}$$

This group is the conformal group for x^α, with the ISp(2) subgroup being the corresponding Poincaré (or Euclidean) subgroup:

$$J_{-\alpha} = -\partial_\alpha \quad , \quad J_{+\alpha} = 2x_\beta{}^2\partial_\alpha + x^\beta M_{\beta\alpha} + x_\alpha d$$
$$J_{\alpha\beta} = x_{(\alpha}\partial_{\beta)} + M_{\alpha\beta} \quad , \quad J_{-+} = -x^\alpha\partial_\alpha + d \quad . \tag{3.3.9}$$

(We define the square of an Sp(2) spinor as $(x_\alpha)^2 \equiv \frac{1}{2}x^\alpha x_\alpha$.) $J_{-\alpha}$ are the translations, $J_{\alpha\beta}$ the Lorentz transformations (rotations), J_{-+} the dilatations, and $J_{+\alpha}$ the conformal boosts. As a result of constraints analogous to (3.3.1a), the translations are realized nonlinearly in (3.3.7) instead of the boosts. This should be compared with the usual conformal group (2.2.4). The action of the generators (3.3.9) have been chosen to have the opposite sign of those of (3.3.8), since it is a coordinate representation instead of a field representation (see sect. 2.2). In later sections we will actually be applying (3.3.7) to coordinates, and hence (3.3.9) should be considered a "zeroth-quantized" formalism.

From the gauge generators $\mathcal{G}_i$, we can define OSp(1,1|2)-invariant generalizations which transform also C and B, as adjoint representations:

$$\widehat{\mathcal{G}}_i = \tfrac{1}{2}\left\{J_-{}^\alpha, \left[J_{-\alpha}, \frac{\partial}{\partial B^i}\right]\right\} = \mathcal{G}_i + C^{j\alpha} f_{ji}{}^k \frac{\partial}{\partial C^{k\alpha}} + B^j f_{ji}{}^k \frac{\partial}{\partial B^k} \quad . \tag{3.3.10}$$

The $\widehat{\mathcal{G}}$'s are the OSp(1,1|2) generalization of the operators (3.2.13).

The OSp(1,1|2) algebra (3.3.7) can be extended to an inhomogeneous algebra IOSp(1,1|2) when one of the generators, which we denote by $\mathcal{G}_0$, is distinguished [3.13]. We then define

$$p_+ = \sqrt{-2i\frac{\partial}{\partial B^0}} \quad ,$$
$$p_\alpha = \frac{1}{p_+} i\left(\frac{\partial}{\partial C^{0\alpha}} + \tfrac{1}{2}C^i{}_\alpha f_{i0}{}^j \frac{\partial}{\partial B^j}\right) \quad ,$$
$$p_- = -\frac{1}{p_+}\left(i\widehat{\mathcal{G}}_0 + p_\alpha{}^2\right) \quad . \tag{3.3.11}$$

(The i indices still include the value 0.) $\widehat{\mathcal{G}}_0$ is then the IOSp(1,1|2) invariant $i\frac{1}{2}(2p_+p_- + p^\alpha p_\alpha)$. This algebra is useful for constructing gauge field theory for closed strings.

OSp(1,1|2) will play a central role in the following chapters: In chapt. 4 it will be used to derive free gauge-invariant actions. A more general form will be derived in the following sections, but the methods of this section will also be used in sect. 8.3 to describe Lorentz-gauge quantization of the string.

3.4. From the light cone

In this section we will derive a general OSp(1,1|2) algebra from the light-cone Poincaré algebra of sect. 2.3, using concepts developed in sect. 2.6. We'll use this general OSp(1,1|2) to derive a general IGL(1), and show how IGL(1) can be extended to include interactions.

The IGL(1) and OSp(1,1|2) algebras of the previous section can be constructed from an arbitrary algebra $\mathcal{G}$, whether first-quantized or second-quantized, and lagrangian or hamiltonian. That already gives 8 different types of BRST formalisms. Furthermore, arbitrary gauges, more general than those obtained by the FP method, and graded algebras (where some of the $\mathcal{G}$'s are anticommuting, as in supersymmetry) can be treated. However, there is a ninth BRST formalism, similar to the BRST1 OSp(1,1|2) hamiltonian formalism, which starts from an IOSp(D,2|2) algebra [2.3] which contains the OSp(1,1|2) as a subgroup. This approach is unique in that, rather than starting from the gauge covariant formalism to derive the BRST algebra, it starts from just the usual Poincaré algebra and derives both the gauge covariant formalism and BRST algebra. In this section, instead of deriving BRST1 from first-quantization, we will describe this special form of BRST1, and give the OSp(1,1|2) subalgebra of which special cases will be found in the following chapters.

The basic idea of the IOSp formalism is to start from the light-cone formalism of the theory with its nonlinear realization of the usual Poincaré group ISO(D-1,1) (with manifest subgroup ISO(D-2)), extend this group to IOSp(D,2|2) (with manifest IOSp(D-1,1|2)) by adding 2 commuting and 2 anticommuting coordinates, and take the ISO(D-1,1)⊗OSp(1,1|2) subgroup, where this ISO(D-1,1) is now manifest and the nonlinear OSp(1,1|2) is interpreted as BRST. Since the BRST operators of BRST1 contain all the field equations, the gauge-invariant action can be derived. Thus, not only can the light-cone formalism be derived from the gauge-invariant formalism, but the converse is also true. Furthermore, for general field theories the light-cone formalism (at least for the free theory) is easier to derive (although more awkward to use), and the IOSp method therefore provides a convenient method to derive the gauge covariant formalism.

We now perform dimensional continuation as in sect. 2.6, but set $x_+ = 0$ as in sect. 2.3. Our fields are now functions of (x_a, x_α, x_-), and have indices corresponding to representations of the spin subgroup OSp(D−1,1|2) in the massless case or OSp(D,1|2) in the massive. Of the full group IOSp(D,2|2) (obtained from extending (2.3.5)) we are now only interested in the subgroup ISO(D−1,1)⊗OSp(1,1|2). The former factor is the usual Poincaré group, acting only in the physical spacetime directions:

$$p_a = i\partial_a \quad , \quad J_{ab} = -ix_{[a}p_{b]} + M_{ab} \quad . \tag{3.4.1}$$

The latter factor is identified as the BRST group, acting in only the unphysical directions:

$$J_{\alpha\beta} = -ix_{(\alpha}p_{\beta)} + M_{\alpha\beta} \quad , \quad J_{-+} = -ix_-p_+ + k \quad , \quad J_{+\alpha} = ix_\alpha p_+ \quad ,$$

$$J_{-\alpha} = -ix_-p_\alpha + \frac{1}{p_+}\left[-ix_\alpha \tfrac{1}{2}(p^bp_b + M^2 + p^\beta p_\beta) + M_\alpha{}^\beta p_\beta + kp_\alpha + \mathcal{Q}_\alpha\right] \quad ,$$

$$\{\mathcal{Q}_\alpha, \mathcal{Q}_\beta\} = -M_{\alpha\beta}(p^ap_a + M^2) \quad ; \tag{3.4.2a}$$

$$\mathcal{Q}_\alpha = M_\alpha{}^bp_b + M_{\alpha m}M \quad . \tag{3.4.2b}$$

We'll generally set $k = 0$.

In order to relate to the BRST1 IGL formalism obtained from ordinary first-quantization (and discussed in the following chapters for the particle and string), we perform an analysis similar to that of (3.3.5,6): Making the (almost) unitary transformation [2.3]

$$ln\ U = (ln\ p_+)\left(c\frac{\partial}{\partial c} + M^3\right) \quad , \tag{3.4.3a}$$

where $x^\alpha = (c, \tilde{c})$, $M^{a\alpha} = (M^{+a}, M^{-a})$, and $M^\alpha{}_m = (M^+{}_m, M^-{}_m)$, we get

$$Q \quad \rightarrow \quad -ic\tfrac{1}{2}(p_a{}^2 + M^2) + M^+i\frac{\partial}{\partial c} + (M^{+a}p_a + M^+{}_mM) + x_-i\frac{\partial}{\partial \tilde{c}} \quad ,$$

$$J^3 = c\frac{\partial}{\partial c} + M^3 - \tilde{c}\frac{\partial}{\partial \tilde{c}} \quad . \tag{3.4.3b}$$

(Cf. (3.2.6a,12a).) As in sect. 3.2, the extra terms in x_- and $\tilde{c}$ (analogous to B^i and $\widetilde{C}^i$) can be dropped in the IGL(1) formalism. After dropping such terms, $J^{3\dagger} = 1 - J^3$. (Or we can subtract $\frac{1}{2}$ to make it simply antihermitian. However, we prefer not to, so that physical states will still have vanishing ghost number.)

Since p_+ is a momentum, this redefinition has a funny effect on reality (but not hermiticity) properties: In particular, c is now a momentum rather than a coordinate (because it has been scaled by p_+, maintaining its hermiticity but making it imaginary in coordinate space). However, we will avoid changing notation or Fourier transforming the fields, in order to simplify comparison to the OSp(1,1|2) formalism. The effect of (3.4.3a) on a field satisfying $\Phi = \Omega\Phi^*$ is that it now satisfies

$$\Phi = (-1)^{c\partial/\partial c + M^3}\Omega\Phi^* \tag{3.4.4}$$

due to the i in $p_+ = i\partial_+$.

These results can be extended to interacting field theory, and we use Yang-Mills as an example [2.3]. Lorentz-covariantizing the light-cone result (2.1.7,2.4.11), we find

$$p_- = \int -\tfrac{1}{4}F^{ij}F_{ji} + \tfrac{1}{2}(p_+A_-)^2 \quad ,$$

$$J_{-a} = \int ix_-(p_a A^i)(p_+A_i) + ix_a\left[-\tfrac{1}{4}F^{ij}F_{ji} + \tfrac{1}{2}(p_+A_-)^2\right] - A_a p_+ A_- \quad ,$$

$$A_- = -\frac{1}{p_+{}^2}[\nabla^i, p_+A_i\} \quad . \tag{3.4.5}$$

When working in the IGL(1) formalism, it's extremely useful to introduce a Lorentz covariant type of second-quantized bracket [3.14]. This bracket can be postulated independently, or derived by covariantization of the light-cone canonical commutator, plus truncation of the $\tilde{c}, x_+$, and x_- coordinates. The latter derivation will prove useful for the derivation of IGL(1) from OSp(1,1|2). Upon covariantization of the canonical light-cone commutator (2.4.5), the arguments of the fields and of the δ-function on the right-hand side are extended accordingly. We now have to truncate. The truncation of x_+ is automatic: Since the original commutator was an equal-time one, there is no x_+ δ-function on the right-hand side, and it therefore suffices to delete the x_+ arguments of the fields. At this stage, in addition to x_- dependence, the fields depend on both c and $\tilde{c}$ and the right-hand side contains both δ-functions. (This commutator may be useful for OSp approaches to field theory.) We now wish to eliminate the $\tilde{c}$ dependence. This cannot be done by straightforward truncation, since expansion of the field in this anticommuting coordinate shows that one cannot eliminate consistently the fields in the $\tilde{c}$ sector. We therefore proceed formally and just delete the $\tilde{c}$ argument from the fields and the corresponding δ-function, obtaining

$$[\Phi^\dagger(1), \Phi(2)]_c = -\frac{1}{2p_{+2}}\delta(x_{2-} - x_{1-})\delta^D(x_2 - x_1)\delta(c_2 - c_1) \quad , \tag{3.4.6}$$

which is a bracket with unusual statistics because of the anticommuting δ-function on the right-hand side. The transformation (3.4.3a) is performed next; its nonunitarity causes the p_+ dependence of (3.4.6) to disappear, enabling one to delete the x_- argument from the fields and the corresponding δ-function to find (using $(c\partial/\partial c)c = c$)

$$[\Phi^\dagger(1), \Phi(2)]_c = -\tfrac{1}{2}\delta^D(x_2 - x_1)\delta(c_2 - c_1) \quad . \tag{3.4.7}$$

This is the covariant bracket. The arguments of the fields are (x_a, c), namely, the usual D bosonic coordinates of covariant theories and the single anticommuting coordinate of the IGL(1) formalism. The corresponding δ-functions appear on the right-hand side. (3.4.6,7) are defined for commuting (scalar) fields, but generalize straightforwardly: For example, for Yang-Mills, where A_i includes both commuting (A_a) and anticommuting (A_α) fields, $[A_i{}^\dagger, A_j]$ has an extra factor of η_{ij}. It might be possible to define the bracket by a commutator $[A,B]_c = A * B - B * A$. Classically it can be defined by a Poisson bracket:

$$[A^\dagger, B]_c = \tfrac{1}{2}\int dz \left(\frac{\delta}{\delta\Phi(z)}A\right)^\dagger \left(\frac{\delta}{\delta\Phi(z)}B\right) \quad , \tag{3.4.8}$$

where z are all the coordinates of Φ (in this case, x_a and c). For $A = B = \Phi$, the result of equation (3.4.7) is reproduced. The above equation implies that the bracket is a derivation:

$$[A, BC]_c = [A,B]_c C + (-1)^{(A+1)B} B[A,C]_c \quad , \tag{3.4.9}$$

where the A's and B's in the exponent of the (-1) are 0 if the corresponding quantity is bosonic and 1 if it's fermionic. This differs from the usual graded Leibnitz rule by a $(-1)^B$ due to the anticommutativity of the dz in the front of (3.4.8), which also gives the bracket the opposite of the usual statistics: We can write $(-1)^{[A,B]_c} = (-1)^{AB+1}$ to indicate that the bracket of 2 bosonic operators is fermionic, etc., a direct consequence of the anticommutativity of the total δ-function in (3.4.7). One can also verify that this bracket satisfies the other properties of a (generalized) Lie bracket:

$$[A,B]_c = (-1)^{AB}[B,A]_c \quad ,$$

$$(-1)^{(A+1)C}[A,[B,C]_c]_c + (-1)^{(B+1)A}[B,[C,A]_c]_c + (-1)^{(C+1)B}[C,[A,B]_c]_c = 0 \quad . \tag{3.4.10}$$

Thus the bracket has the opposite of the usual graded symmetry, being antisymmetric for objects of odd statistics and symmetric otherwise. This property follows from the hermiticity condition (3.4.4): $(-1)^{c\partial/\partial c}$ gives $(-1)^{-(\partial/\partial c)c} = -(-1)^{c\partial/\partial c}$ upon integration by parts, which gives the effect of using an antisymmetric metric. The Jacobi identity has the same extra signs as in (3.4.9). These properties are sufficient to perform the manipulations analogous to those used in the light cone.

Before applying this bracket, we make some general considerations concerning the derivation of interacting IGL(1) from OSp(1,1|2). We start with the original untransformed generators J^3 and $J_-{}^c = Q$. The first step is to restrict our attention to just the fields at $\tilde{c} = 0$. Killing all the fields at linear order in $\tilde{c}$ is consistent with the transformation laws, since the transformations of the latter fields include no terms which involve only $\tilde{c} = 0$ fields. Since the linear-in-$\tilde{c}$ fields are canonically conjugate to the $\tilde{c} = 0$ fields, the only terms in the generators which could spoil this property would themselves have to depend on only $\tilde{c} = 0$ fields, which, because of the $d\tilde{c}$ $(= \partial/\partial\tilde{c})$ integration, would require explicit $\tilde{c}$-dependence. However, from (2.4.15), since $\eta^{cc} = C^{cc} = 0$, we see that $J_-{}^c$ anticommutes with p^c, and thus has no explicit $\tilde{c}$-dependence at either the free or interacting levels. (The only explicit coordinate dependence in Q is from a c term.)

The procedure of restricting to $\tilde{c} = 0$ fields can then be implemented very simply by dropping all $p^c(= -\partial/\partial\tilde{c})$'s in the generators. As a consequence, we also lose all explicit x_- terms in Q. (This follows from $[J_-{}^c, p_+] = -p^c$.) Since $\tilde{c}$ and $\partial/\partial\tilde{c}$ now occur nowhere explicitly, we can also kill all implicit dependence on $\tilde{c}$: All fields are evaluated at $\tilde{c} = 0$, the $d\tilde{c}$ is removed from the integral in the generators, and the $\delta(\tilde{c}_2 - \tilde{c}_1)$ is removed from the canonical commutator, producing (3.4.6). In the case of Yang-Mills fields $A^i = (A^a, A^\alpha) = (A^a, A^c, A^{\tilde{c}})$, the BRST generator at this point is given by

$$Q = \int ic[-\tfrac{1}{4}F^{ij}F_{ji} + \tfrac{1}{2}(p_+A_-)^2] - A^c p_+ A_- \quad , \tag{3.4.11}$$

where the integrals are now over just x_a, x_-, and c, and some of the field strengths simplify:

$$F^{cc} = 2(A^c)^2 \quad , \quad other\ F^{ic} = [\nabla^i, A^c\} \quad . \tag{3.4.12}$$

Before performing the transformations which eliminate p_+ dependence, it's now convenient to expand the fields over c as

$$\begin{aligned} A^a &= A^a + c\chi^a \quad , \\ A^c &= iC + cB \quad , \\ A^{\tilde{c}} &= i\tilde{C} + cD \quad , \end{aligned} \tag{3.4.13}$$

where the fields on the right-hand sides are x^α-independent. (The i's have been chosen in accordance with (3.4.4) to make the final fields real.) We next perform the dc integration, and then perform as the first transformation the first-quantized one (3.4.3a), $\Phi \to p_+{}^{-J^3}\Phi$ (using the first-quantized $J^3 = c\partial/\partial c + M^3$), which gives

$$-iQ = \int \tfrac{1}{4}F_{ab}{}^2 - \tfrac{1}{2}\left(2B + \frac{1}{p_+}i[\nabla^a, p_+A_a] - \{C, \tilde{C}\} - \frac{2}{p_+}\left\{p_+{}^2C, \frac{1}{p_+}\tilde{C}\right\}\right)^2$$

$$- \left[2D + \tilde{C}^2 - 2p_+{}^2 \left(\frac{1}{p_+}\tilde{C}\right)^2\right] C^2$$

$$+ \left(2\chi^a - i[\nabla^a, \tilde{C}] - 2i\left[p_+A^a, \frac{1}{p_+}\tilde{C}\right]\right)[\nabla_a, C] \quad . \qquad (3.4.14)$$

This transformation also replaces (3.4.6) with (3.4.7), with an extra factor of η_{ij} for $[A_i{}^\dagger, A_j]$, but still with the x_- δ-function. Expanding the bracket over the c's,

$$[\chi_a, A_b]_c = \tfrac{1}{2}\eta_{ab} \quad , \quad [D, C]_c = \tfrac{1}{2} \quad , \quad [B, \tilde{C}]_c = -\tfrac{1}{2} \quad , \qquad (3.4.15)$$

where we have left off all the δ-function factors (now in commuting coordinates only). Note that, by (3.4.10), all these brackets are *symmetric*.

We might also define a second-quantized

$$J^3 = \int A^i p_+ \left(c\frac{\partial}{\partial c} + M^3\right) A_i \quad , \qquad (3.4.16a)$$

but this form automatically keeps just the antihermitian part of the first-quantized operator $c\partial/\partial c = \frac{1}{2}[c, \partial/\partial c] + \frac{1}{2}$: Doing the c integration and transformation (3.4.3a),

$$J^3 = \int \chi^a A_a - \tilde{C}B - 3CD \quad . \qquad (3.4.16b)$$

As a result, the terms in Q of different orders in the fields have different second-quantized ghost number. Therefore, we use only the first-quantized ghost operator (or second-quantize it in functional form).

As can be seen in the above equations, despite the rescaling of the fields by suitable powers of p_+ there remains a fairly complicated dependence on p_+. There is no explicit x_- dependence anywhere but, of course, the fields have x_- as an argument. It would seem that there should be a simple prescription to get rid of the p_+'s in the transformations. Setting $p_+ = constant$ does not work, since it violates the Leibnitz rule for derivatives ($p_+\phi = a\phi$ implies that $p_+\phi^2 = 2a\phi^2$ and not $a\phi^2$). Even setting $p_+\phi_i = \lambda_i\phi_i$ does not work. An attempt that comes very close is the following: Give the fields some specific x_- dependence in such a way that the p_+ factors can be evaluated and that afterwards such dependence can be canceled between the right-hand side and left-hand side of the transformations. In the above case it seems that the only possibility is to set every field proportional to $(x_-)^0$ but then it is hard to define $1/p_+$ and p_+. One then tries setting each field proportional to $(x_-)^\epsilon$ and then let $\epsilon \to 0$ at the end. In fact this prescription gives the correct answer for the quadratic terms of the Yang-Mills BRST transformations. Unfortunately it does not give the correct cubic terms.

It might be possible to eliminate p_+-dependence simply by applying $J_{-+} = 0$ as a constraint. However, this would require resolving some ambiguities in the evaluation of the nonlocal (in x_-) operator p_+ in the interaction vertices.

We therefore remove the explicit p_+-dependence by use of an explicit transformation. In the Yang-Mills case, this transformation can be completely determined by choosing it to be the one which redefines the auxiliary field B in a way which eliminates interaction terms in Q involving it, thus making $B + i\frac{1}{2}p \cdot A$ BRST-invariant. The resulting transformation [3.14] redefines only the BRST auxiliary fields:

$$Q \to e^{L_\Delta}Q \quad , \quad L_A B \equiv [A, B]_c \quad , \quad L_\Delta{}^2 = 0 \quad ,$$

$$\Delta = \int \tilde{C}\frac{1}{p_+}i[A^a, p_+ A_a] - \tilde{C}^2 C + 2\left(\frac{1}{p_+}\tilde{C}\right)^2 (p_+{}^2 C) \quad , \tag{3.4.17}$$

simply redefines the auxiliary fields to absorb the awkward interaction terms in (3.4.14). (We can also eliminate the free terms added to B and χ by adding a term $\int \tilde{C} i p^a A_a$ to Δ to make the first term $\tilde{C}(1/p_+)i[\nabla^a, p_+ A_a]$.) We then find for the transformed BRST operator

$$-iQ = \int \tfrac{1}{4}F_{ab}{}^2 - \tfrac{1}{4}(2B + ip \cdot A)^2 - 2DC^2 + (2\chi^a - ip^a\widetilde{C})[\nabla_a, C] \quad . \tag{3.4.18}$$

The resulting transformations are then

$$\begin{aligned}
QA_a &= -i[\nabla_a, C] \quad , \\
Q\chi_a &= i\tfrac{1}{2}[\nabla^b, F_{ba}] + i\{C, \chi_a - i\tfrac{1}{2}p_a\widetilde{C}\} + p_a(B + i\tfrac{1}{2}p \cdot A) \quad , \\
Q\widetilde{C} &= -2i(B + i\tfrac{1}{2}p \cdot A) \quad , \\
QD &= -i\left[\nabla^a, \chi_a - i\tfrac{1}{2}p_a\widetilde{C}\right] + i[C, D] \quad , \\
QC &= iC^2 \quad , \\
QB &= -\tfrac{1}{2}p^a[\nabla_a, C] \quad .
\end{aligned} \tag{3.4.19}$$

Since all the p_+'s have been eliminated, we can now drop all x_- dependence from the fields, integration, and δ-functions. On the fields A_a, C, $\widetilde{C}$, B of the usual BRST2 formalism, this result agrees with the corresponding transformations (3.2.8), where this $B = \frac{1}{2}\widetilde{B}$ of (3.2.11). By working with the second-quantized operator form of Q (and of the redefinition Δ), we have automatically obtained a form which makes $Q\Phi$ integrable in Φ, or equivalently makes the vertices which follow from this operator cyclic in all the fields (or symmetric, if one takes group-theory indices into account). The significance of this property will be described in the next chapter.

This extended-light-cone form of the OSp(1,1|2) algebra will be used to derive free gauge-invariant actions in the next chapter. The specific form of the generators for the case of the free open string will be given in sect. 8.2, and the generalization to the free closed string in sect. 11.1. A partial analysis of the interacting string along these lines will be given in sect. 12.1.

3.5. Fermions

These results can be extended to fermions [3.15]. This requires a slight modification of the formalism, since the Sp(2) representations resulting from the above analysis for spinors don't include singlets. This modification is analogous to the addition of the $B\partial/\partial\tilde{C}$ terms to Q in (3.2.6a). We can think of the OSp(1,1|2) generators of (3.4.2) as "orbital" generators, and add "spin" generators which themselves generate OSp(1,1|2). In particular, since we are here considering spinors, we choose the spin generators to be those for the simplest spinor representation, the graded generalization of a Dirac spinor, whose generators can be expressed in terms of graded Dirac "matrices":

$$\{\tilde{\gamma}^A,\tilde{\gamma}^B] = 2\eta^{AB} \quad , \quad S_{AB} = \tfrac{1}{4}[\tilde{\gamma}_A,\tilde{\gamma}_B\} \quad , \quad J_{AB}{}' = J_{AB} + S_{AB} \quad , \tag{3.5.1}$$

where { ,] is the opposite of [, }. These $\tilde{\gamma}$ matrices are not to be confused with the "ordinary" γ matrices which appear in M_{ij} from the dimensional continuation of the true spin operators. The $\tilde{\gamma}^A$, like γ^i, are hermitian. (The hermiticity of γ^i in the light-cone formalism follows from $(\gamma^i)^2 = 1$ for each i and the fact that all states in the light-cone formalism have nonnegative norm, since they're physical.) The choice of whether the $\tilde{\gamma}$'s (and also the graded γ^i's) commute or anticommute with other operators (which could be arbitrarily changed by a Klein transformation) follows from the index structure as usual (bosonic for indices $\pm$, fermionic for α). (Thus, as usual, the ordinary γ matrices γ^a commute with other operators, although they anticommute with each other.)

In order to put the OSp(1,1|2) generators in a form more similar to (3.4.2), we need to perform unitary transformations which eliminate the new terms in J_{-+} and $J_{+\alpha}$ (while not affecting $J_{\alpha\beta}$, although changing $J_{-\alpha}$). In general, the appropriate transformations $J_{AB}{}' = UJ_{AB}U^{-1}$ to eliminate such terms are:

$$\ln U = -(\ln p_+)S_{-+} \tag{3.5.2a}$$

to first eliminate the S_{-+} term from J_{-+}, and then

$$ln\ U = S_{+}{}^{\alpha}p_{\alpha} \tag{3.5.2b}$$

to do the same for $J_{+\alpha}$. The general result is

$$J_{-+} = -ix_{-}p_{+} + k \quad , \quad J_{+\alpha} = ix_{\alpha}p_{+} \quad , \quad J_{\alpha\beta} = -ix_{(\alpha}p_{\beta)} + \widehat{M}_{\alpha\beta} \quad ,$$

$$J_{-\alpha} = -ix_{-}p_{\alpha} + \frac{1}{p_{+}}\left[-ix_{\alpha}\tfrac{1}{2}(p^{b}p_{b} + M^{2} + p^{\beta}p_{\beta}) + \widehat{M}_{\alpha}{}^{\beta}p_{\beta} + kp_{\alpha} + \widehat{Q}_{\alpha}\right] \quad ; \tag{3.5.3a}$$

$$\widehat{M}_{\alpha\beta} = M_{\alpha\beta} + S_{\alpha\beta} \quad ,$$

$$\widehat{Q}_{\alpha} = (M_{\alpha}{}^{b}p_{b} + M_{\alpha m}M) + \left[S_{-\alpha} + S_{+\alpha}\tfrac{1}{2}(p^{b}p_{b} + M^{2})\right] \quad . \tag{3.5.3b}$$

(3.5.3a) is the same as (3.4.2a), but with $M_{\alpha\beta}$ and Q_{α} replaced by $\widehat{M}_{\alpha\beta}$ and $\widehat{Q}_{\alpha}$. In this case the last term in $\widehat{Q}_{\alpha}$ is

$$S_{-\alpha} + S_{+\alpha}\tfrac{1}{2}(p^{b}p_{b} + M^{2}) = -\tfrac{1}{2}\tilde{\gamma}_{\alpha}\left[\tilde{\gamma}_{-} + \tilde{\gamma}_{+}\tfrac{1}{2}(p^{b}p_{b} + M^{2})\right] \quad . \tag{3.5.4}$$

We can again choose $k = 0$.

This algebra will be used to derive free gauge-invariant actions for fermions in sect. 4.5. The generalization to fermionic strings follows from the representation of the Poincaré algebra given in sect. 7.2.

3.6. More dimensions

In the previous section we saw that fermions could be treated in a way similar to bosons by including an OSp(1,1|2) Clifford algebra. In the case of the Dirac spinor, there is already an OSp(D−1,1|2) Clifford algebra (or OSp(D,1|2) in the massive case) obtained by adding 2+2 dimensions to the light-cone γ-matrices, in terms of which M_{ij} (and therefore the OSp(1,1|2) algebra) is defined. Including the additional γ-matrices makes the spinor a representation of an OSp(D,2|4) Clifford algebra (OSp(D+1,2|4) for massive), and is thus equivalent to adding 4+4 dimensions to the original light-cone spinor instead of 2+2, ignoring the extra spacetime coordinates. This suggests another way of treating fermions which allows bosons to be treated identically, and should thus allow a straightforward generalization to supersymmetric theories [3.16].

We proceed similarly to the 2+2 case: Begin by adding 4+4 dimensions to the light-cone Poincaré algebra (2.3.5). Truncate the resulting IOSp(D+1,3|4) algebra to ISO(D−1,1)⊗IOSp(2,2|4). IOSp(2,2|4) contains (in particular) 2 inequivalent truncations to IOSp(1,1|2), which can be described by (defining-representation direct-product) factorization of the OSp(2,2|4) metric into the OSp(1,1|2) metric times the metric of either SO(2) (U(1)) or SO(1,1) (GL(1)):

$$\eta_{\mathcal{AB}} = \eta_{AB}\eta_{\hat{a}\hat{b}} \quad , \quad \mathcal{A} = A\hat{a}$$

$$\rightarrow \quad J_{AB} = \eta^{\hat{b}\hat{a}} J_{A\hat{a},B\hat{b}} \quad , \quad \epsilon_{\hat{a}\hat{b}}\Delta = \eta^{BA} J_{A\hat{a},B\hat{b}} \quad , \qquad (3.6.1)$$

where Δ is the generator of the $U(1)$ or $GL(1)$ and $\eta_{\hat{a}\hat{b}} = I$ or σ_1. These 2 OSp(1,1|2)'s are Wick rotations of each other. We'll treat the 2 cases separately.

The GL(1) case corresponds to first taking the GL(2|2) (=SL(2|2)⊗GL(1)) subgroup of OSp(2,2|4) (as SU(N)⊂SO(2N), or GL(1|1)⊂OSp(1,1|2)), keeping also half of the inhomogeneous generators to get IGL(2|2). Then taking the OSp(1,1|2) subgroup of the SL(2|2) (in the same way as SO(N)⊂SU(N)), we get IOSp(1,1|2)⊗GL(1), which is like the Poincaré group in (1,1|2) dimensions plus dilatations. (There is also an SL(1|2)=OSp(1,1|2) subgroup of SL(2|2), but this turns out not to be useful.) The advantage of breaking down to GL(2|2) is that for this subgroup the coordinates of the string (sect. 8.3) can be redefined in such a way that the extra zero-modes are separated out in a natural way while leaving the generators local in σ. This GL(2|2) subgroup can be described by writing the OSp(2,2|4) metric as

$$\eta_{\mathcal{AB}} = \begin{pmatrix} 0 & \eta_{AB'} \\ \eta_{A'B} & 0 \end{pmatrix} , \quad \eta_{AB'} = (-1)^{AB}\eta_{B'A} \quad , \quad \mathcal{A} = (A, A') \qquad (3.6.2a)$$

$$\rightarrow \quad J_{\mathcal{AB}} = \begin{pmatrix} \tilde{J}_{AB} & \tilde{J}_{AB'} \\ \tilde{J}_{A'B} & \tilde{J}_{A'B'} \end{pmatrix} \quad , \quad \tilde{J}_{AB'} = -(-1)^{AB}\tilde{J}_{B'A} \quad , \qquad (3.6.2b)$$

where $\tilde{J}_{A'B}$ are the GL(2|2) generators, to which we add the $\tilde{p}_{A'}$ half of $p_{\mathcal{A}}$ to form IGL(2|2). (The metric $\eta_{AB'}$ can be used to eliminate primed indices, leaving covariant and contravariant unprimed indices.) In this notation, the original $\pm$ indices of the light cone are now $+'$ and $-$ (whereas $+$ and $-'$ are "transverse"). To reduce to the IOSp(1,1|2)⊗GL(1) subgroup we identify primed and unprimed indices; i.e., we choose the subgroup which transforms them in the same way:

$$J_{AB} = \tilde{J}_{A'B} + \tilde{J}_{AB'} , \quad \check{p}_A = \tilde{p}_{A'} \quad , \quad \Delta = \eta^{BA'}\tilde{J}_{A'B} \quad . \qquad (3.6.3)$$

We distinguish the momenta $\check{p}_A$ and their conjugate coordinates $\check{x}_A$, which we wish to eliminate, from $p_A = \tilde{p}_A$ and their conjugates x_A, which we'll keep as the usual ones of OSp(1,1|2) (including the nonlinear p_-). At this point these generators take the explicit form

$$\Delta = i\check{x}^A\check{p}_A - ix_-p_+ - ix^\alpha p_\alpha + M_{-'+} + C^{\beta\alpha}M_{\alpha'\beta} \quad ,$$

$$J_{+\alpha} = i\breve{p}_+\breve{x}_\alpha - i\breve{x}_+\breve{p}_\alpha + ip_+x_\alpha + M_{+\alpha'} \quad ,$$
$$J_{-+} = -i\breve{x}_-\breve{p}_+ + i\breve{x}_+\breve{p}_- - ix_-p_+ + M_{-'+} \quad ,$$
$$J_{\alpha\beta} = -i\breve{x}_{(\alpha}\breve{p}_{\beta)} - ix_{(\alpha}p_{\beta)} + M_{\alpha\beta'} + M_{\alpha'\beta} \quad ,$$
$$J_{-\alpha} = -i\breve{x}_-\breve{p}_\alpha + i\breve{x}_\alpha\breve{p}_- - ix_-p_\alpha + ix_\alpha p_- + M_{-'\alpha} + \frac{1}{\breve{p}_+}Q_{\alpha'} \quad ,$$
$$p_- = -\frac{1}{2\breve{p}_+}({p_a}^2 + M^2 + 2p_+\breve{p}_- + 2p^\alpha\breve{p}_\alpha) \quad ,$$
$$Q_{\alpha'} = M_{\alpha'}{}^a p_a + M_{\alpha' m}M + M_{\alpha'+}\breve{p}_- + M_{\alpha'-'}p_+ - M_{\alpha'\beta}\breve{p}^\beta - M_{\alpha'\beta'}p^\beta \quad . \tag{3.6.4}$$

(All the $\breve{p}$'s are linear, being unconstrained so far.) We now use p and Δ to eliminate the extra zero-modes. We apply the constraints and corresponding gauge conditions

$$\Delta = 0 \quad \rightarrow \quad \breve{x}_- = \frac{1}{\breve{p}_+}(x_-p_+ + x^\alpha p_\alpha - \breve{x}_+\breve{p}_- - \breve{x}^\alpha\breve{p}_\alpha + iM_{-'+} + iC^{\beta\alpha}M_{\alpha'\beta}) \quad ,$$
$$gauge \quad \breve{p}_+ = 1 \quad ;$$
$$\breve{p}^A\breve{p}_A = 0 \quad \rightarrow \quad \breve{p}_- = -\tfrac{1}{2}\breve{p}^\alpha\breve{p}_\alpha \quad ,$$
$$gauge \quad \breve{x}_+ = 0 \quad . \tag{3.6.5}$$

These constraints are directly analogous to (2.2.3), which were used to obtain the usual coordinate representation of the conformal group SO(D,2) from the usual coordinate representation (with 2 more coordinates) of the same group as a Lorentz group. In fact, after making a unitary transformation of the type (3.5.2b),

$$U = e^{-M_{+\alpha'}\breve{p}^\alpha} \quad , \tag{3.6.6}$$

the remaining unwanted coordinates $\breve{x}_\alpha$ completely decouple:

$$UJ_{AB}U^{-1} = \overset{\circ}{J}_{AB} + J'_{AB} \quad , \tag{3.6.7a}$$

$$\overset{\circ}{J}_{+\alpha} = i\breve{x}_\alpha \quad , \quad \overset{\circ}{J}_{-+} = i\breve{x}^\alpha\breve{p}_\alpha \quad , \quad \overset{\circ}{J}_{\alpha\beta} = -i\breve{x}_{(\alpha}\breve{p}_{\beta)} \quad , \quad \overset{\circ}{J}_{-\alpha} = -i\breve{x}_\alpha\breve{p}^\beta\breve{p}_\beta \quad ; \tag{3.6.7b}$$

$$J'_{+\alpha} = 2(ip_+x_\alpha + M_{+\alpha'}) \quad , \quad J'_{-+} = 2(-ix_-p_+ + M_{-'+}) + (-ix^\alpha p_\alpha + C^{\beta\alpha}M_{\alpha'\beta}) \quad ,$$
$$J'_{\alpha\beta} = -ix_{(\alpha}p_{\beta)} + (M_{\alpha'\beta} + M_{\alpha\beta'}) \quad ,$$
$$J'_{-\alpha} = (-ix_-p_\alpha + M_{-'\alpha})$$
$$+ \left[-ix_\alpha\tfrac{1}{2}({p_a}^2 + M^2) + (M_{\alpha'}{}^a p_a + M_{\alpha' m}M + M_{\alpha'-'}p_+ - M_{\alpha'\beta'}p^\beta)\right] \quad , \tag{3.6.7c}$$

where $\overset{\circ}{J}$ are the generators of the conformal group in 2 anticommuting dimensions ((3.3.9), after switching coordinates and momenta), and J' are the desired OSp(1,1|2) generators.

To eliminate zero-modes, it's convenient to transform these OSp(1,1|2) generators to the canonical form (3.4.2a). This is performed [3.7] by the redefinition

$$p_+ \to \tfrac{1}{2}p_+{}^2 \quad , \tag{3.6.8a}$$

followed by the unitary transformations

$$U_1 = p_+{}^{-\left(-i\frac{1}{2}[x^\alpha,p_\alpha]+2M_{-'+}+C^{\beta\alpha}M_{\alpha'\beta}\right)} \quad ,$$

$$U_2 = e^{-2M_{+\alpha'}p^{\dot\alpha}} \quad . \tag{3.6.8b}$$

Since p_+ is imaginary (though hermitian) in (x_-) coordinate space, U_1 changes reality conditions accordingly (an i for each p_+). The generators are then

$$J_{+\alpha} = ix_\alpha p_+ , \quad J_{-+} = -ix_- p_+ \quad , \quad J_{\alpha\beta} = -ix_{(\alpha}p_{\beta)} + \widehat{M}_{\alpha\beta} \quad ,$$

$$J_{-\alpha} = -ix_- p_\alpha + \frac{1}{p_+}\left[-ix_\alpha\tfrac{1}{2}(p_a{}^2 + M^2 + p^\beta p_\beta) + \widehat{M}_\alpha{}^\beta p_\beta + \widehat{Q}_\alpha\right] \quad ; \tag{3.6.9a}$$

$$\widehat{M}_{\alpha\beta} = M_{\alpha\beta'} + M_{\alpha'\beta} \quad ,$$

$$\widehat{Q}_\alpha = M_{-'\alpha} - \tfrac{1}{2}M_{-'\alpha'} + (M_{\alpha'}{}^a p_a + M_{\alpha'm}M) + M_{+\alpha'}(p_a{}^2 + M^2) \quad . \tag{3.6.9b}$$

For the U(1) case the derivation is a little more straightforward. It corresponds to first taking the U(1,1|1,1) (=SU(1,1|1,1)⊗ U(1)) subgroup of OSp(2,2|4). From (3.6.1), instead of (3.6.2a,3) we now have

$$\eta_{\mathcal{AB}} = \begin{pmatrix} \eta_{AB} & 0 \\ 0 & \eta_{A'B'} \end{pmatrix} \quad , \quad \eta_{AB} = \eta_{A'B'} \quad , \quad \mathcal{A} = (A, A') \quad ,$$

$$J_{AB} = \tilde{J}_{AB} + \tilde{J}_{A'B'} \quad , \quad \check{p}_A = \tilde{p}_{A'} \quad , \quad \Delta = \eta^{BA}\tilde{J}_{A'B} \quad . \tag{3.6.10}$$

The original light-cone ± are now still ± (no primes), so the unwanted zero-modes can be eliminated by the constraints and gauge choices

$$\check{p}_A = 0 \quad \to \quad gauge \quad \check{x}_A = 0 \quad . \tag{3.6.11}$$

Alternatively, we could include $\check{p}_A$ among the generators, using IOSp(1,1|2) as the group (as for the usual closed string: see sects. 7.1, 11.1). (The same result can be obtained by replacing (3.6.11) with the constraints $\Delta = 0$ and $\epsilon^{\hat{b}a} p_{Aa} p_{B\hat{b}} \sim p_{[A}\check{p}_{B)} = 0$.) The OSp(1,1|2) generators are now

$$J_{+\alpha} = ix_\alpha p_+ + M_{+'\alpha'} \quad , \quad J_{-+} = -ix_- p_+ + M_{-'+'} \quad ,$$

$$J_{\alpha\beta} = -ix_{(\alpha} p_{\beta)} + M_{\alpha\beta} + M_{\alpha'\beta'} \quad ,$$

$$J_{-\alpha} = -ix_- p_\alpha - ix_\alpha \frac{1}{2p_+}({p_a}^2 + M^2 + p^\beta p_\beta) + \frac{1}{p_+}({M_\alpha}^a p_a + M_{\alpha m} M + {M_\alpha}^\beta p_\beta) + M_{-'\alpha'} \quad , \tag{3.6.12a}$$

or, in other words (symbols), these OSp(1,1|2) generators are just the usual ones plus the spin of a second OSp(1,1|2), with the same representation as the spin of the first OSp(1,1|2):

$$J_{AB} = \tilde{J}_{AB} + M_{A'B'} \quad . \tag{3.6.12b}$$

(However, for the string $M_{\alpha m}M$ will contain oscillators from both sets of 2+2 dimensions, so these sets of oscillators won't decouple, even though $\tilde{J}_{AB}$ commutes with $M_{A'B'}$.) To simplify the form of J_{-+} and $J_{-\alpha}$, we make the consecutive unitary transformations (3.5.2):

$$U_1 = {p_+}^{-M_{-'+'}} \quad , \quad U_2 = e^{M_{+'}{}^{\alpha'} p_\alpha} \quad , \tag{3.6.13}$$

after which the generators again take the canonical form:

$$J_{+\alpha} = ix_\alpha p_+ \quad , \quad J_{-+} = -ix_- p_+ \quad , J_{\alpha\beta} = -ix_{(\alpha} p_{\beta)} + \widehat{M}_{\alpha\beta} \quad ,$$

$$J_{-\alpha} = -ix_- p_\alpha + \frac{1}{p_+}\left[-ix_\alpha \tfrac{1}{2}({p_a}^2 + M^2 + p^\beta p_\beta) + {\widehat{M}_\alpha}{}^\beta p_\beta + \widehat{Q}_\alpha\right] \quad ; \tag{3.6.14a}$$

$$\widehat{M}_{\alpha\beta} = M_{\alpha\beta} + M_{\alpha'\beta'} \quad ,$$

$$\widehat{Q}_\alpha = M_{-'\alpha'} + ({M_\alpha}^b p_b + M_{\alpha m} M) + M_{+'\alpha'} \tfrac{1}{2}({p_a}^2 + M^2) \quad . \tag{3.6.14b}$$

Because of U_1, formerly real fields now satisfy $\phi^\dagger = (-1)^{M_{-'+'}} \phi$.

Examples and actions of this 4+4-extended OSp(1,1|2) will be considered in sect. 4.1, its application to supersymmetry in sect. 5.5, and its application to strings in sect. 8.3.

Exercises

(1) Derive the time derivative of (3.1.6) from (3.1.7).

(2) Derive (3.1.12). Compare with the usual derivation of the Noether current in field theory.

(3) Derive (3.1.15,17).

(4) Show that Q of (3.2.6a) is nilpotent. Show this directly for (3.2.8).

(5) Derive (3.3.1b).

(6) Use (3.3.6) to rederive (3.2.6a,12a).

(7) Use (3.3.2,7) to derive the OSp(1,1|2) algebra for Yang-Mills in terms of the explicit independent fields (in analogy to (3.2.8)).

(8) Perform the transformation (3.4.3a) to obtain (3.4.3b). Choose the Dirac spinor representation of the spin operators (in terms of γ-matrices). Compare with (3.2.6), and identify the field equations $\mathcal{G}$ and ghosts C.

(9) Check that the algebra of $\widehat{Q}_\alpha$ and $\widehat{M}_{\alpha\beta}$ closes for (3.5.3), (3.6.9), and (3.6.14), and compare with (3.4.2).

4. GENERAL GAUGE THEORIES

4.1. OSp(1,1|2)

In this chapter we will use the results of sects. 3.4-6 to derive free gauge-invariant actions for arbitrary field theories, and discuss some preliminary results for the extension to interacting theories.

The (free) gauge covariant theory for arbitrary representations of the Poincaré group (except perhaps for those satisfying self-duality conditions) can be constructed from the BRST1 OSp(1,1|2) generators [2.3]. For the fields described in sect. 3.4 which are representations of OSp(D,2|2), consider the gauge invariance generated by OSp(1,1|2) and the obvious (but unusual) corresponding gauge-invariant action:

$$\delta\Phi = \tfrac{1}{2} J^{BA} \Lambda_{AB} \quad \rightarrow \quad S = \int d^D x_a dx_- d^2 x_\alpha \ \tfrac{1}{2} \Phi^\dagger p_+ \delta(J_{AB}) \Phi \quad , \tag{4.1.1}$$

where J_{AB} for $A = (+, -, \alpha)$ (graded antisymmetric in its indices) are the generators of OSp(1,1|2), and we have set $k = 0$, so that the p_+ factor is the Hilbert space metric. In particular, the J_{-+} and $J_{+\alpha}$ transformations allow all dependence on the unphysical coordinates to be gauged away:

$$\delta\Phi = -ix_- p_+ \Lambda_{-+} + ix^\alpha p_+ \Lambda_{-\alpha} \tag{4.1.2}$$

implies that only the part of Φ at $x_- = x_\alpha = 0$ can be gauge invariant. A more explicit form of $\delta(J_{AB})$ is given by

$$\begin{aligned} p_+ \delta(J_{AB}) &= p_+ \delta(J_{\alpha\beta}{}^2) i \delta(J_{-+}) \delta^2(J_{+\alpha}) \delta^2(J_{-\alpha}) \\ &= \delta(x_-) \delta^2(x_\alpha) \delta(M_{\alpha\beta}{}^2) p_+{}^2 J_{-\alpha}{}^2 \quad , \end{aligned} \tag{4.1.3}$$

where we have used

$$J_{-+}\delta(J_{-+}) = \delta(J_{-+}) J_{-+} = 0 \quad \rightarrow \quad \delta(J_{-+}) = i \frac{1}{p_+} \delta(x_-) \quad , \tag{4.1.4}$$

since $p_+ \neq 0$ in light-cone formalisms. The gauge invariance of the kinetic operator follows from the fact that the δ-functions can be reordered fairly freely: $\delta(J_{\alpha\beta}{}^2)$ (which is really a Kronecker δ) commutes with all the others, while

$$\delta(J_{-+}+a)\delta^2(J_{\pm\alpha}) = \delta^2(J_{\pm\alpha})\delta(J_{-+}+a\mp 2) \quad \rightarrow \quad [\delta(J_{-+}),\delta^2(J_{+\alpha})\delta^2(J_{-\alpha})] = 0 \quad ,$$

$$[\delta^2(J_{-\alpha}),\delta^2(J_{+\beta})] = 2J_{-+} + (C^{\alpha\beta}J_{-+} + J^{\alpha\beta})\tfrac{1}{2}[J_{-\alpha},J_{+\beta}] \quad , \tag{4.1.5}$$

where the J_{-+} and $J^{\alpha\beta}$ each can be freely moved to either side of the $[J_{-\alpha},J_{+\beta}]$. After integration of the action over the trivial coordinate dependence on x_- and x_α, (4.1.1) reduces to (using (3.4.2,4.1.3))

$$S = \int d^D x_a \; \tfrac{1}{2}\phi^\dagger \delta(M_{\alpha\beta}{}^2)(\Box - M^2 + Q^2)\phi \quad , \quad \delta\phi = -i\tfrac{1}{2}Q_\alpha \Lambda^\alpha + \tfrac{1}{2}M_{\alpha\beta}\Lambda^{\alpha\beta} \; , \tag{4.1.6}$$

where ϕ now depends only on the usual spacetime coordinates x_a, and for irreducible Poincaré representations ϕ has indices which are the result of starting with an irreducible representation of OSp(D−1,1|2) in the massless case, or OSp(D,1|2) in the massive case, and then truncating to the Sp(2) singlets. (This type of action was first proposed for the string [4.1,2].) Λ^α is the remaining part of the $J_{-\alpha}$ transformations after using up the transformations of (4.1.2) (and absorbing a $1/\partial_+$), and contains the usual component gauge transformations, while $\Lambda^{\alpha\beta}$ just gauges away the Sp(2) nonsinglets. We have thus derived a general gauge-covariant action by adding 2+2 dimensions to the light-cone theory. In sect. 4.4 we'll show that gauge-fixing to the light cone gives back the original light-cone theory, proving the consistency of this method.

In the BRST formalism the field contains not only physical polarizations, but also auxiliary fields (nonpropagating fields needed to make the action local, such as the trace of the metric tensor for the graviton), ghosts (including antighosts, ghosts of ghosts, etc.), and Stueckelberg fields (gauge degrees of freedom, such as the gauge part of Higgs fields, which allow more renormalizable and less singular formalisms for massive fields). All of these but the ghosts appear in the gauge-invariant action. For example, for a massless vector we start with $A_i = (A_a, A_\alpha)$, which appears in the field ϕ as

$$\big|\phi\big\rangle = \big|{}^i\big\rangle A_i \quad , \quad \big\langle{}^i\big|{}^j\big\rangle = \eta^{ij} \quad . \tag{4.1.7}$$

Reducing to Sp(2) singlets, we can truncate to just A_a. Using the relations

$$M^{ij}\big|{}^k\big\rangle = \big|{}^{[i}\big\rangle\eta^{j)k} \quad \rightarrow \quad M^{\alpha a}\big|{}^b\big\rangle = \eta^{ab}\big|{}^\alpha\big\rangle \quad , \quad M^{\alpha a}\big|{}^\beta\big\rangle = -C^{\alpha\beta}\big|{}^a\big\rangle \quad , \tag{4.1.8}$$

where [ ) is graded antisymmetrization, we find

$$\tfrac{1}{2}M^{\alpha b}M_\alpha{}^c\Big|{}^a\Big\rangle = \tfrac{1}{2}M^{\alpha b}\eta^{ac}\Big|{}_\alpha\Big\rangle = \eta^{ac}\Big|{}^b\Big\rangle \quad , \tag{4.1.9}$$

and thus the lagrangian

$$\mathcal{L} = \tfrac{1}{2}\Big\langle\phi\Big|\delta(M_{\alpha\beta}{}^2)[\Box - (M_\alpha{}^b\partial_b)^2]\Big|\phi\Big\rangle = \tfrac{1}{2}A^a(\Box A_a - \partial_a\partial^b A_b) \quad . \tag{4.1.10}$$

Similarly, for the gauge transformation

$$\Big|\Lambda^\alpha\Big\rangle = \Big|{}^i\Big\rangle\Lambda_i{}^\alpha \quad \to \quad \delta A_a = \Big\langle{}_a\Big|\tfrac{1}{2}M_\alpha{}^b\partial_b\Big|\Lambda^\alpha\Big\rangle = -\tfrac{1}{2}\partial_a\Lambda_\alpha{}^\alpha \quad . \tag{4.1.11}$$

As a result of the $\delta(M_{\alpha\beta}{}^2)$ acting on $Q_\alpha\Lambda^\alpha$, the only part of Λ which survives is the part which is an overall singlet in the matrix indices and explicit α index: in this case, $\Lambda_i{}^\alpha = \delta_i{}^\alpha\lambda \to \delta A_a = -\partial_a\lambda$. Note that the $\phi Q^2\phi$ term can be written as a $(Q\phi)^2$ term: This corresponds to subtracting out a "gauge-fixing" term from the "gauge-fixed" lagrangian $\phi(\Box - M^2)\phi$. (See the discussion of gauge fixing in sect. 4.4.)

For a massless antisymmetric tensor we start with $A_{[ij)} = (A_{[ab]}, A_{a\alpha}, A_{(\alpha\beta)})$ appearing as

$$\Big|\phi\Big\rangle = -\tfrac{1}{2}\Big|{}^{ij}\Big\rangle_A A_{ji} \quad , \quad \Big|{}^{ij}\Big\rangle_A = \frac{1}{\sqrt{2}}\Big|{}^{[i}\Big\rangle \otimes \Big|{}^{j)}\Big\rangle \tag{4.1.12}$$

(and similarly for $|\Lambda^\alpha\rangle$), and truncate to just $A_{[ab]}$. Then, from (4.1.8),

$$\begin{aligned}\tfrac{1}{2}M^{\alpha c}M_\alpha{}^d\Big|{}^a\Big\rangle\Big|{}^b\Big\rangle &= \tfrac{1}{2}M^{\alpha c}\left(\eta^{da}\Big|{}_\alpha\Big\rangle\Big|{}^b\Big\rangle + \eta^{db}\Big|{}^a\Big\rangle\Big|{}_\alpha\Big\rangle\right) \\ &= \left(\eta^{da}\Big|{}^c\Big\rangle\Big|{}^b\Big\rangle + \eta^{db}\Big|{}^a\Big\rangle\Big|{}^c\Big\rangle\right) + \eta^{d(a}\eta^{b)c}\tfrac{1}{2}\Big|{}^\alpha\Big\rangle\Big|{}_\alpha\Big\rangle \quad , \end{aligned} \tag{4.1.13}$$

and we have

$$\mathcal{L} = \tfrac{1}{4}A^{ab}(\Box A_{ab} + \partial^c\partial_{[a}A_{b]c}) \quad , \quad \delta A_{ab} = \tfrac{1}{2}\partial_{[a}\Lambda_{b]\alpha}{}^\alpha \quad . \tag{4.1.14}$$

For a massless traceless symmetric tensor we start with $h_{(ij]} = (h_{(ab)}, h_{\alpha b}, h_{[\alpha\beta]})$ satisfying $h^i{}_i = h^a{}_a + h^\alpha{}_\alpha = 0$, appearing as

$$\Big|\phi\Big\rangle = \tfrac{1}{2}\Big|{}^{ij}\Big\rangle_S h_{ji} \quad , \quad \Big|{}^{ij}\Big\rangle_S = \frac{1}{\sqrt{2}}\Big|{}^{(i}\Big\rangle\Big|{}^{j]}\Big\rangle \quad , \tag{4.1.15}$$

and truncate to $(h_{(ab)}, h_{[\alpha\beta]})$, where $h_{[\alpha\beta]} = \tfrac{1}{2}C_{\alpha\beta}\eta^{ab}h_{(ab)}$, leaving just an unconstrained symmetric tensor. Then, using (4.1.13), as well as

$$\tfrac{1}{2}M^{\gamma a}M_\gamma{}^b\Big|{}^\alpha\Big\rangle\Big|{}^\beta\Big\rangle = \tfrac{1}{2}C^{\alpha\beta}\Big|{}^{(a}\Big\rangle\Big|{}^{b)}\Big\rangle - \eta^{ab}\Big|{}^\alpha\Big\rangle\Big|{}^\beta\Big\rangle \quad , \tag{4.1.16}$$

and using the condition $h^\alpha{}_\alpha = -h^a{}_a$, we find

$$\mathcal{L} = \tfrac{1}{4}h^{ab}\Box h_{ab} - \tfrac{1}{2}h^{ab}\partial_b\partial_c h_a{}^c + \tfrac{1}{2}h^c{}_c\partial_a\partial_b h^{ab} - \tfrac{1}{4}h^a{}_a\Box h^b{}_b \quad ,$$

$$\delta h_{ab} = -\tfrac{1}{2}\partial_{(a}\Lambda_{b)\alpha}{}^\alpha \quad . \qquad (4.1.17)$$

This is the linearized Einstein-Hilbert action for gravity.

The massive cases can be obtained by the dimensional reduction technique, as in (2.2.9), since that's how it was done for this entire procedure, from the light-cone Poincaré algebra down to (4.1.6). (For the string, the OSp generators are represented in terms of harmonic oscillators, and $M_{\alpha m}M$ is cubic in those oscillators instead of quadratic, so the oscillator expressions for the generators don't follow from dimensional reduction, and (4.1.6) must be used directly with the $M_{\alpha m}M$ terms.) Technically, $p_m = m$ makes sense only for complex fields. However, at least for free theories, the resulting i's that appear in the $p_a p_m$ crossterms can be removed by appropriate redefinitions for the complex fields, after which they can be chosen real. (See the discussion below (2.2.10).) For example, for the massive vector we replace $A_m \to iA_m$ (and then take A_m real) to obtain

$$\big|\phi\big\rangle = \big|^a\big\rangle A_a + i\big|_m\big\rangle A_m + \big|^\alpha\big\rangle A_\alpha \quad , \quad \big\langle\phi\big| = A^a\big\langle_a\big| - iA_m\big\langle_m\big| + A^\alpha\big\langle_\alpha\big| \quad . \qquad (4.1.18)$$

The lagrangian and invariance then become

$$\begin{aligned}\mathcal{L} &= \tfrac{1}{2}A^a[(\Box - m^2)A_a - \partial_a\partial^b A_b] + \tfrac{1}{2}A_m\Box A_m + mA_m\partial^a A_a \\ &= \tfrac{1}{4}F_{ab}{}^2 - \tfrac{1}{2}(mA_a + \partial_a A_m)^2 \quad ,\end{aligned}$$

$$\delta A_a = -\partial_a\lambda \quad , \quad \delta A_m = m\lambda \quad . \qquad (4.1.19)$$

This gives a Stueckelberg formalism for a massive vector.

Other examples reproduce all the special cases of higher-spin fields proposed earlier [4.3] (as well as cases that hadn't been obtained previously). For example, for totally symmetric tensors, the usual "double-tracelessness" condition is automatic: Starting from the light cone with a totally symmetric and traceless tensor (in transverse indices), extending $i \to (a,\alpha)$ and restricting to Sp(2) singlets, directly gives a totally symmetric and traceless tensor (in D-dimensional indices) of the same rank, and one of rank 2 lower (but no lower than that, due to the total antisymmetry in the Sp(2) indices).

The most important feature of the BRST method of deriving gauge-invariant actions from light-cone (unitary) representations of the Poincaré group is that it automatically includes exactly the right number of auxiliary fields to make the action local. In the case of Yang-Mills, the auxiliary field (A_-) was obvious, since it results directly from adding just 2 commuting dimensions (and not 2 anticommuting) to the light cone, i.e., from making $D-2$-dimensional indices D-dimensional. Furthermore, the necessity of this field for locality doesn't occur until interactions are included (see sect. 2.1). A less trivial example is the graviton: Naively, a traceless symmetric D-dimensional tensor would be enough, since this would automatically include the analog of A_-. However, the BRST method automatically includes the trace of this tensor. In general, the extra auxiliary fields with anticommuting "ghost-valued" Lorentz indices are necessary for gauge-covariant, local formulations of field theories [4.4,5]. In order to study this phenomenon in more detail, and because the discussion will be useful later in the 2D case for strings, we now give a brief discussion of general relativity.

General relativity is the gauge theory of the Poincaré group. Since local translations (i.e., general coordinate transformations) include the orbital part of Lorentz transformations (as translation by an amount linear in x), we choose as the group generators ∂_m and the Lorentz spin M_{ab}. Treating M_{ab} as second-quantized operators, we indicate how they act by writing explicit "spin" vector indices $a, b, \ldots$ (or spinor indices) on the fields, while using $m, n, \ldots$ for "orbital" vector indices on which M_{ab} doesn't act, as on ∂_m. (The action of the second-quantized M_{ab} follows from that of the first-quantized: E.g., from (4.1.8), (2.2.5), and the fact that $(M^{ij})^\dagger = -M^{ij}$, we have $M_{ab}A_c = -\eta_{c[a}A_{b]}$.) The spin indices (but not the orbital ones) can be contracted with the usual constant tensors of the Lorentz group (the Lorentz metric and γ matrices). The (antihermitian) generators of gauge transformations are thus

$$\lambda = \lambda^m(x)\partial_m + \tfrac{1}{2}\lambda^{ab}(x)M_{ba} \quad , \tag{4.1.20}$$

and the covariant derivatives are

$$\nabla_a = e_a{}^m\partial_m + \tfrac{1}{2}\omega_a{}^{bc}M_{cb} \quad , \tag{4.1.21}$$

where we have absorbed the usual derivative term, since derivatives are themselves generators, and to make the covariant derivative transform covariantly under the gauge transformations

$$\nabla_a{}' = e^{\lambda}\nabla_a e^{-\lambda} \quad . \tag{4.1.22}$$

Covariant field strengths are defined, as usual, by commutators of covariant derivatives,

$$[\nabla_a, \nabla_b] = T_{ab}{}^c\nabla_c + \tfrac{1}{2}R_{ab}{}^{cd}M_{dc} \quad , \tag{4.1.23}$$

since that automatically makes them transform covariantly (i.e., by a similarity transformation, as in (4.1.22)), as a consequence of the transformation law (4.1.22) of the covariant derivatives themselves. Without loss of generality, we can choose

$$T_{ab}{}^{c} = 0 \quad , \tag{4.1.24}$$

since this just determines $\omega_{ab}{}^{c}$ in terms of $e_a{}^m$, and any other ω can always be written as this ω plus a tensor that is a function of just T. (The theory could then always be rewritten in terms of the $T = 0$ ∇ and T itself, making T an arbitrary extra tensor with no special geometric significance.) To solve this constraint we first define

$$e_a = e_a{}^m \partial_m \quad ,$$
$$[e_a, e_b] = c_{ab}{}^{c} e_c \quad . \tag{4.1.25}$$

$c_{ab}{}^{c}$ can then be expressed in terms of $e_a{}^m$, the matrix inverse $e_m{}^a$,

$$e_a{}^m e_m{}^b = \delta_a{}^b \quad , \quad e_m{}^a e_a{}^n = \delta_m{}^n \quad , \tag{4.1.26}$$

and their derivatives. The solution to (4.1.24) is then

$$\omega_{abc} = \tfrac{1}{2}(c_{bca} - c_{a[bc]}) \quad . \tag{4.1.27}$$

The usual global Lorentz transformations, which include orbital and spin pieces together in a specific way, are a symmetry of the vacuum, defined by

$$\langle \nabla_a \rangle = \kappa \partial_a \quad \leftrightarrow \quad \langle e_a{}^m \rangle = \kappa \delta_a{}^m \quad . \tag{4.1.28}$$

κ is an arbitrary constant, which we can choose to be a unit of length, so that ∇ is dimensionless. (In $D = 4$ it's just the usual gravitational coupling constant, proportional to the square root of Newton's gravitational constant.) As a result of general coordinate invariance, any covariant object (i.e., a covariant derivative or tensor with only spin indices uncontracted) will then also be dimensionless. The subgroup of the original gauge group which leaves the vacuum (4.1.28) invariant is just the usual (global) Poincaré group, which treats orbital and spin indices in the same way. We can also treat these indices in a similar way with respect to the full gauge group by using the "vielbein" $e_a{}^m$ and its inverse to convert between spin and orbital indices. In particular, the orbital indices on all fields except the vielbein itself can be converted into spin ones. Also, since integration measures are antisymmetric, converting dx^m into $\Omega^a = dx^m e_m{}^a$ converts $d^D x$ into $\Omega^D = d^D x\, \mathrm{e}^{-1}$, where $e = det(e_a{}^m)$. On such covariant fields, ∇ always acts covariantly. On the other hand, in the absence of spinors, all indices can be converted into orbital ones. In particular, instead of the vielbein we could work with the metric tensor and its inverse:

$$g_{mn} = \eta_{ab} e_m{}^a e_n{}^b \quad , \quad g^{mn} = \eta^{ab} e_a{}^m e_b{}^n \quad . \tag{4.1.29}$$

Then, instead of ∇, we would need a covariant derivative which knows how to treat uncontracted orbital indices covariantly.

The action for gravity can be written as

$$S = -\tfrac{1}{2} \int d^D x \, e^{-1} R \quad , \quad R = \tfrac{1}{2} R_{ab}{}^{ba} \quad . \tag{4.1.30}$$

This can be rewritten in terms of c_{abc} as

$$e^{-1} R = -\partial_m (e^{-1} e^{am} c_a{}^b{}_b) + e^{-1} \left[-\tfrac{1}{2} (c^{ab}{}_b)^2 + \tfrac{1}{8} c^{abc} c_{abc} - \tfrac{1}{4} c^{abc} c_{bca} \right] \tag{4.1.31}$$

using

$$e^{-1} e^a f_a = \partial_m (e^{-1} e^{am} f_a) + e^{-1} c^{ab}{}_b f_a \quad . \tag{4.1.32}$$

Expanding about the vacuum,

$$e_a{}^m = \kappa \delta_a{}^m + \kappa^{D/2} h_a{}^m \quad , \tag{4.1.33}$$

where we can choose e^{am} (and thus h^{am}) to be symmetric by the λ_{ab} transformation, the linearized action is just (4.1.17). As an alternative form for the action, we can consider making the field redefinition

$$e_a{}^m \to \phi^{-2/(D-2)} e_a{}^m \quad , \tag{4.1.34}$$

which introduces the new gauge invariance of (Weyl) local scale transformations

$$e_a{}^{m\prime} = e^{\zeta} e_a{}^m \quad , \quad \phi' = e^{(D-2)\zeta/2} \phi \quad . \tag{4.1.35}$$

(The gauge choice $\phi = constant$ returns the original fields.) Under the field redefinition (4.1.34), the action (4.1.30) becomes

$$S \to \int d^D x \, e^{-1} \left[2 \tfrac{D-1}{D-2} (\nabla_a \phi)^2 - \tfrac{1}{2} R \phi^2 \right] \quad . \tag{4.1.36}$$

We have actually started from (4.1.30) without the total-derivative term of (4.1.31), which is then a function of just $e_a{}^m$ and its first derivatives, and thus correct even at boundaries. (We also dropped a total-derivative term $-\partial_m (\tfrac{1}{2} \phi^2 e^{-1} e^{am} c_a{}^b{}_b)$ in

(4.1.36), which will be irrelevant for the following discussion.) If we eliminate ϕ by its field equation, but keep surface terms, this becomes

$$\begin{aligned} S &\to \int d^D x\, e^{-1}\, 2\tfrac{D-1}{D-2}\nabla\cdot(\phi\nabla\phi) \\ &= \int d^D x\, e^{-1}\, \tfrac{D-1}{D-2}\Box\phi^2 \\ &= \int d^D x\, e^{-1}\, \tfrac{D-1}{D-2}\Box\left[\langle\phi\rangle^2 + 2\langle\phi\rangle(\phi-\langle\phi\rangle) + (\phi-\langle\phi\rangle)^2\right] \end{aligned} \tag{4.1.37}$$

We can solve the ϕ field equation as

$$\phi = \langle\phi\rangle\left(1 - \frac{1}{4\frac{D-1}{D-2}\Box + R}R\right) \tag{4.1.38}$$

(We can choose $\langle\phi\rangle = 1$, or take the κ out of (4.1.28) and introduce it instead through $\langle\phi\rangle = \kappa^{-(D-2)/2}$ by a global ζ transformation.) Assuming ϕ falls off to $\langle\phi\rangle$ fast enough at ∞, the last term in (4.1.37) can be dropped, and, using (4.1.38), the action becomes [4.6]

$$S \to -\tfrac{1}{2}\int d^D x\, e^{-1}\left(R - R\frac{1}{4\frac{D-1}{D-2}\Box + R}R\right) \tag{4.1.39}$$

Since this action has the invariance (4.1.35), we can gauge away the trace of h or, equivalently, gauge the determinant of $e_a{}^m$ to 1. In fact, the same action results from (4.1.30) if we eliminate this determinant by its equation of motion.

Thus, we see that, although gauge-covariant, Lorentz-covariant formulations are possible without the extra auxiliary fields, they are nonlocal. Furthermore, the nonlocalities become more complicated when coupling to nonconformal matter (such as massive fields), in a way reminiscent of Coulomb terms or the nonlocalities in light-cone gauges. Thus, the construction of actions in such a formalism is not straightforward, and requires the use of Weyl invariance in a way analogous to the use of Lorentz invariance in light-cone gauges. Another alternative is to eliminate the trace of the metric from the Einstein action by a coordinate choice, but the remaining constrained (volume-preserving) coordinate invariance causes difficulties in quantization [4.7].

We have also seen that some properties of gravity (the ones relating to conformal transformations) become more transparent when the scale compensator ϕ is introduced. (This is particularly true for supergravity.) Introducing such fields into the OSp formalism requires introducing new degrees of freedom, to make the representation larger (at least in terms of gauge degrees of freedom). Although such invariances are hard to recognize at the free level, the extensions of sect. 3.6 show signs of performing such generalizations. However, while the U(1)-type extension can be applied to arbitrary Poincaré representations, the GL(1)-type has difficulty with fermions. We'll first discuss this difficulty, then show how the 2 types differ for bosons even for the vector, and finally look again at gravity.

The U(1) case of spin 1/2 reproduces the algebra of sect. 3.5, since $M_{A'B'}$ of (3.6.12b) is exactly the extra term of (3.5.1):

$$M_{ij} = \tfrac{1}{2}\gamma_{ij} \quad \rightarrow$$

$$\begin{aligned} \bar{\phi}\delta(\widehat{M}_{\alpha\beta}{}^2)(\Box + \widehat{Q}^2)\phi &= \bar{\phi}\delta(\;)e^{i\gamma_{+'}\not{p}/2} i\tfrac{1}{2}\not{p}\gamma_{-'}e^{i\gamma_{+'}\not{p}/2}\phi \\ &= i\tfrac{1}{4}\overline{(\gamma_{-'}\phi + i\tfrac{1}{2}\gamma_{-'}\gamma_{+'}\not{p}\phi)}\delta(\;)\gamma_{+'}\not{p}(\gamma_{-'}\phi + i\tfrac{1}{2}\gamma_{-'}\gamma_{+'}\not{p}\phi) \\ &= i\tfrac{1}{2}\hat{\bar{\phi}}\not{p}\gamma_{-'}\hat{\phi} \quad , \end{aligned} \tag{4.1.40}$$

where $\gamma_{ij} = \frac{1}{2}[\gamma_i, \gamma_j\}$, and we have used

$$0 = \tfrac{1}{8}(\gamma^{\alpha\beta} + \gamma^{\alpha'\beta'})(\gamma_{\alpha\beta} + \gamma_{\alpha'\beta'}) = (\gamma^\alpha\gamma_{\alpha'})^2 + 4 \quad \rightarrow \quad \gamma^\alpha\gamma_{\alpha'} = 2i \quad . \tag{4.1.41}$$

(We could equally well have chosen the other sign. This choice, with our conventions, corresponds to harmonic-oscillator boundary conditions: See sect. 4.5.) After eliminating $\gamma_{+'}\phi$ by gauge choice or, equivalently, by absorbing it into $\gamma_{-'}\phi$ by field redefinition, this becomes just $\bar{\varphi}\not{p}\varphi$. However, in the GL(1) case, the analog to (4.1.41) is

$$\begin{aligned} 0 = \tfrac{1}{8}(\{\gamma^\alpha, \gamma^{\beta'}\} + \{\gamma^{\alpha'}, \gamma^\beta\})(\{\gamma_\alpha, \gamma_{\beta'}\} + \{\gamma_{\alpha'}, \gamma_\beta\}) = (\gamma^\alpha\gamma_{\alpha'})(\gamma^{\beta'}\gamma_\beta) \quad , \\ \gamma^\alpha\gamma_{\alpha'} + \gamma^{\alpha'}\gamma_\alpha = -4 \quad , \end{aligned} \tag{4.1.42}$$

and to (4.1.40) is

$$\Box + \widehat{Q}^2 = -p^2 + \tfrac{1}{8}(\gamma^\alpha\gamma_{\alpha'})\gamma_{-'}(\not{p} - \gamma_+ p^2) + \tfrac{1}{8}(\gamma^{\alpha'}\gamma_\alpha)(\not{p} - \gamma_+ p^2)\gamma_{-'} \quad . \tag{4.1.43}$$

Unfortunately, ϕ and $\bar{\phi}$ must have opposite boundary conditions $\gamma^\alpha\gamma_{\alpha'} = 0$ or $\gamma^{\alpha'}\gamma_\alpha = 0$ in order to contribute in the presence of $\delta(\widehat{M}_{\alpha\beta}{}^2)$, as is evidenced by the asymmetric form of (4.1.43) for either choice. Consequently, the parts of ϕ and $\bar{\phi}$ that survive are not hermitian conjugates of each other, and the action is not unitary. (Properly speaking, if we choose consistent boundary conditions for both ϕ and $\bar{\phi}$, the action vanishes.) Thus, the GL(1)-type OSp(1,1|2) is unsuitable for spinors unless further modified. In any case, such a modification would not treat bosons and fermions symmetrically, which is necessary for treating supersymmetry. (Fermions in the usual OSp formalism will be discussed in more detail in sect. 4.5.)

For the case of spin 1 (generalizing the light-cone Hilbert space, as in (4.1.7-8)), we expand

$$\phi = |^a\rangle A_a + i\,|_{-'}\rangle A_+ + i\,|_+\rangle A_- \quad , \quad \phi^\dagger = A^a \langle_a| - iA_+ \langle_{-'}| - iA_- \langle_+| \quad , \tag{4.1.44}$$

for the GL(1) case, and the same for U(1) with $|_+\rangle \rightarrow |_{+'}\rangle$ (Sp(2)-spinor fields again drop out of the full ϕ). We find for GL(1) [3.7]

$$L = -\tfrac{1}{4}{F_{ab}}^2 - \tfrac{1}{2}(A_- + \partial \cdot A)^2 = -\tfrac{1}{4}{F_{ab}}^2 - \tfrac{1}{2}{\hat{A}_-}^2 \quad , \tag{4.1.45a}$$

where $F_{ab} = \partial_{[a}A_{b]}$, and for U(1)

$$L = -\tfrac{1}{4}{F_{ab}}^2 + \tfrac{1}{2}(A_- + \tfrac{1}{2}\Box A_+)^2 = -\tfrac{1}{4}{F_{ab}}^2 + \tfrac{1}{2}{\hat{A}_-}^2 \quad . \tag{4.1.45b}$$

In both cases A_+ can be gauged away, and A_- is auxiliary. However, the sign for U(1)-type OSp(1,1|2) is the same as for auxiliary fields in supersymmetry (for off-shell irreducible multiplets), whereas the sign for GL(1) is opposite. The sign difference is not surprising, considering the U(1) and GL(1) types are Wick rotations of each other: This auxiliary-field term, together with the auxiliary component of A_a (the light-cone A_-), appear with the metric $\eta_{\hat{a}\hat{b}}$ of (3.6.1), and thus with the same sign for SO(2) (U(1)). In fact, (4.1.45b) is just the part of the 4D N=1 super-Yang-Mills lagrangian for fields which are R-symmetry invariant: A_- can be identified with the usual auxiliary field, and A_+ with the $\theta = 0$ component of the superfield. Similarly, $\gamma_{+'}\phi$ for spin 1/2 can be identified with the linear-in-θ part of this superfield. This close analogy strongly suggests that the nonminimal fields of this formalism may be necessary for treating supersymmetry. Note also that for GL(1) the auxiliary automatically mixes with the spin-1 "gauge-fixing" function, like a Nakanishi-Lautrup field, while for U(1) there is a kind of "parity" symmetry of the OSp(1,1|2) generators, $|_{A'}\rangle \rightarrow -\,|_{A'}\rangle$, which prevents such mixing, and can be included in the usual parity transformations to strengthen the identification with supersymmetry.

For spin 2, for U(1) we define

$$\begin{aligned}\phi = \tfrac{1}{2}h^{ab}\frac{1}{\sqrt{2}}\,|_{(a}\rangle\,|_{b)}\rangle + iA^a{}_+\frac{1}{\sqrt{2}}\,|_{(a}\rangle\,|_{-')}\rangle + iA^a{}_-\frac{1}{\sqrt{2}}\,|_{(a}\rangle\,|_{+')}\rangle \\ + \varphi_{++}\,|_{-'}\rangle\,|_{-'}\rangle + \varphi_{+-}\frac{1}{\sqrt{2}}\,|_{(+'}\rangle\,|_{-')}\rangle + \varphi_{--}\,|_{+'}\rangle\,|_{+'}\rangle \\ + \varphi\frac{1}{\sqrt{2}}\,|^\alpha\rangle\,|_\alpha\rangle + \varphi'\tfrac{1}{2}\left(\left|^\alpha\right\rangle\left|_{\alpha'}\right\rangle + \left|^{\alpha'}\right\rangle\left|_\alpha\right\rangle\right) + \varphi''\frac{1}{\sqrt{2}}\left|^{\alpha'}\right\rangle\left|_{\alpha'}\right\rangle \quad ,\end{aligned} \tag{4.1.46a}$$

subject to the tracelessness condition ($h^i{}_i = 0$)

$$\tfrac{1}{2}h^a{}_a + \varphi_{+-} - \varphi - \varphi'' = 0 \quad , \tag{4.1.46b}$$

and find the lagrangian

$$\begin{aligned} L = & -\tfrac{1}{4}h^{ab}(p^2 h_{ab} - 2p_a p^c h_{cb} + 2p_a p_b h^c{}_c - \eta_{ab}p^2 h^c{}_c) \\ & + \tfrac{1}{2}(A_{a-} - \tfrac{1}{2}p^2 A_{a+} + p_a p^b A_{b+} - i\sqrt{2}p_a\varphi')^2 \\ & + (\varphi_{+-} - \varphi'')(\sqrt{2}\varphi_{--} + p^2\varphi_{+-} - 2p^2\varphi'' + \frac{1}{2\sqrt{2}}p^4\varphi_{++} + p^2 h^b{}_b - p^b p^c h_{bc}) \\ = & \text{ “R”} + \tfrac{1}{2}\hat{A}_{a-}{}^2 + \sqrt{2}\hat{\varphi}_{+-}\hat{\varphi}_{--} \quad . \end{aligned} \tag{4.1.47}$$

The second term is the square of an auxiliary “axial” vector (which again appears with sign opposite to that in GL(1) [3.7]), which resembles the axial vector auxiliary field of supergravity (including terms which can be absorbed, as for spin 1). In the last term, the redefinition $\varphi_{--} \to \hat{\varphi}_{--}$ involves the (linearized) Ricci scalar. Although it's difficult to tell from the free theory, it may also be possible to identify some of the gauge degrees of freedom with conformal compensators: φ' with the compensator for local R-symmetry, and φ_{+-} (or φ or φ''; one is eliminated by the tracelessness condition and one is auxiliary) with the local scale compensator.

A simple expression for interacting actions in terms of just the OSp(1,1|2) group generators has not yet been found. (However, this is not the case for IGL(1): See the following section.) The usual gauge-invariant interacting field equations can be derived by imposing $J_{\alpha\beta}\phi = J_{-\alpha}\phi = 0$, which are required in a (anti)BRST formalism, and finding the equations satisfied by the $x_- = x_\alpha = 0$ sector. However, this requires use of the other sectors as auxiliary fields, whereas in the approach described here they would be gauge degrees of freedom.

These results for gauge-invariant actions from OSp(1,1|2) will be applied to the special case of the string in chapt. 11.

4.2. IGL(1)

We now derive the corresponding gauge-invariant action in the IGL(1) formalism and compare with the OSp(1,1|2) results. We begin with the form of the generators (3.4.3b) obtained from the transformation (3.4.3a). For the IGL(1) formalism we can then drop the zero-modes x_- and $\tilde{c}$, and the action and invariance then are (using $\delta(Q) = Q$)

$$S = -\int d^D x dc\; \Phi^\dagger\, iQ\delta(J^3)\,\Phi \quad , \quad \delta\Phi = -iQ\Lambda + J^3\hat{\Lambda} \quad . \tag{4.2.1}$$

This is the IGL(1) analog of (4.1.1). (This action also was first proposed for the string [4.8,9].) The $\delta(J^3)$ kills the sign factor in (3.4.4). However, even though some unphysical coordinates have been eliminated, the field is still a representation of the spin group OSp(D-1,1|2) (or OSp(D,1|2)), and thus there is still a "hidden" Sp(2) symmetry broken by this action (but only by auxiliary fields: see below).

To obtain the analog of (4.1.6), we first expand the field in the single ghost zero-mode c:

$$\Phi = \phi + ic\psi \quad . \tag{4.2.2}$$

ϕ is the field of the OSp(1,1|2) formalism after elimination of all its gauge zero-modes, and ψ is an auxiliary field (identified with the Nakanishi-Lautrup auxiliary fields in the gauge-fixed formalism [4.4,5]). If we expand the action (4.2.1) in c, using (3.4.3b), and the reality condition on the field to combine crossterms, we obtain, with $Q^\alpha = (Q^+, Q^-)$,

$$\begin{aligned}\mathcal{L} &= -\int dc\ \Phi^\dagger iQ\delta(J^3)\Phi \\ &= \tfrac{1}{2}\phi^\dagger(\Box - M^2)\delta(M^3)\phi - \psi^\dagger M^+\delta(M^3+1)\psi + 2\psi^\dagger Q^+\delta(M^3)\phi \ .\end{aligned} \tag{4.2.3}$$

As an example of this action, we again consider a massless vector. In analogy to (4.1.7),

$$\Phi = |^i\rangle\, \phi_i + ic\, |^i\rangle\, \psi_i \quad . \tag{4.2.4a}$$

After the $\delta(J^3)$ projection, the only surviving fields are

$$\Phi = \big|^a\big\rangle A_a + ic\big|^-\big\rangle B \quad , \tag{4.2.4b}$$

where B is the auxiliary field. Then, using the relations (from (4.1.8))

$$M^3\big|^a\big\rangle = 0 \quad , \quad M^3\big|^\pm\big\rangle = \pm\big|^\pm\big\rangle \quad ;$$

$$M^+\big|^a\big\rangle = M^+\big|^+\big\rangle = 0 \quad , \quad M^+\big|^-\big\rangle = 2i\big|^+\big\rangle \quad ;$$

$$M^{+a}\big|^b\big\rangle = \eta^{ab}\big|^+\big\rangle \quad , \quad M^{+a}\big|^+\big\rangle = 0 \quad , \quad M^{+a}\big|^-\big\rangle = -i\big|^a\big\rangle \quad ; \tag{4.2.5}$$

we find the lagrangian and invariance

$$\mathcal{L} = \tfrac{1}{2}A^a\Box A_a - 2B^2 + 2B\partial^a A_a \quad ;$$

$$\delta A_a = \partial_a\lambda \quad , \quad \delta B = \tfrac{1}{2}\Box\lambda \quad ; \tag{4.2.6}$$

which yields the usual result after elimination of B by its equation of motion. This is the same lagrangian, including signs and auxiliary-field redefinitions, as for the GL(1)-type 4+4-extended OSp(1,1|2), (4.1.45a).

Any IGL(1) action can be obtained from a corresponding OSp(1,1|2) action, and vice versa [3.13]. Eliminating ψ from (4.2.3) by its equation of motion,

$$0 = \frac{\delta S}{\delta \psi^\dagger} \sim M^+ \delta(M^3 + 1)\psi - \mathcal{Q}^+ \delta(M^3)\phi$$

$$\rightarrow \quad \mathcal{L}' = \tfrac{1}{2}\phi^\dagger(\Box - M^2 - 2\mathcal{Q}^+ M^{+-1}\mathcal{Q}^+)\delta(M^3)\phi \quad , \tag{4.2.7}$$

the OSp(1,1|2) action (4.1.6) is obtained:

$$\begin{aligned}
(\Box - M^2 &- 2\mathcal{Q}^+ M^{+-1}\mathcal{Q}^+)\left[\delta(M^3) - \delta({M_{\alpha\beta}}^2)\right] \\
&= (\Box - M^2 - 2\mathcal{Q}^+ M^{+-1}\mathcal{Q}^+) M^+ M^{+-1}\delta(M^3) \\
&= \left[(\Box - M^2)M^+ - 2\mathcal{Q}^+ M^{+-1} M^+ \mathcal{Q}^+\right] M^{+-1}\delta(M^3) \\
&= \left[(\Box - M^2)M^+ - 2\mathcal{Q}^{+2}\right] M^{+-1}\delta(M^3) \\
&= 0
\end{aligned}$$

$$\begin{aligned}
\rightarrow \quad \mathcal{L}' &= \tfrac{1}{2}\phi^\dagger(\Box - M^2 - 2\mathcal{Q}^+ M^{+-1}\mathcal{Q}^+)\delta({M_{\alpha\beta}}^2)\phi \\
&= \tfrac{1}{2}\phi^\dagger(\Box - M^2 + {\mathcal{Q}_\alpha}^2)\delta({M_{\alpha\beta}}^2)\phi \quad .
\end{aligned} \tag{4.2.8}$$

We have also used $\mathcal{Q}^{+2} = (\Box - M^2)M^+$, which follows from the OSp commutation relations, or from $Q^2 = 0$. M^{+-1} is an Sp(2) lowering operator normalized so that it is the inverse of the raising operator M^+, except that it vanishes on states where M^3 takes its minimum value [4.1]. It isn't an inverse in the strict sense, since M^+ vanishes on certain states, but it's sufficient for it to satisfy

$$M^+ M^{+-1} M^+ = M^+ \quad . \tag{4.2.9}$$

We can obtain an explicit expression for M^{+-1} using familiar properties of SO(3) (SU(2)). The Sp(2) operators are related to the conventionally normalized SO(3) operators by $(M^3, M^\pm) = 2(T^3, T^\pm)$. However, these are really SO(2,1) operators, and so have unusual hermiticity conditions: T^+ and T^- are each hermitian, while T^3 is antihermitian. Since for any SU(2) algebra $\vec{T}$ the commutation relations

$$[T^3, T^\pm] = \pm T^\pm \quad , \quad [T^+, T^-] = 2T^3 \tag{4.2.10}$$

imply

$$\vec{T}^2 = (T^3)^2 + \tfrac{1}{2}\{T^+, T^-\} = T(T+1) \quad \to \quad T^\mp T^\pm = (T \mp T^3)(T \pm T^3 + 1) \quad , \tag{4.2.11}$$

we can write

$$T^{+-1} = \frac{1}{T^- T^+} T^- = \frac{1 - \delta_{T^3,T}}{(T - T^3)(T + T^3 + 1)} T^- \quad . \tag{4.2.12}$$

We can then verify (4.2.9), as well as the identities

$$T^{+-1} T^+ T^{+-1} = T^{+-1} \quad ,$$

$$T^{+-1} T^+ = 1 - \delta_{T^3,T} \quad , \quad T^+ T^{+-1} = 1 - \delta_{T^3,-T} \quad . \tag{4.2.13}$$

Conversely, the IGL(1) action can be obtained by partial gauge-fixing of the OSp(1,1|2) action, by writing $\mathcal{L}$ of (4.2.3) as $\mathcal{L}'$ of (4.2.7) plus a pure BRST variation. Using the covariantly second-quantized BRST operator of sect. 3.4, we can write

$$\mathcal{L} = \mathcal{L}' + \left[Q, -\phi^\dagger M^{+-1}[Q, \delta(M^3)\phi]_c\right]_c \quad . \tag{4.2.14}$$

Alternatively, we can use functional notation, defining the operator

$$\mathcal{Q} = -\int dx dc \, (Q\Phi) \frac{\delta}{\delta \Phi} \quad . \tag{4.2.15}$$

In terms of $J_+{}^\alpha = (R, \widetilde{R})$, the extra terms fix the invariance generated by R, which had allowed c to be gauged away. This also breaks the Sp(2) down to GL(1), and breaks the antiBRST invariance. Another way to understand this is by reformulating the IGL(1) in terms of a field which has all the zero-modes of the OSp(1,1|2) field Φ. Consider the action

$$S = \int d^D x d^2 x_\alpha dx_- \; \Phi^\dagger \, p_+ \delta(J^3) i \delta(J_{-+}) \delta(\widetilde{R}) \delta(Q) \, \Phi \quad . \tag{4.2.16}$$

The gauge invariance is now given by the 4 generators appearing as arguments of the δ functions, and is reduced from the OSp(1,1|2) case by the elimination of the generators $(J^\pm, R, \widetilde{Q})$. This algebra is the algebra GL(1|1) of $N = 2$ supersymmetric quantum mechanics (also appearing in the IGL(1) formalism for the closed string [4.10]: see sect. 8.2): The 2 fermionic generators are the "supersymmetries," $J^3 + J_{-+}$ is the O(2) generator which scales them, and $J^3 - J_{-+}$ is the "momentum." If the gauge coordinates x_- and $\tilde{c}$ are integrated out, the action (4.2.1) is obtained, as can be seen with the aid of (3.4.3).

In contrast to the light cone, where the hamiltonian operator H $(= -p_-)$ is essentially the action ((2.4.4)), we find that with the new covariant, second-quantized bracket of (3.4.7) the covariant action is the BRST operator: Because the action (2.4.8) of a generator (2.4.7) on $\Phi^\dagger$ is equivalent to the generator's functional derivative (because of (3.4.7)), the gauge-invariant action now thought of as an operator satisfies

$$[S, \Phi^\dagger]_c = \tfrac{1}{2}\frac{\delta S}{\delta \Phi} \quad . \tag{4.2.17}$$

Furthermore, since the gauge-covariant equations of motion of the theory are given by the BRST transformations generated by the operator Q, one has

$$\frac{\delta S}{\delta \Phi} = -2i[Q, \Phi^\dagger]_c \tag{4.2.18}$$

$$\rightarrow \quad S = -iQ \quad . \tag{4.2.19}$$

(Strictly speaking, S and Q may differ by an irrelevant Φ-independent term.) This statement can be applied to any formalism with field equations that follow from the BRST operator, independent of whether it originates from the light-cone, and it holds in interacting theories as well as free ones. In particular, for the case of interacting Yang-Mills, the action follows directly from (3.4.18). After restricting the fields to $J^3 = 0$, this gives the interacting generalization of the example of (4.2.6). The action can also be written as $S = -2i \int d\Phi \ Q\Phi$, where $\int d\Phi \ \Phi^n \equiv \frac{1}{n+1} \int \Phi^\dagger \Phi^n$.

This operator formalism is also useful for deriving the gauge invariances of the interacting theory, in either the IGL(1) or OSp(1,1|2) formalisms (although the corresponding interacting action is known in this form only for IGL(1), (4.2.19).) Just as the global BRST invariances can be written as a unitary transformation (in the notation of (3.4.17))

$$U = e^{iL_G} \quad , \quad G = \epsilon \mathcal{O} \quad , \quad \epsilon = constant \quad , \tag{4.2.20}$$

where $\mathcal{O}$ is any IGL(1) (or OSp(1,1|2)) operator (in covariant second-quantized form), the gauge transformations can be written similarly but with

$$G = [f, \mathcal{O}]_c \quad , \tag{4.2.21}$$

where f is linear in Φ ($f = \int \Lambda\Phi$) for the usual gauge transformation (and f's higher-order in Φ may give field-dependent gauge transformations). Thus, $\Phi' = U\Phi U^{-1}$, and $g(\Phi)' = Ug(\Phi)U^{-1}$ for any functional g of Φ. In the free case, this reproduces (4.1.1,4.2.1).

This relation between OSp(1,1|2) and IGL(1) formalisms is important for relating different first-quantizations of the string, as will be discussed in sect. 8.2.

4.3. Extra modes

As discussed in sect. 3.2, extra sets of unphysical modes can be added to BRST formalisms, such as those which Lorentz gauges have with respect to temporal gauges, or those in the 4+4-extended formalisms of sect. 3.6. We now prove the equivalence of the OSp(1,1|2) actions of formulations with and without such modes [3.13]. Given that IGL(1) actions and equations of motion can be reduced to OSp forms, it's sufficient to show the equivalence of the IGL actions with and without extra modes. The BRST and ghost-number operators with extra modes, after the redefinition of (3.3.6), differ from those without by the addition of abelian terms. We'll prove that the addition of these terms changes the IGL action (4.2.1) only by adding auxiliary and gauge degrees of freedom. To prove this, we consider adding such terms 2 sets of modes at a time (an even number of additional ghost modes is required to maintain the fermionic statistics of the integration measure):

$$Q = Q_0 + (\mathbf{b}^\dagger\mathbf{f} - \mathbf{f}^\dagger\mathbf{b}) \quad ,$$

$$[\mathbf{b},\mathbf{g}^\dagger] = [\mathbf{g},\mathbf{b}^\dagger] = \{\mathbf{c},\mathbf{f}^\dagger\} = \{\mathbf{f},\mathbf{c}^\dagger\} = 1 \quad , \tag{4.3.1}$$

in terms of the "old" BRST operator Q_0 and the 2 new sets of modes **b**, **g**, **c**, and **f**, and their hermitian conjugates. We also assume boundary conditions in the new coordinates implied by the harmonic-oscillator notation. (Otherwise, additional unphysical fields appear, and the new action isn't equivalent to the original one: see below.) By an explicit expansion of the new field over all the new oscillators,

$$\Phi = \sum_{m,n=0}^{\infty} (A_{mn} + iB_{mn}\mathbf{c}^\dagger + iC_{mn}\mathbf{f}^\dagger + iD_{mn}\mathbf{f}^\dagger\mathbf{c}^\dagger)\frac{1}{\sqrt{m!n!}}(\mathbf{b}^\dagger)^m(\mathbf{g}^\dagger)^n\,|0\rangle \quad , \tag{4.3.2}$$

we find

$$\begin{aligned}\Phi^\dagger Q\Phi &= (A^\dagger Q_0 A + 2B^\dagger Q_0 C - D^\dagger Q_0 D) + 2B^\dagger(\mathbf{b}^\dagger D - i\mathbf{b}A) \\ &= \sum_{m,n=0}^{\infty}\Big[(A^\dagger{}_{mn}Q_0 A_{nm} + 2B^\dagger{}_{mn}Q_0 C_{nm} - D^\dagger{}_{mn}Q_0 D_{nm}) \\ &\qquad + 2B^\dagger{}_{mn}\Big(\sqrt{n}D_{n-1,m} - i\frac{1}{\sqrt{m+1}}A_{n,m+1}\Big)\Big] \quad .\end{aligned} \tag{4.3.3}$$

(The ground state in (4.3.2) and the matrix elements evaluated in (4.3.3) are with respect to only the new oscillators.) We can therefore shift A_{mn} by a $Q_0 C_{m,n-1}$ term to cancel the $B^\dagger Q_0 C$ term (using $Q_0{}^2 = 0$), and then B_{mn} by a $Q_0 D_{m,n-1}$ term to cancel the $D^\dagger Q_0 D$ term. We can then shift A_{mn} by $D_{m-1,n-1}$ to cancel the $B^\dagger D$ term, leaving only the $A^\dagger Q_0 A$ and $B^\dagger A$ terms. (These redefinitions are equivalent to the gauge choices $C = D = 0$ using the usual invariance $\delta\Phi = Q\Lambda$.) Finally, we can eliminate the Lagrange multipliers B by their equations of motion, which eliminate all of A except A_{m0}, and from the form of the remaining $A^\dagger Q_0 A$ term we find that all the remaining components of A except A_{00} drop out (i.e., are pure gauge). This leaves only the term $A^\dagger{}_{00} Q_0 A_{00}$. Thus, all the components except the ground state with respect to the new oscillators can be eliminated as auxiliary or gauge degrees of freedom. The net result is that all the new oscillators are eliminated from the fields and operators in the action (4.2.1), with Q thus replaced by Q_0 (and similarly for J^3). (A similar analysis can be performed directly on the equations of motion $Q\Phi = 0$, giving this general result for the cohomology of Q even in cases when the action is not given by (4.2.1).) This elimination of new modes required that the creation operators in (4.3.3) be left-invertible:

$$a^{\dagger -1} a^\dagger = 1 \quad \to \quad a^{\dagger -1} = \frac{1}{a^\dagger a + 1} a = \frac{1}{N+1} a \quad \to \quad N \geq 0 \quad , \qquad (4.3.4)$$

implying that all states must be expressible as creation operators acting on a ground state, as in (4.3.2) (the usual boundary conditions on harmonic oscillator wave functions, except that here **b** and **g** correspond to a space of indefinite metric). This proves the equivalence of the IGL(1) actions, and thus, by the previous argument, also the OSp(1,1|2) actions, with and without extra modes, and that the extra modes simply introduce more gauge and auxiliary degrees of freedom.

Such extra modes, although redundant in free theories, may be useful in formulating larger gauge invariances which simplify the form of interacting theories (as, e.g., in nonlinear σ models). The use in string theories of such extra modes corresponding to the world-sheet metric will be discussed in sect. 8.3.

4.4. Gauge fixing

We now consider gauge fixing of these gauge-invariant actions using the BRST algebra from the light cone, and relate this method to the standard second-quantized BRST methods described in sects. 3.1-3 [4.1]. We will find that the first-quantized BRST transformations of the fields in the usual gauge-fixed action are *identical* to the second-quantized BRST transformations, but the first-quantized BRST formalism has a larger set of fields, some of which drop out of the usual gauge-fixed action.

(E.g., see (3.4.19). However, gauges exist where these fields also appear.) Even in the IGL(1) formalism, although all the "propagating" fields appear, only a subset of the BRST "auxiliary" fields appear, since the two sets are equal in number in the first-quantized IGL(1) but the BRST auxiliaries are fewer in the usual second-quantized formalism. We will also consider gauge fixing to the light-cone gauge, and reobtain the original light-cone theories to which 2+2 dimensions were added.

For covariant gauge fixing we will work primarily within the IGL(1) formalism, but similar methods apply to OSp(1,1|2). Since the entire "hamiltonian" $\square - M^2$ vanishes under the constraint $Q = 0$ (acting on the field), the free gauge-fixed action of the field theory consists of only a "gauge-fixing" term:

$$\begin{aligned} S &= i\left[Q, \int dx dc \; \tfrac{1}{2}\Phi^\dagger \mathcal{O}\Phi\right]_c \\ &= \int dx dc \; \tfrac{1}{2}\Phi^\dagger [\mathcal{O}, iQ\}\Phi \\ &= \int dx dc \; \tfrac{1}{2}\Phi^\dagger K\Phi \quad , \end{aligned} \tag{4.4.1}$$

for some operator $\mathcal{O}$, where the first Q, appearing in the covariant bracket, is understood to be the second-quantized one. In order to get $\square - M^2$ as the kinetic operator for part of Φ, we choose

$$\mathcal{O} = -\left[c, \frac{\partial}{\partial c}\right] \quad \rightarrow \quad K = c(\square - M^2) - 2\frac{\partial}{\partial c}M^+ \ . \tag{4.4.2}$$

When expanding the field in c, $\square - M^2$ is the kinetic operator for the piece containing all physical and ghost fields. Explicitly, (3.4.3b), when substituted into the lagrangian $L = \frac{1}{2}\Phi^\dagger K\Phi$ and integrated over c, gives

$$\frac{\partial}{\partial c}L = \tfrac{1}{2}\phi^\dagger(\square - M^2)\phi + \psi^\dagger M^+\psi \ , \tag{4.4.3}$$

and in the BRST transformations $\delta\Phi = i\epsilon Q\Phi$ gives

$$\delta\phi = i\epsilon(Q^+\phi - M^+\psi) \quad , \quad \delta\psi = i\epsilon\left[Q^+\psi - \tfrac{1}{2}(\square - M^2)\phi)\right] \quad . \tag{4.4.4}$$

ϕ contains propagating fields and ψ contains BRST auxiliary fields. Although the propagating fields are completely gauge-fixed, the BRST auxiliary fields have the gauge invariance

$$\delta\psi = \lambda \quad , \quad M^+\lambda = 0 \quad . \tag{4.4.5}$$

The simplest case is the scalar $\Phi = \varphi(x)$. In this case, all of ψ can be gauged away by (4.4.5), since $M^+ = 0$. The lagrangian is just $\frac{1}{2}\varphi(\square - m^2)\varphi$. For the massless vector (cf. (4.2.4b)),

$$\Phi = \left|^i\right\rangle A_i + ic\left|^-\right\rangle B \quad , \tag{4.4.6}$$

where we have again used (4.4.5). By comparing (4.4.3) with (4.2.3), we see that the ϕ^2 term is extended to all J^3, the ψ^2 term has the opposite sign, and the crossterm is dropped. We thus find

$$\mathcal{L} = \tfrac{1}{2}(A^i)^\dagger \Box A_i + 2B^2 = \tfrac{1}{2}A^a \Box A_a - i\widetilde{C}\Box C + 2B^2 \quad , \tag{4.4.7}$$

where we have written $A^\alpha = i(C, \widetilde{C})$, due to (3.4.4,13). This agrees with the result (3.2.11) in the gauge $\zeta = 1$, where this $B = \frac{1}{2}\widetilde{B}$. The BRST transformations (4.4.4), using (4.2.5), are

$$\begin{aligned} \delta A_a &= i\epsilon\partial_a C \quad , \\ \delta C &= 0 \quad , \\ \delta\widetilde{C} &= \epsilon(2B - \partial\cdot A) \quad , \\ \delta B &= i\tfrac{1}{2}\epsilon\Box C \quad , \end{aligned} \tag{4.4.8}$$

which agrees with the linearized case of (3.2.8).

We next prove the equivalence of this form of gauge fixing with the usual approach, described in sect. 3.2 [4.1] (as we have just proven for the case of the massless vector). The steps are: (1) Add terms to the original BRST auxiliary fields, which vanish on shell, to make them BRST invariant, as they are in the usual BRST formulation of field theory. (In Yang-Mills, this is the redefinition $B \to \widetilde{B}$ in (3.2.11).) (2) Use the BRST transformations to identify the physical fields (which may include auxiliary components). We can then reobtain the gauge-invariant action by dropping all other fields from the lagrangian, with the gauge transformations given by replacing the ghosts in the BRST transformations by gauge parameters.

In the lagrangian (4.4.3) only the part of the BRST auxiliary field ψ which appears in $M^+\psi$ occurs in the action; the rest of ψ is pure gauge and drops out of the action. Thus, we only require that the shifted $M^+\psi$ be BRST-invariant:

$$\psi = \widetilde{\psi} + A\phi \quad , \quad \delta M^+\widetilde{\psi} = 0 \quad . \tag{4.4.9}$$

Using the BRST transformations (4.4.4) and the identities (from (3.4.3b))

$$\begin{aligned} Q^2 = 0 \quad \to \quad [\Box - M^2, M^+] = [\Box - M^2, Q^+] = [M^+, Q^+] = 0 \quad , \\ Q^{+2} = \tfrac{1}{2}(\Box - M^2)M^+ \quad , \end{aligned} \tag{4.4.10}$$

we obtain the conditions on A:

$$(Q^+ - M^+A)Q^+ = (Q^+ - M^+A)M^+ = 0 \quad . \tag{4.4.11}$$

The solution to these equations is

$$A = M^{+-1}Q^+ . \tag{4.4.12}$$

Performing the shift (4.4.9), the gauge-fixed lagrangian takes the form

$$\frac{\partial}{\partial c}L = \tfrac{1}{2}\phi^\dagger \hat{K}\phi + 2\widetilde{\psi}^\dagger(1-\delta_{M^3,-2T})Q^+\phi + \widetilde{\psi}^\dagger M^+\widetilde{\psi} \quad , \tag{4.4.13a}$$

$$\hat{K} = \Box - M^2 - 2Q^+M^{+-1}Q^+ \quad , \tag{4.4.13b}$$

where T is the "isospin," as in (4.2.11), for $M_{\alpha\beta}$. The BRST transformations can now be written as

$$\delta\delta_{M^3,-2T}\phi = \epsilon\delta_{M^3,-2T}Q^+\phi \quad , \quad \delta(1-\delta_{M^3,-2T})\phi = -\epsilon M^+\widetilde{\psi} \quad ,$$

$$\delta\widetilde{\psi} = \epsilon\delta_{M^3,2T}(Q^+\widetilde{\psi} - \tfrac{1}{2}\hat{K}\phi) \quad . \tag{4.4.14}$$

The BRST transformation of $\widetilde{\psi}$ is pure gauge, and can be dropped. (In some of the manipulations we have used the fact that Q, Q^+, and $\Box - M^2$ are symmetric, i.e., even under integration by parts, while M^+ is antisymmetric, and Q and $\mathcal{Q}$ are antihermitian while M^+ and $\Box - M^2$ are hermitian. In a coordinate representation, particularly for c, all symmetry generators, such as Q, Q^+, and M^+, would be antisymmetric, since the fields would be real.)

We can now throw away the BRST-invariant BRST auxiliary fields $\widetilde{\psi}$, but we must also separate the ghost fields in ϕ from the physical ones. According to the usual BRST procedure, the physical modes of a theory are those which are both BRST-invariant and have vanishing ghost number (as well as satisfy the field equations). In particular, physical fields may transform into ghosts (corresponding to gauge transformations, since the gauge pieces are unphysical), but never transform into BRST auxiliary fields. Therefore, from (4.4.14) we must require that the physical fields have $M^3 = -2T$ to avoid transforming into BRST auxiliary fields, but we also require vanishing ghost number $M^3 = 0$. Hence, the physical fields (located in ϕ) are selected by requiring the simple condition of vanishing isospin $T = 0$. If we project out the ghosts with the projection operator δ_{T0} and use the identity (4.2.8), we obtain a lagrangian containing only physical fields:

$$\mathcal{L}_1 = \tfrac{1}{2}\phi^\dagger(\Box - M^2 + Q^2)\delta_{T0}\phi \tag{4.4.15}$$

Its gauge invariance is obtained from the BRST transformations by replacing the ghosts (the part of ϕ appearing on the right-hand side of the transformation law) with the gauge parameter (the reverse of the usual BRST quantization procedure), and we add gauge transformations to gauge away the part of ϕ with $T \neq 0$:

$$\delta\phi = -i\tfrac{1}{2}\mathcal{Q}_\alpha\Lambda^\alpha + \tfrac{1}{2}M_{\alpha\beta}\Lambda^{\alpha\beta} \quad , \qquad (4.4.16)$$

where we have obtained the $\mathcal{Q}^-$ term from closure of $\mathcal{Q}^+$ with $M_{\alpha\beta}$. The invariance of (4.4.15) under (4.4.16) can be verified using the above identities. This action and invariance are just the original ones of the OSp(1,1|2) gauge-invariant formalism (or the IGL(1) one, after eliminating the NL auxiliary fields). The gauge-fixing functions for the Λ transformations are also given by the BRST transformations: They are the transformations of ghosts into physical fields:

$$F_{GF} = \mathcal{Q}^+\delta_{T0}\phi = p^a(M^+{}_a\delta_{T0}\phi) + M(M^+{}_m\delta_{T0}\phi) \quad . \qquad (4.4.17)$$

(The first term is the usual Lorentz-gauge gauge-fixing function for massless fields, the second term the usual addition for Stueckelberg/Higgs fields.) The gauge-fixed lagrangian (physical fields only) is thus

$$\mathcal{L}_{GF} = \mathcal{L}_1 - \tfrac{1}{4}F_{GF}{}^\dagger M^- F_{GF} = \tfrac{1}{2}\phi^\dagger(\Box - M^2)\delta_{T0}\phi \quad , \qquad (4.4.18)$$

in agreement with (4.4.3).

In summary, we see that this first-quantized gauge-fixing procedure is identical to the second-quantized one with regard to (1) the physical gauge fields, their gauge transformations, and the gauge-invariant action, (2) the BRST transformations of the physical fields, (3) the closure of the BRST algebra, (4) the BRST invariance of the gauge-fixed action, and (5) the invertibility of the kinetic operator after elimination of the NL fields. (1) implies that the two theories are physically the same, (2) and (3) imply that the BRST operators are the same, up to additional modes as in sect. 4.3, (4) implies that both are correctly gauge fixed (but perhaps in different gauges), and (5) implies that all gauge invariances have been fixed, including those of ghosts. Concerning the extra modes, from the $\Box - M^2$ form of the gauge-fixed kinetic operator we see that they are exactly the ones necessary to give good high-energy behavior of the propagator, and that we have chosen a generalized Fermi-Feynman gauge. Also, note the fact that the $c = 0$ (or $x^\alpha = 0$) part of the field contains exactly the right set of ghost fields, as was manifest by the arguments of sect. 2.6, whereas in the usual second-quantized formalism one begins with just the physical fields manifest, and the ghosts and their ghosts, etc., must be found by a step-by-step procedure. Thus we see that the OSp from the light cone not only gives a straightforward way for deriving general free gauge-invariant actions, but also gives a method for gauge fixing which is equivalent to, but more direct than, the usual methods.

We now consider gauge fixing to the light cone. In this gauge the gauge theory reduces back to the original light-cone theory from which it was heuristically obtained by adding 2+2 dimensions in sects. 3.4, 4.1. This proves a general "no-ghost theorem," that the OSp(1,1|2) (and IGL(1)) gauge theory is equivalent on-shell to the corresponding light-cone theory, for any Poincaré representation (including strings as a special case).

Consider an arbitrary bosonic gauge field theory, with action (4.1.6). (Fermions will be considered in the following section.) Without loss of generality, we can choose $M^2 = 0$, since the massive action can be obtained by dimensional reduction. The light-cone gauge is then described by the gauge-fixed field equations

$$p^2\phi = 0 \tag{4.4.19a}$$

subject to the gauge conditions, in the Lorentz frame $p_a = \delta_a{}^+ p_+$,

$$M_{\alpha\beta}\phi = 0 \quad , \quad \mathcal{Q}_\alpha\phi = M_{\alpha-}p_+\phi = 0 \quad , \tag{4.4.19b}$$

with the residual part of the gauge invariance

$$\delta\phi = -i\tfrac{1}{2}\mathcal{Q}_\alpha\Lambda^\alpha \sim M_{\alpha-}\Lambda^\alpha \quad , \tag{4.4.19c}$$

where $\pm$ now refer to the usual "longitudinal" Lorentz indices. (The light-cone gauge is thus a further gauge-fixing of the Landau gauge, which uses only (4.4.19ab).) (4.4.19bc) imply that the only surviving fields are singlets of the new OSp(1,1|2) algebra generated by $M_{\alpha\beta}$, $M_{\alpha\pm}$, M_{-+} (with longitudinal Lorentz indices $\pm$): i.e., those which satisfy $M_{AB}\phi = 0$ and can't be gauged away by $\delta\phi = M^{BA}\Lambda_{AB}$.

We therefore need to consider the subgroup SO(D−2)⊗OSp(1,1|2) of OSp(D−1,1|2) (the spin group obtained by adding 2+2 dimensions to the original SO(D−2)), and determine which parts of an irreducible OSp(D−1,12) representation are OSp(1,1|2) singlets. This is done most simply by considering the corresponding Young tableaux (which is also the most convenient method for adding 2+2 dimensions to the original representation of the light-cone SO(D−2)). This means considering tensor products of the vector (defining) representation with various graded symmetrizations and antisymmetrizations, and (graded) tracelessness conditions on all the indices. The obvious OSp(1,1|2) singlet is given by allowing all the vector indices to take only SO(D−2) values. However, the resulting SO(D−2) representation is reducible, since it is not SO(D−2)-traceless. The OSp(D−1,1|2)-tracelessness condition equates its SO(D−2)-traces to OSp(1,1|2)-traces of representations which differ only by replacing the traced SO(D−2) indices with traced OSp(1,1|2) ones. However, OSp(1,1|2) (or OSp(2N|2N) more generally) has the unusual property that its traces are not true singlets. The simplest example [4.11] (and one which we'll show is sufficient to treat the general case) is the graviton of (4.1.15). Considering just the OSp(1,1|2) values of the indices, there are 2 states which are singlets under the bosonic subgroup GL(2) generated by $M_{\alpha\beta}$, M_{-+}, namely $|_{(+}\rangle\, |_{-)}\rangle$, $|^{\alpha}\rangle\, |_{\alpha}\rangle$. However, of these two states, one linear combination is pure gauge and one is pure auxiliary:

$$\begin{aligned}
\left|\phi_1\right\rangle = \left|^A\right\rangle\left|_A\right\rangle \quad &\rightarrow \quad M_{AB}\left|\phi_1\right\rangle = 0 \quad , \quad \textit{but} \\
&\qquad \left|\phi_1\right\rangle = -\tfrac{1}{2}M_{\pm}{}^{\alpha}\left|_{(\mp}\right\rangle\left|_{\alpha)}\right\rangle \quad , \\
\left|\phi_2\right\rangle = \left|_{(+}\right\rangle\left|_{-)}\right\rangle - \left|^{\alpha}\right\rangle\left|_{\alpha}\right\rangle \quad &\rightarrow \quad \left|\phi_2\right\rangle \neq M^{AB}\left|\Lambda_{BA}\right\rangle \quad , \quad \textit{but} \\
&\qquad M_{\pm\alpha}\left|\phi_2\right\rangle = -2\left|_{(\pm}\right\rangle\left|_{\alpha)}\right\rangle \neq 0 \quad . \qquad (4.4.20)
\end{aligned}$$

This result is due basically to the fact that the graded trace can't be separated out in the usual way with the metric because of the identity

$$\eta^{AB}\eta_{BA} = \eta^{A}{}_{A} = (-1)^{A}\delta_{A}{}^{A} = 2 - 2 = 0 \quad . \qquad (4.4.21)$$

Similarly, the reducible OSp(1,1|2) representation which consists of the unsymmetrized direct product of an arbitrary number of vector representations will contain no singlets, since any one trace reduces to the case just considered, and thus the representations which result from graded symmetrizations and antisymmetrizations will also contain none. Thus, no SO(D−2)-traces of the original OSp(D−1,1|2) representation need be considered, since they are equated to OSp(1,1|2)-nonsinglet traces by the OSp(D−1,1|2)-tracelessness condition. Hence, the only surviving SO(D−2) representation is the original irreducible light-cone one, obtained by restricting all vector indices to their SO(D−2) values and imposing SO(D−2)-tracelessness.

These methods apply directly to open strings. The modification for closed strings will be discussed in sect. 11.1.

4.5. Fermions

The results of this section are based on the OSp(1,1|2) algebra of sect. 3.5.

The action and invariances are again given by (4.1.1), with the modified J_{AB} of sect. 3.5, and (4.1.3) is unchanged except for $M_{\alpha\beta} \to \widehat{M}_{\alpha\beta}$. We also allow for the inclusion of a matrix factor in the Hilbert-space metric to maintain the (pseudo)hermiticity of the spin operators (e.g., γ^0 for a Dirac spinor, so $\Phi^\dagger \to \bar{\Phi} = \Phi^\dagger\gamma^0$). Under the action of the δ functions, we can make the replacement

$$p_+{}^2 J_-{}_\alpha{}^2 \to -\tfrac{3}{4}(p^2+M^2) + (M_\alpha{}^b p_b + M_{\alpha m}M)^2$$
$$-\tfrac{1}{2}\tilde{\gamma}^\alpha(M_\alpha{}^b p_b + M_{\alpha m}M)\left[\tilde{\gamma}_- + \tilde{\gamma}_+\tfrac{1}{2}(p^2+M^2)\right] \quad , \qquad (4.5.1)$$

where $p^2 \equiv p^a p_a$. Under the action of the same δ functions, the gauge transformation generated by $J_{-\alpha}$ is replaced with

$$\delta\Phi = J_-{}^\alpha\Lambda_\alpha \quad \to \quad \left\{(M^{\alpha b}p_b + M^\alpha{}_m M) - \tfrac{1}{2}\tilde{\gamma}^\alpha\left[\tilde{\gamma}_- + \tilde{\gamma}_+\tfrac{1}{2}(p^2+M^2)\right]\right\}\Lambda_\alpha \quad . \qquad (4.5.2)$$

Choosing $\Lambda_\alpha = \tilde{\gamma}_\alpha\Lambda$, the $\tilde{\gamma}_-$ part of this gauge transformation can be used to choose the partial gauge

$$\tilde{\gamma}_+\Phi = 0 \quad . \qquad (4.5.3)$$

The action then becomes

$$S = \int d^D x\, \bar{\phi}\, \delta(\widehat{M}_{\alpha\beta}{}^2) i\tilde{\gamma}^\alpha(M_\alpha{}^b p_b + M_{\alpha m}M)\,\phi \quad , \qquad (4.5.4)$$

where we have reduced Φ to the half represented by ϕ by using (4.5.3). (The $\tilde{\gamma}_\pm$ can be represented as 2×2 matrices.) The remaining part of (4.5.2), together with the $J_{\alpha\beta}$ transformation, can be written in terms of $\tilde{\gamma}_\pm$-independent parameters as

$$\delta\phi = (M^{\alpha b}p_b + M^\alpha{}_m M)(\Lambda_\alpha + \tfrac{1}{2}\tilde{\gamma}_\alpha\tilde{\gamma}^\beta\Lambda_\beta) + \left[-\tfrac{1}{4}(p^2+M^2) + (M_\alpha{}^b p_b + M_{\alpha m}M)^2\right]B$$
$$+ \tfrac{1}{2}\widehat{M}^{\alpha\beta}\Lambda_{\alpha\beta} \quad . \qquad (4.5.5)$$

One way to get general irreducible spinor representations of orthogonal groups (except for chirality conditions) is to take the direct product of a Dirac spinor with an irreducible tensor representation, and then constrain it by setting to zero the result of multiplying by a γ matrix and contracting vector indices. Since the OSp representations used here are obtained by dimensional continuation, this means we use the same constraints, with the vector indices i running over all commuting and anticommuting values (including m, if we choose to define M_{im} by dimensional reduction from one extra dimension). The OSp spin operators can then be written as

$$M_{ij} = \check{M}_{ij} + \tfrac{1}{4}[\gamma_i, \gamma_j\} \quad , \tag{4.5.6}$$

where $\check{M}$ are the spin operators for some tensor representation and γ_i are the OSp Dirac matrices, satisfying the OSp Clifford algebra

$$\{\gamma_i, \gamma_j] = 2\eta_{ij} \quad . \tag{4.5.7}$$

We choose similar relations between γ's and $\tilde{\gamma}$'s:

$$\{\gamma_i, \tilde{\gamma}_B] = 0 \quad . \tag{4.5.8}$$

Then, noting that $\frac{1}{2}(\gamma^\alpha + i\tilde{\gamma}^\alpha)$ and $\frac{1}{2}(\gamma_\alpha - i\tilde{\gamma}_\alpha)$ satisfy the same commutation relations as creation and annihilation operators, respectively, we define

$$\gamma_\alpha = a_\alpha + a^\dagger{}_\alpha \quad , \quad \tilde{\gamma}_\alpha = i\left(a_\alpha - a^\dagger{}_\alpha\right) \quad ; \quad \left[a_\alpha, a^{\dagger\beta}\right] = \delta_\alpha{}^\beta \quad . \tag{4.5.9}$$

We also find

$$\widehat{M}_{\alpha\beta} = \check{M}_{\alpha\beta} + a^\dagger{}_{(\alpha} a_{\beta)} \quad . \tag{4.5.10}$$

This means that an arbitrary representation $\psi_{(\alpha\cdots\beta)}$ of the part of the Sp(2) generated by $\check{M}_{\alpha\beta}$ that is also a singlet under the full Sp(2) generated by $\widehat{M}_{\alpha\beta}$ can be written as

$$\psi_{(\alpha\cdots\beta)} = a^\dagger{}_\alpha \cdots a^\dagger{}_\beta \psi \quad , \quad a_\alpha \psi = 0 \quad . \tag{4.5.11}$$

In particular, for a Dirac spinor $\check{M} = 0$, so the action (4.5.4) becomes simply (see also (4.1.40))

$$S = \int d^D x \; \bar{\phi}(\not{p} + \not{M})\phi \quad , \tag{4.5.12}$$

where $\not{p} \equiv \gamma^a p_a$, $\not{M} \equiv \gamma_m M$ (γ_m is like γ_5), all dependence on γ^α and $\tilde{\gamma}^\alpha$ has been eliminated, and the gauge transformation (4.5.5) vanishes. (The transformation $\phi \to e^{i\gamma_m \pi/4}\phi$ takes $\not{p} + \not{M} \to \not{p} + iM$.)

In the case of the gravitino, we start with $\phi = |^i\rangle\, \phi_i$, where $|^i\rangle$ is a basis for the representation of $\check{M}$ (only). ϕ must satisfy not only $\widehat{M}_{\alpha\beta}\phi = 0$ but also the irreducibility condition

$$\gamma^i \phi_i = 0 \quad \to \quad \phi_\alpha = \tfrac{1}{2}\gamma_\alpha \gamma^a \phi_a \quad , \quad a_\alpha \phi^a = 0 \quad . \tag{4.5.13}$$

($a^\dagger{}_\alpha$ or $i\tilde{\gamma}_\alpha$ could be used in place of γ_α in the solution for ϕ_α.) Then using (4.1.8) for $\check{M}$ on ϕ, straightforward algebra gives the action

$$\begin{aligned} S &= \int d^D x \; \bar{\phi}_a (\eta^{ab} \not{p} - p^{(a}\gamma^{b)} + \gamma^a \not{p} \gamma^b)\phi_b \\ &= \int d^D x \; \bar{\phi}_a \gamma^{abc} p_b \phi_c \quad , \end{aligned} \tag{4.5.14}$$

where $\gamma^{abc} = (1/3!)\gamma^{[a}\gamma^{b}\gamma^{c]}$, giving the usual gravitino action for arbitrary D. The gauge invariance remaining in (4.5.5) after using (4.5.13-14), and making suitable redefinitions, reduces to the usual

$$\delta\phi_a = \partial_a\lambda \quad . \tag{4.5.15}$$

We now derive an alternative form of the fermionic action which corresponds to actions given in the literature for fermionic strings. Instead of explicitly solving the constraint $\widehat{M}_{\alpha\beta} = 0$ as in (4.5.11), we use the $\widehat{M}_{\alpha\beta}$ gauge invariance of (4.5.5) to "rotate" the $a^{\alpha\dagger}$'s. For example, writing $a^\alpha = (a^+, a^-)$, we can rotate them so they all point in the "+" direction: Then we need consider only ϕ's of the form $\phi(a^{+\dagger})\,|0\rangle$. (The + value of α should not be confused with the + index on p_+.) The $\delta(\widehat{M}_{\alpha\beta}{}^2)$ then picks out the piece of the form (4.5.11). (It "smears" over directions in Sp(2). This use of a^α is similar to the "harmonic coordinates" of harmonic superspace [4.12].) We can also pick

$$\phi = e^{-ia^{+\dagger}a^{-\dagger}}\phi(a^{+\dagger})\,|0\rangle \quad , \tag{4.5.16a}$$

since the exponential (after $\delta(\widehat{M})$ projection) just redefines some components by shifting by components of lower $\check{M}$-spin. In this gauge, writing $\gamma^\alpha = (s, u)$, $\tilde{\gamma}^\alpha = (\tilde{s}, \tilde{u})$, we can rewrite ϕ as

$$|\tilde{0}\rangle = e^{-ia^{+\dagger}a^{-\dagger}}\,|0\rangle \quad \leftrightarrow \quad \tilde{s}\,|\tilde{0}\rangle = u\,|\tilde{0}\rangle = 0 \quad ,$$
$$\phi = \phi(s)\,|\tilde{0}\rangle \quad \leftrightarrow \quad \tilde{s}\phi = 0 \quad . \tag{4.5.16b}$$

By using an appropriate (indefinite-metric) coherent-state formalism, we can choose s to be a coordinate (and $u = -2i\partial/\partial s$). We next make the replacement

$$\delta(\widehat{M}_{\alpha\beta}{}^2) \quad \to \quad \int ds\ \mu(s)\ \delta(2\check{M}^+ + s^2)\delta\left(\check{M}^3 + s\frac{\partial}{\partial s}\right) \tag{4.5.17}$$

after pushing it to the right in (4.5.4) so it hits the ϕ, where we have just replaced projection onto $\widehat{M}_{\alpha\beta}{}^2 = 0$ with $\widehat{M}^+ = \widehat{M}^3 = 0$ (which implies $\widehat{M}^- = 0$). The first δ function factor is a Dirac δ function, while the second is a Kronecker δ. $\mu(s)$ is an appropriate measure factor; instead of determining it by an explicit use of coherent states, we fix it by comparison with the simplest case of the Dirac spinor: Using

$$\int ds\ \epsilon(s)\delta(s^2 + r^2)(a + bs) = b \quad , \tag{4.5.18a}$$

we find

$$\mu(s) \sim \epsilon(s) \quad . \tag{4.5.18b}$$

Then only the $\tilde{u}$ term of (4.5.4) contributes in this gauge, and we obtain (the $\tilde{u}$ itself having already been absorbed into the measure (4.5.18b))

$$S = -2\int d^Dx\; ds\; \epsilon(s)\; \bar{\phi}\; Q^+\delta(2\check{M}^+ + s^2)\delta\left(\check{M}^3 + s\frac{\partial}{\partial s}\right)\phi \quad . \tag{4.5.19}$$

(All dependence on γ^α and $\tilde{\gamma}^\alpha$ has been reduced to ϕ being a function of just s. This action was first proposed for the string [4.13].) Since the first δ function can be used to replace any s^2 with a $-2\check{M}^+$, we can perform all such replacements, or equivalently choose the gauge

$$\phi(s) = \phi_0 + s\phi_1 \quad . \tag{4.5.20}$$

(An equivalent procedure was performed for the string in [4.14].) For the Dirac spinor, after integration over s (including that in $Q^+ = \frac{1}{2}s(\not{p} + \not{M})$), the Dirac action is easily found (ϕ_1 drops out). For general spinor representations, Q^+ has an additional $\check{Q}^+$ term, and s integration gives the lagrangian

$$\mathcal{L} = \bar{\phi}_0(\not{p} + \not{M})\delta(\check{M}^3)\phi_0 + 2\left[\bar{\phi}_1\check{Q}^+\delta(\check{M}^3)\phi_0 - \bar{\phi}_0\check{Q}^+\delta(\check{M}^3 + 1)\phi_1\right] \quad . \tag{4.5.21}$$

However, γ-matrix trace constraints (such as (4.5.13)) must still be solved to relate the components.

The explicit form of the OSp(1,1|2) operators for the fermionic string to use with these results follows from the light-cone Poincaré generators which will be derived in sect. 7.2. The s dependence of Q^+ is then slightly more complicated (it also has a $\partial/\partial s$ term). (The resulting action first appeared in [4.14,15].)

The proof of equivalence to the light cone is similar to that for bosons in the previous section. Again considering the massless case, the basic difference is that we now have to use, from (3.5.3b),

$$\widehat{M}_{\alpha\beta} = M_{\alpha\beta} + S_{\alpha\beta} \quad , \quad \widehat{Q}_\alpha = M_{\alpha-}p_+ + S_{-\alpha} \quad , \tag{4.5.22}$$

and other corresponding generators, as generating the new OSp(1,1|2). This is just the diagonal OSp(1,1|2) obtained from S_{AB} and the one used in the bosonic case. In analogy to the bosonic case, we consider reducible OSp(D−1,1|2) representations corresponding to direct products of arbitrary numbers of vector representations with one spinor representation (represented by graded γ-matrices). We then take the direct product of this with the S_{AB} representation, which is an OSp(1,1|2) spinor but an SO(D−2) scalar. Since the direct product of 2 OSp(1,1|2) spinors gives the direct sum of all (graded) antisymmetric tensor representations, each once (by the usual γ-matrix decomposition), from the bosonic result we see that the only way to get an OSp(1,1|2) singlet is if all vector indices again take only their SO(D−2) values. The OSp(D−1,1|2) spinor is the direct product of an SO(D−2) spinor with an OSp(1,1|2) spinor, so the net result is the original light-cone one. In the bosonic case traces in OSp(1,1|2) vector indices did not give singlets because of (4.4.21); a similar result holds for γ-matrix traces because of

$$\gamma^A \gamma_A = \eta^A{}_A = 0 \quad . \tag{4.5.23}$$

More general representations for S_{AB} could be considered, e.g., as in sect. 3.6. The action can then be rewritten as (4.1.6), but with $M_{\alpha\beta}$ and Q_α replaced by $\widehat{M}_{\alpha\beta}$ and $\widehat{Q}_\alpha$ of (3.5.3b). In analogy to (4.5.2,3), the $S_{-\alpha}$ part of the $J_{-\alpha}$ transformation can be used to choose the gauge $S_{+\alpha}\Phi = 0$. Then, depending on whether the representation allows application of the "lowering" operators $S_{-\alpha}$ 0,1, or 2 times, only the terms of zeroth, first, or second order in $S_{-\alpha}$, respectively, can contribute in the kinetic operator. Since these terms are respectively second, first, and zeroth order in derivatives, they can be used to describe bosons, fermions, and auxiliary fields.

The argument for equivalence to the light cone directly generalizes to the U(1)-type 4+4-extended OSp(1,1|2) of sect. 3.6. Then $\widehat{M}_{AB} = M_{AB} + S_{AB}$ has a singlet only when M_{AB} and S_{AB} are both singlets (for bosons) or both Dirac spinors (for fermions).

Exercises

(1) Derive (4.1.5).

(2) Derive (4.1.6).

(3) Find the gauge-invariant theory resulting from the light-cone theory of a totally symmetric, traceless tensor of arbitrary rank.

(4) Find the explicit infinitesimal gauge transformations of $e_a{}^m$, $e_m{}^a$, e^{-1}, g^{mn}, g_{mn}, and ω_{abc} from (4.1.20-22). Linearize, and show the gauge $e^{[am]} = 0$ can be obtained with λ_{ab}. Find the transformation for a covariant vector A_a (from a similarity transformation, like (4.1.22)).

(5) Write c_{abc} explicitly in terms of $e_a{}^m$. Find T_{abc} and R_{abcd} in terms of c_{abc} and ω_{abc}. Derive (4.1.31). Linearize to get (4.1.17).

(6) Find an expression for ω_{abc} when (4.1.24) is not imposed, in terms of T_{abc} and the ω of (4.1.27).

(7) Derive global Poincaré transformations by finding the subgroup of (4.1.20) which leaves (4.1.28) invariant.

(8) Find the field equation for ϕ from (4.1.36), and show that (4.1.38) satisfies it.

(9) Derive the gauge-covariant action for gravity in the GL(1)-type 4+4-extension of OSp(1,1|2), and compare with the U(1) result, (4.1.47).

(10) Find the BRST transformations for the IGL(1) formalism of sect. 4.2 (BRST1, derived from the light cone) for free gravity. Find those for the usual IGL(1) formalism of sect. 3.2 (BRST2, derived from second-quantizing the gauge-invariant field theory). After suitable redefinitions of the BRST1 fields (including auxiliaries and ghosts), show that a subset of these fields that corresponds to the complete set of fields in the BRST2 formalism has identical BRST transformations.

(11) Formulate ϕ^3 theory as in (4.2.19), using the bracket of (3.4.7).

(12) Derive the gauge transformations for interacting Yang-Mills by the covariant second-quantized operator method of (4.2.21), in both the IGL(1) and OSp(1,1|2) formalisms.

(13) Find the free gauge-invariant action for gravity in the IGL(1) formalism, and compare with the OSp(1,1|2) result (4.1.17). Find the gauge-fixed action by (4.4.1-5).

(14) Perform IGL(1) gauge-fixing, as in sect. 4.4, for a second-rank antisymmetric tensor gauge field. Perform the analogous gauge fixing by the method of sect. 3.2, and compare. Note that there are scalar commuting ghosts which can be interpreted as the ghosts for the gauge invariances of the vector ghosts ("ghosts for ghosts").

(15) Derive (4.5.14,15).

5. PARTICLE

5.1. Bosonic

If coordinates are considered as fields, and their arguments as the coordinates of small spacetimes, then the mechanics of particles and strings can be considered as 1- and 2-dimensional field theories, respectively (see sect. 1.1). (However, to avoid confusion, we will avoid referring to mechanics theories as field theories.) Thus, the particle is a useful analog of the string in one lower "dimension", and we here review its properties that will be found useful below for the string.

As described in sect. 3.1, the mechanics action for any relativistic particle is completely determined by the constraints it satisfies, which are equivalent to the free equations of motion of the corresponding field theory. The first-order (hamiltonian) form ((3.1.10)) is more convenient than the second-order one because (1) it makes canonical conjugates explicit, (2) the inverse propagator (and, in more general cases, all other operator equations of motion) can be directly read off as the hamiltonian, (3) path-integral quantization is easier, and (4) treatment of the supersymmetric case is clearer. The simplest example is a massless, spinless particle, whose only constraint is the Klein-Gordon equation $p^2 = 0$. From (3.1.10), the action [5.1] can thus be written in first-order form as

$$S = \int d\tau\, \mathcal{L} \quad , \quad \mathcal{L} = \dot{x} \cdot p - g\tfrac{1}{2}p^2 \quad , \tag{5.1.1}$$

where τ is the proper time, of which the position x, momentum p, and 1-dimensional metric g are functions, and $\dot{} = \partial/\partial\tau$. The action is invariant under Poincaré transformations in the higher-dimensional spacetime described by x, as well as 1D general coordinate transformations (τ-reparametrizations). The latter can be obtained from (3.1.11):

$$\delta x = \zeta p \quad , \quad \delta p = 0 \quad , \quad \delta g = \dot{\zeta} \quad . \tag{5.1.2}$$

These differ from the usual transformations by terms which vanish on shell: In general, any action with more than one field is invariant under $\delta\phi_i = \lambda_{ij}\delta S/\delta\phi_j$, where λ_{ij} is antisymmetric. Such invariances may be necessary for off-shell closure of the above algebra, but are irrelevant for obtaining the field theory from the classical mechanics. (In fact, in the component formalism for supergravity, gauge invariance is more easily proven using a first-order formalism with the type of transformations in (5.1.2) rather than the usual transformations which follow from the second-order formalism [5.2].) In this case, if we add the transformations

$$\delta' x = \epsilon \frac{\delta S}{\delta p} \quad , \quad \delta' p = -\epsilon \frac{\delta S}{\delta x} \tag{5.1.3}$$

to (5.1.2), and choose $\epsilon = g^{-1}\zeta$, we obtain the usual general coordinate transformations (see sect. 4.1)

$$\delta'' x = \epsilon \dot{x} \quad , \quad \delta'' p = \epsilon \dot{p} \quad , \quad \delta'' g = (\epsilon g)^{\bullet} \quad . \tag{5.1.4}$$

The second-order form is obtained by eliminating p:

$$\mathcal{L} = g^{-1} \tfrac{1}{2} \dot{x}^2 \quad . \tag{5.1.5}$$

The transformations (5.1.4) for x and g also follow directly from (5.1.2) upon eliminating p by its equation of motion. The massive case is obtained by replacing p^2 with $p^2 + m^2$ in (5.1.1). When the additional term is carried over to (5.1.5), we get

$$\mathcal{L} = \tfrac{1}{2} g^{-1} \dot{x}^2 - \tfrac{1}{2} g m^2 \quad . \tag{5.1.6}$$

g can now also be eliminated by its equation of motion, producing

$$S = -m \int d\tau \, \sqrt{-\dot{x}^2} = -m \int \sqrt{-dx^2} \quad , \tag{5.1.7}$$

which is the length of the world line.

Besides the 1D invariance of (5.1.1) under reparametrization of τ, it also has the discrete invariance of τ reversal. If we choose $x(\tau) \to x(-\tau)$ under this reversal, then $p(\tau) \to -p(-\tau)$, and thus this proper-time reversal can be identified as the classical mechanical analog of charge (or complex) conjugation in field theory [5.3], where $\phi(x) \to \phi^*(x)$ implies $\phi(p) \to \phi^*(-p)$ for the fourier transform. (Also, the electromagnetic coupling $q \int d\tau \; \dot{x} \cdot A(x)$ when added to (5.1.1) requires the charge $q \to -q$.)

There are two standard gauges for quantizing (5.1.1). In the light-cone formalism the gauge is completely fixed (for $p_+ \neq 0$, up to global transformations, which are eliminated by boundary conditions) by

$$x_+ = p_+ \tau \quad . \tag{5.1.8}$$

We then eliminate p_- as a lagrange multiplier with field equation $g = 1$. The lagrangian then simplifies to

$$\mathcal{L} = \dot{x}_- p_+ + \dot{x}_i p_i - \tfrac{1}{2}(p_i{}^2 + m^2) \quad , \tag{5.1.9}$$

with (retarded) propagator

$$-i\Theta(\tau) e^{i\tau\frac{1}{2}(p_i{}^2+m^2)} \tag{5.1.10a}$$

(where $\Theta(\tau) = 1$ for $\tau > 0$ and 0 otherwise) or, fourier transforming with respect to τ,

$$\frac{1}{i\frac{\partial}{\partial\tau} + \frac{1}{2}(p_i{}^2 + m^2) + i\epsilon} = \frac{1}{p_+ p_- + \frac{1}{2}(p_i{}^2 + m^2) + i\epsilon} = \frac{1}{\frac{1}{2}(p^2 + m^2) + i\epsilon} \ . \tag{5.1.10b}$$

For interacting particles, it's preferable to choose

$$x_+ = \tau \quad , \tag{5.1.11}$$

so that the same τ coordinate can be used for all particles. Then $g = 1/p_+$, so the hamiltonian $\frac{1}{2}(p_i{}^2 + m^2)$ gets replaced with $(p_i{}^2 + m^2)/2p_+$, which more closely resembles the nonrelativistic case. If we also use the remaining (global) invariance of τ reparametrizations (generated by p^2), we can choose the gauge $x_+ = 0$, which is the same as choosing the Schrödinger picture.

Alternatively, in the covariant formalism one chooses the gauge

$$g = constant \quad , \tag{5.1.12}$$

where g can't be completely gauge-fixed because of the invariance of the 1D volume $T = \int d\tau\, g$. The functional integral over g is thus replaced by an ordinary integral over T [5.4], and the propagator is [5.3,5]

$$-i\int_0^\infty dT\ \Theta(T) e^{iT\frac{1}{2}(p^2+m^2)} = \frac{1}{\frac{1}{2}(p^2 + m^2) + i\epsilon} \ . \tag{5.1.13}$$

The use of the mechanics approach to the particle is somewhat pointless for the free theory, since it contains no information except the constraints (from which it was derived), and it requires treatment of the irrelevant "off-shell" behavior in the "coordinate" τ. However, the proper-time is useful in interacting theories for studying certain classical limits and various properties of perturbation theory. In particular, the form of the propagator given in (5.1.13) (with Wick-rotated τ: see sect. 2.5) is the most convenient for doing loop integrals using dimensional regularization: The momentum integrations become simple Gaussian integrals, which can be trivially evaluated in arbitrary dimensions by analytic continuation from integer ones:

$$\int d^D p\, e^{-ap^2} = \left(\int d^2 p\, e^{-ap^2} \right)^{D/2} = \left(\frac{\pi}{a} \right)^{D/2} \tag{5.1.14}$$

(The former integral factors into 1-dimensional ones; the latter is easily performed in polar coordinates.) The Schwinger parameters τ are then converted into the usual Feynman parameters α by inserting unity as $\int_0^\infty d\lambda\ \delta(\lambda - \sum \tau)$, rescaling $\tau_i = \lambda\alpha_i$, and integrating out λ, which now appears in standard Γ-function integrals, to get the usual Feynman denominators. An identical procedure is applied in string theory, but writing the parameters as $x = e^{-\tau}$, $w = e^{-\lambda}$. (See (9.1.10).) By not converting the τ's into α's, the high-energy behavior of scattering amplitudes can be analyzed more easily [5.6]. Also, the singularities in an amplitude correspond to classical paths of the particles, and this identification can be seen to be simply the identification of the τ parameters with the classical proper-time [5.7]. 1-loop calculations can be performed by introducing external fields (see also sect. 9.1) and treating the path of the particle in spacetime as closed [5.5,8]. Such calculations can treat arbitrary numbers of external lines (or nonperturbative external fields) for certain external field configurations (such as constant gauge-field strengths). Finally, the introduction of such expressions for propagators in external fields allows the study of classical limits of quantum field theories in which some quantum fields (represented by the external field) become classical fields, as in the usual classical limit, while other fields (represented by the particles described by the mechanics action) become classical particles [5.9].

This classical mechanics analysis will be applied to the string in chapt. 6.

5.2. BRST

In this section we'll apply the methods of sect. 3.2-3 to study BRST quantization of particle mechanics, and find results equivalent to those obtained by more general methods in sect. 3.4.

In the case of particle mechanics (according to sect. 3.1), for the action of the previous section we have $\mathcal{G} = -i\frac{1}{2}(p^2+m^2)$, and thus [4.4], for the "temporal" gauge $g = 1$, from (3.2.6)

$$Q = -ic\tfrac{1}{2}(p^2 + m^2) \quad , \tag{5.2.1}$$

which agrees with the general result (3.4.3b). We could write $c = \partial/\partial \mathbf{C}$ so that in the classical field theory which follows from the quantum mechanics the field

$\phi(x, \mathrm{C})$ could be real (see sect. 3.4). This also follows from the fact that the (τ-reparametrization) gauge-parameter corresponding to c carries a (proper-)time index (it's a 1D vector), and thus changes sign under τ-reversal (mechanics' equivalent of field theory's complex conjugation), and so c is a momentum ($\phi(x, \mathrm{C}) = \phi^*(x, \mathrm{C})$, $\phi(p, c) = \phi^*(-p, -c)$).

We now consider extending IGL(1) to OSp(1,1|2) [3.7]. By (3.3.2),

$$Q^\alpha = -ix^\alpha \tfrac{1}{2}(p^2 + m^2) - b\partial^\alpha \ . \tag{5.2.2}$$

In order to compare with sect. 3.4, we make the redefinitions (see (3.6.8))

$$b = i\frac{\partial}{\partial g} \ , \quad g = \tfrac{1}{2}{p_+}^2 \ , \tag{5.2.3a}$$

(where g is the world-line metric) and the unitary transformation

$$ln\ U = -(ln\ p_+)\tfrac{1}{2}[x^\alpha, \partial_\alpha] \ , \tag{5.2.3b}$$

finding

$$UQ^\alpha U^{-1} = -ix^\alpha \frac{1}{2p_+}(p^2 + m^2 + p^\alpha p_\alpha) - ix_- p_\alpha \ , \tag{5.2.4}$$

which agrees with the expression given in (3.4.2) for the generators $J_{-\alpha}$ for the case of the spinless particle, as does the rest of the OSp(1,1|2) obtained from (3.3.7).

In a lagrangian formalism, for the action (5.1.6) with invariance (5.1.4), (3.3.2) gives the BRST transformation laws

$$\begin{aligned} Q^\alpha x^a &= x^\alpha \dot{x}^a \ , \\ Q^\alpha g &= (x^\alpha g)^{\bullet} \ , \\ Q^\alpha x^\beta &= \tfrac{1}{2}x^{(\alpha}\dot{x}^{\beta)} - C^{\alpha\beta}b \ , \\ Q^\alpha b &= \tfrac{1}{2}(x^\alpha \dot{b} - b\dot{x}^\alpha) - \tfrac{1}{4}(\dot{x}_\beta{}^2 x^\alpha + x_\beta{}^2 \ddot{x}^\alpha) \ . \end{aligned} \tag{5.2.5}$$

We first make the redefinition

$$\tilde{b} = b - \tfrac{1}{2}({x_\alpha}^2)^{\bullet} \tag{5.2.6}$$

to simplify the transformation law of x^α and thus b:

$$\begin{aligned} Q^\alpha x^\beta &= x^\alpha \dot{x}^\beta - C^{\alpha\beta}\tilde{b} \ , \\ Q^\alpha \tilde{b} &= x^\alpha \dot{\tilde{b}} \ . \end{aligned} \tag{5.2.7}$$

We then make further redefinitions

$$x^\alpha \rightarrow g^{-1}x^\alpha \quad , \quad \bar{b} \rightarrow g^{-1}\left[b - 2(g^{-1}\dot{x}_\alpha{}^2)\right] \quad , \tag{5.2.8}$$

which simplify the g transformation, allowing a further simplification for b:

$$\begin{aligned} Q^\alpha x^a &= x^\alpha g^{-1}\dot{x}^a = x^\alpha p^a \quad , \\ Q^\alpha g &= \dot{x}^\alpha \quad , \\ Q^\alpha x^\beta &= -C^{\alpha\beta}b \quad , \\ Q^\alpha b &= 0 \quad . \end{aligned} \tag{5.2.9}$$

To get just a BRST operator (as for the IGL(1) formalism), we can restrict the Sp(2) indices in (5.2.9) to just one value. Then x^α for the other value of α (the antighost) and b can be dropped. (They form an independent IGL(1) multiplet, as described in sect. 3.2.)

To get the OSp(1,1|2) formalism, we choose a "Lorentz" gauge. We then quantize with the ISp(2)-invariant gauge-fixing term

$$\mathcal{L}_1 = -Q_\alpha{}^2 f(g) = f''(g)(\dot{g}b - \dot{x}_\alpha{}^2) \tag{5.2.10}$$

for some arbitrary function f such that $f'' \neq 0$. Canonically quantizing (where $f'(g)$ is conjugate to b), and using the equations of motion, we find Q^α from its Noether current (which in $D = 1$ is also the charge) to be given by (5.2.2). For an IGL(1) formalism, we can use the temporal gauge (writing $x^\alpha = (c, \tilde{c})$)

$$\mathcal{L}_1 = iQ[\tilde{c}f(g)] = bf(g) - if'(g)\tilde{c}\dot{c} \quad . \tag{5.2.11}$$

(Compare the discussions of gauge choices in sect. 3.2-3.)

Although Lorentz-gauge quantization gave a result equivalent to that obtained from the light cone in sect. 3.4, we'll find in sect. 8.3 for the string a result equivalent to that obtained from the light cone in sect. 3.6.

5.3. Spinning

The mechanics of a relativistic spin-1/2 particle [5.1] is obtained by symmetrizing the particle action for a spinless particle with respect to *one-dimensional* (local) supersymmetry. We thus generalize $x(\tau) \rightarrow X(\tau, \theta)$, etc., where θ is a single, real, anticommuting coordinate. We first define global supersymmetry by the generators

$$q = \frac{\partial}{\partial\theta} - \theta i\partial \quad , \quad i\partial \equiv i\frac{\partial}{\partial\tau} = -q^2 \quad , \tag{5.3.1a}$$

which leave invariant the derivatives

$$\mathrm{d} = \frac{\partial}{\partial\theta} + \theta i\partial \quad , \quad \partial = -i\mathrm{d}^2 \quad . \tag{5.3.1b}$$

The local invariances are then generated by (expanding covariantly)

$$K = \kappa i\mathrm{d} + ki\partial \quad , \tag{5.3.2a}$$

which act covariantly (i.e., as $(\ )' = e^{iK}(\ )e^{-iK}$) on the derivatives

$$\mathcal{D} = G\mathrm{d} + \mathfrak{G}\partial \quad , \quad -i\mathcal{D}^2 \quad . \tag{5.3.2b}$$

This gives the infinitesimal transformation

$$\delta\mathcal{D} = i[K, \mathcal{D}] = i(KG - \mathcal{D}\kappa)\mathrm{d} + i(K\mathfrak{G} - \mathcal{D}k)\partial + 2i\kappa G\partial \quad . \tag{5.3.2c}$$

We next use κ by the last part of this transformation to choose the gauge

$$\mathfrak{G} = 0 \rightarrow \kappa = \tfrac{1}{2}\mathrm{d}k \quad . \tag{5.3.3}$$

The action (5.1.1) becomes

$$\begin{aligned} S &= \int d\tau\, d\theta\, G^{-1}\left[(-i\mathcal{D}^2 X)\cdot P - \tfrac{1}{2}P\cdot\mathcal{D}P\right] \\ &= \int d\tau\, d\theta\, \left[-iG(\mathrm{d}X)\cdot(\mathrm{d}P) - \tfrac{1}{2}P\cdot \mathrm{d}P\right] \quad . \end{aligned} \tag{5.3.4}$$

When expanded in components by $\int d\theta \rightarrow \mathrm{d}$, and defining

$$\begin{aligned} X = x \quad &, \quad \mathcal{D}X = i\gamma \quad ; \\ P = \zeta \quad &, \quad \mathcal{D}P = p \quad ; \\ G = g^{-1/2} \quad &, \quad \mathrm{d}G = ig^{-1}\psi \quad ; \end{aligned} \tag{5.3.5}$$

when evaluated at $\theta = 0$ (in analogy to sect. 3.2), we find

$$S = \int d\tau\, \left(\dot{x}\cdot p + i\psi\gamma\cdot p - i\gamma\cdot\dot{\zeta} - g\tfrac{1}{2}p^2 + \tfrac{1}{2}i\zeta\cdot\dot{\zeta}\right) \quad . \tag{5.3.6}$$

The (g, x, p) sector works as for the bosonic case. In the (ψ, γ, ζ) sector we see that the quantity $i(\gamma - \frac{1}{2}\zeta)$ is canonically conjugate to ζ, and thus

$$\gamma = \frac{\partial}{\partial\zeta} + \tfrac{1}{2}\zeta \quad , \tag{5.3.7a}$$

which has γ-matrix type commutation relations. It anticommutes with

$$\hat{\gamma} = \gamma - \zeta = \frac{\partial}{\partial \zeta} - \tfrac{1}{2}\zeta \quad , \qquad (5.3.7b)$$

which is an independent set of γ-matrices. However, it is γ which appears in the Dirac equation, obtained by varying ψ.

In a light-cone formalism, we again eliminate all auxiliary "−" components by their equations of motion, and use the gauge invariance (5.3.2-3) to fix the "+" components

$$X_+ = p_+\tau \rightarrow \quad x_+ = p_+\tau \ , \quad \gamma_+ = 0 \quad . \qquad (5.3.8)$$

We then find $G = 1$, and (5.3.6) reduces to

$$\mathcal{L} = \dot{x}_- p_+ + \dot{x}_i p_i - \tfrac{1}{2}p_i{}^2 - i(\gamma_- - \zeta_-)\dot{\zeta}_+ - i\gamma_i\dot{\zeta}_i + \tfrac{1}{2}i\zeta_i\dot{\zeta}_i \quad . \qquad (5.3.9)$$

In order to obtain the usual spinor field, it's necessary to add a lagrange multiplier term to the action constraining $\hat{\gamma} = 0$. This constraint can either be solved classically (but only for even spacetime dimension D) by determining half of the ζ's to be the canonical conjugates of the other half (consider $\zeta_1 + i\zeta_2$ vs. $\zeta_1 - i\zeta_2$, etc.), or by imposing it quantum mechanically on the field Gupta-Bleuler style. The former approach sacrifices manifest Lorentz invariance in the coordinate approach; however, if the γ's are considered simply as operators (without reference to their coordinate representation), then the field is the usual spinor representation, and both can be represented in the usual matrix notation. This constrained action is equivalent to the second-order action

$$\begin{aligned} S &= \int d\tau\, d\theta\, \tfrac{1}{2}G^{-1}(\mathcal{D}^2 X)\cdot(\mathcal{D}X) \\ &= \int d\tau\, d\theta\, \tfrac{1}{2}(G\mathrm{d}X)\mathrm{d}(G\mathrm{d}X) \quad , \\ &= \int d\tau\, (\tfrac{1}{2}g^{-1}\dot{x}^2 + ig^{-1}\psi\gamma\cdot\dot{x} - \tfrac{1}{2}i\gamma\cdot\dot{\gamma}) \quad , \end{aligned} \qquad (5.3.10a)$$

or, in first-order form for x only,

$$S = \int d\tau\, (\dot{x}\cdot p - g\tfrac{1}{2}p^2 + \tfrac{1}{2}i\dot{\gamma}\cdot\gamma + i\psi\gamma\cdot p) \quad . \qquad (5.3.10b)$$

The constraint $\gamma\cdot p = 0$ (the Dirac equation) is just a factorized form of the constraint (2.2.8) for this particular representation of the Lorentz group.

A further constraint is necessary to get an irreducible Poincaré representation in even D. Since any function of an anticommuting coordinate contains bosonic and fermionic terms as the coefficients of even and odd powers of that coordinate, we need the constraint $\gamma_D = \pm 1$ on the field (where γ_D means just the product of all the γ's) to pick out a field of just one statistics (in this case, a Weyl spinor: notice that D is even in order for the previous constraint to be applied). In the OSp approach this Weyl chirality condition can also be obtained by an extension of the algebra [4.10]: OSp(1,1|2)⊗U(1), where the U(1) is chiral transformations, results in an extra Kronecker δ which is just the usual chirality projector. This U(1) generator (for at least the special case of a Dirac spinor or Ramond string) can also be derived as a constraint from first-quantization: The classical mechanics action for a Dirac spinor, under the global transformation $\delta\gamma_a \sim \epsilon_{abc\cdots d}\gamma^b\gamma^c\cdots\gamma^d$, varies by a boundary term $\sim \int d\tau \frac{\partial}{\partial\tau}\gamma_D$, where as usual $\gamma_D \sim \epsilon_{abc\cdots d}\gamma^a\gamma^b\gamma^c\cdots\gamma^d$. By adding a lagrange multiplier term for $\gamma_D \pm 1$, this symmetry becomes a local one, gauged by the lagrange multiplier (as for the other equations of motion). By 1D supersymmetrization, there is also a lagrange multiplier for $\epsilon_{abc\cdots d}p^a\gamma^b\gamma^c\cdots\gamma^d$. The action then describes a Weyl spinor.

Many supersymmetric gauges are possible for g and ψ. The simplest sets both to constants ("temporal" gauge $G = 1$), but this gauge doesn't allow an OSp(1,1|2) algebra. The next simplest gauge, $\mathbf{d}G = 0$, does the same to ψ but sets the τ derivative of g to vanish, making it an extra coordinate in the field theory (related to x_-, or p_+), giving the generators of (3.4.2). However, the gauge which also keeps ψ as a coordinate (and as a partner to g) is $\dot{G} = 0$. In order to get the maximal coordinates (or at least zero-modes, for the string) we choose an OSp(1,1|2) which keeps ψ (related to $\tilde{\gamma}_\pm$, and the corresponding extra ghost, related to $\tilde{\gamma}_\alpha$). This gives the modified BRST algebra of (3.5.1).

An *"isospinning" particle* [5.10] can be described similarly. By dropping the ψ term in (5.3.6,10b) it's possible to have a different symmetry on the indices of (γ, ζ) than on those of (x, p). In fact, even the range of the indices and the metric can be different. Thus, spin separates from orbital angular momentum and becomes isospin. There is no longer an anticommuting gauge invariance, but with a positive definite metric on the isospinor indices it's no longer necessary to have one to maintain unitarity. If we use the constraint $\hat{\gamma} = 0$ we get an isospinor, but if we don't we get a matrix, with the γ's acting on one side and the $\hat{\gamma}$'s on the other. Noting that τ reversal switches γ with $-\hat{\gamma}$, we see that the matrix gets transposed. Therefore, the complex conjugation that is the quantum-mechanical analog of τ reversal is actually *hermitian* conjugation, particularly on a field which is a matrix representation of some group. (When $\hat{\gamma}$ is constrained to vanish, τ reversal is not

an invariance.) By combining these anticommuting variables with the previous ones we get an isospinning spinning particle.

At this point we take a slight diversion to discuss properties of spinors in arbitrary dimensions with arbitrary spacetime signature. This will complete our discussion of spinors in this section, and will be useful in the following section, where representations of supersymmetry, which is itself described by a spinor generator, will be found to depend qualitatively on the dimension. The analysis of spinors in Euclidean space (i.e, the usual spinor representations of SO(D)) can be obtained by the usual group theoretical methods (see, e.g., [5.11]), using either Dynkin diagrams or an explicit representation of the γ-matrices. The properties of spinors in SO(D_+,D_-) can then be obtained by Wick rotation of D_- space directions into time ones. (Of particular interest are the D-dimensional Lorentz group SO(D−1,1) and the D-dimensional conformal group SO(D,2).) This affects the spinors with respect to only complex conjugation properties. A useful notation to classify spinors and their properties is: Denote a fundamental spinor ("spin 1/2") as ψ_α, and its hermitian conjugate as $-\bar{\psi}_{\dot{\alpha}}$. Denote another spinor ψ^α such that the contraction $\psi^\alpha\psi_\alpha$ is invariant under the group, and its hermitian conjugate $\bar{\psi}^{\dot{\alpha}}$. The representation of the group on these various spinors is then related by taking complex conjugates and inverses of the matrices representing the group on the fundamental one. For SO(D_+,D_-) there are always some of these representations that are equivalent, since SO(2N) has only 2 inequivalent spinor representations and SO(2N+1) just 1. (In γ-matrix language, the Dirac spinor can be reduced to 2 inequivalent Weyl spinors by projection with $\frac{1}{2}(1 \pm \gamma_D)$ in even dimensions.) In cases where there is another fundamental spinor representation not included in this set, we also introduce a $\psi_{\alpha'}$ and the corresponding 3 other spinors. (However, in that case all 4 in each set will be equivalent, since there are at most 2 inequivalent altogether.) Many properties of the spinor representations can be described by classifying the index structure of: (1) the inequivalent spinors, (2) the bispinor invariant tensors, or "metrics," which are just the matrices relating the equivalent spinors in the sets of 4, and (3) the σ-matrices (γ-matrices for D odd, but in even D the matrices half as big which remain after Weyl projection), which are simply the Clebsch-Gordan coefficients for relating spinor⊗spinor to vector. In the latter 2 cases, we also classify the symmetry in the 2 spinor indices, where appropriate.

The metrics are of 3 types (along with their complex conjugates and inverses): (1) $M_\alpha{}^{\dot{\beta}}$, which gives charge conjugation for (pseudo)real representations, and is re-

lated to complex conjugation properties of γ-matrices, (2) $M_{\alpha\dot{\beta}}$, which is the matrix which relates the Dirac spinors Ψ and $\bar{\Psi}$, if it commutes with Weyl projection, and is related to hermitian conjugation properties of γ-matrices, and (3) $M_{\alpha\beta}$, which is the Clebsch-Gordan coefficients for spinor⊗same-representation spinor to scalar, and is related to transposition properties of γ-matrices. For all of these it's important to know whether the metric is symmetric or antisymmetric; in particular, for the first type we get either real or pseudoreal representations, respectively. In γ-matrix language, this charge conjugation matrix is straightforwardly constructed in the representation where the γ-matrices are expressed in terms of direct products of the Pauli matrices for the 2-dimensional subspaces. Upon Wick rotation of 1 direction each in any number of pairs corresponding to these 2-dimensional subspaces, the corresponding Pauli matrix factor in the charge conjugation matrix must be dropped (with perhaps some change in the choice of Pauli matrix factors for the other subspaces). It then follows that (pseudo)reality is the same in SO(D_++1,D_-+1) as in SO(D_+,D_-), so all cases follow from the Euclidean case. For the second type of metric, $\bar{\Psi} = \Psi^\dagger$ in the Euclidean case, so $M_{\alpha\dot{\beta}}$ is just the identity matrix (i.e., the spinor representations are unitary). After Wick rotation, this matrix becomes the product of all the γ-matrices in the Wick rotated directions, since those γ-matrices got factors of i in the Wick rotation, and thus need this extra factor to preserve the reality of the tensors $\bar{\Psi}\gamma\cdots\gamma\Psi$. The symmetry properties of this metric then follow from those of the γ-matrices. Also, because of the signature of the γ-matrices, it follows that this metric, except in the Euclidean case, has half its eigenvalues +1 and half −1. The last type of metric has only undotted indices and thus has nothing to do with complex conjugation, so its properties are unchanged by Wick rotation. It's identical to the first type in Euclidean space (since the second type is the identity there; in general, if 2 of the metrics exist, the third is just their product), which thus determines it in the general case. Various types of groups are defined by these metrics alone (real, unitary, orthogonal, symplectic, etc.), with the SO(D_+,D_-) group as a subgroup. (In fact, these metrics *completely* determine the SO(D_+,D_-) group, up to abelian factors, in $D \equiv D_+ + D_- \leq 6$, and allow all vector indices to be replaced by pairs of spinor indices. They also determine the group in $D = 8$ for D_- even, due to "triality," the discrete symmetry which permutes the vector representation with the 2 spinors.) We also classify the σ-matrices by their symmetry properties only when its 2 spinor indices are for equivalent representations, so they are unrelated to complex conjugation (both indices undotted), and thus their symmetry is determined by the Euclidean case.

We now summarize the results obtained by the methods sketched above for spinors ψ, metrics η (symmetric) and Ω (antisymmetric), and σ-matrices, in terms of D mod 8 and D_- mod 4:

D_- D	0 Euclidean	1 Lorentz	2 conformal	3
0	$\psi_\alpha\ \psi_{\alpha'}$ $\eta_{\alpha\beta}\ \eta_\alpha{}^{\dot\beta}\ \eta_{\alpha\dot\beta}$ $\sigma_{\alpha\beta'}$	$\psi_\alpha\ \psi_{\dot\alpha}$ $\eta_{\alpha\beta}$ $\sigma_{\alpha\dot\beta}$	$\psi_\alpha\ \psi_{\alpha'}$ $\eta_{\alpha\beta}\ \Omega_\alpha{}^{\dot\beta}\ \Omega_{\alpha\dot\beta}$ $\sigma_{\alpha\beta'}$	$\psi_\alpha\ \psi_{\dot\alpha}$ $\eta_{\alpha\beta}$ $\sigma_{\alpha\dot\beta}$
1	ψ_α $\eta_{\alpha\beta}\ \eta_\alpha{}^{\dot\beta}\ \eta_{\alpha\dot\beta}$ $\sigma_{(\alpha\beta)}$	ψ_α $\eta_{\alpha\beta}\ \eta_\alpha{}^{\dot\beta}\ \eta_{\alpha\dot\beta}$ $\sigma_{(\alpha\beta)}$	ψ_α $\eta_{\alpha\beta}\ \Omega_\alpha{}^{\dot\beta}\ \Omega_{\alpha\dot\beta}$ $\sigma_{(\alpha\beta)}$	ψ_α $\eta_{\alpha\beta}\ \Omega_\alpha{}^{\dot\beta}\ \Omega_{\alpha\dot\beta}$ $\sigma_{(\alpha\beta)}$
2	$\psi_\alpha\ \psi^\alpha$ $\eta_{\alpha\dot\beta}$ $\sigma_{(\alpha\beta)}\ \sigma^{(\alpha\beta)}$	$\psi_\alpha\ \psi^\alpha$ $\eta_\alpha{}^{\dot\beta}$ $\sigma_{(\alpha\beta)}\ \sigma^{(\alpha\beta)}$	$\psi_\alpha\ \psi^\alpha$ $\Omega_{\alpha\dot\beta}$ $\sigma_{(\alpha\beta)}\ \sigma^{(\alpha\beta)}$	$\psi_\alpha\ \psi^\alpha$ $\Omega_\alpha{}^{\dot\beta}$ $\sigma_{(\alpha\beta)}\ \sigma^{(\alpha\beta)}$
3	ψ_α $\Omega_{\alpha\beta}\ \Omega_\alpha{}^{\dot\beta}\ \eta_{\alpha\dot\beta}$ $\sigma_{(\alpha\beta)}$	ψ_α $\Omega_{\alpha\beta}\ \eta_\alpha{}^{\dot\beta}\ \Omega_{\alpha\dot\beta}$ $\sigma_{(\alpha\beta)}$	ψ_α $\Omega_{\alpha\beta}\ \eta_\alpha{}^{\dot\beta}\ \Omega_{\alpha\dot\beta}$ $\sigma_{(\alpha\beta)}$	ψ_α $\Omega_{\alpha\beta}\ \Omega_\alpha{}^{\dot\beta}\ \eta_{\alpha\dot\beta}$ $\sigma_{(\alpha\beta)}$
4	$\psi_\alpha\ \psi_{\alpha'}$ $\Omega_{\alpha\beta}\ \Omega_\alpha{}^{\dot\beta}\ \eta_{\alpha\dot\beta}$ $\sigma_{\alpha\beta'}$	$\psi_\alpha\ \psi_{\dot\alpha}$ $\Omega_{\alpha\beta}$ $\sigma_{\alpha\dot\beta}$	$\psi_\alpha\ \psi_{\alpha'}$ $\Omega_{\alpha\beta}\ \eta_\alpha{}^{\dot\beta}\ \Omega_{\alpha\dot\beta}$ $\sigma_{\alpha\beta'}$	$\psi_\alpha\ \psi_{\dot\alpha}$ $\Omega_{\alpha\beta}$ $\sigma_{\alpha\dot\beta}$
5	ψ_α $\Omega_{\alpha\beta}\ \Omega_\alpha{}^{\dot\beta}\ \eta_{\alpha\dot\beta}$ $\sigma_{[\alpha\beta]}$	ψ_α $\Omega_{\alpha\beta}\ \Omega_\alpha{}^{\dot\beta}\ \eta_{\alpha\dot\beta}$ $\sigma_{[\alpha\beta]}$	ψ_α $\Omega_{\alpha\beta}\ \eta_\alpha{}^{\dot\beta}\ \Omega_{\alpha\dot\beta}$ $\sigma_{[\alpha\beta]}$	ψ_α $\Omega_{\alpha\beta}\ \eta_\alpha{}^{\dot\beta}\ \Omega_{\alpha\dot\beta}$ $\sigma_{[\alpha\beta]}$
6	$\psi_\alpha\ \psi^\alpha$ $\eta_{\alpha\dot\beta}$ $\sigma_{[\alpha\beta]}\ \sigma^{[\alpha\beta]}$	$\psi_\alpha\ \psi^\alpha$ $\Omega_\alpha{}^{\dot\beta}$ $\sigma_{[\alpha\beta]}\ \sigma^{[\alpha\beta]}$	$\psi_\alpha\ \psi^\alpha$ $\Omega_{\alpha\dot\beta}$ $\sigma_{[\alpha\beta]}\ \sigma^{[\alpha\beta]}$	$\psi_\alpha\ \psi^\alpha$ $\eta_\alpha{}^{\dot\beta}$ $\sigma_{[\alpha\beta]}\ \sigma^{[\alpha\beta]}$
7	ψ_α $\eta_{\alpha\beta}\ \eta_\alpha{}^{\dot\beta}\ \eta_{\alpha\dot\beta}$ $\sigma_{[\alpha\beta]}$	ψ_α $\eta_{\alpha\beta}\ \Omega_\alpha{}^{\dot\beta}\ \Omega_{\alpha\dot\beta}$ $\sigma_{[\alpha\beta]}$	ψ_α $\eta_{\alpha\beta}\ \Omega_\alpha{}^{\dot\beta}\ \Omega_{\alpha\dot\beta}$ $\sigma_{[\alpha\beta]}$	ψ_α $\eta_{\alpha\beta}\ \eta_\alpha{}^{\dot\beta}\ \eta_{\alpha\dot\beta}$ $\sigma_{[\alpha\beta]}$

(We have omitted the vector indices on the σ-matrices. We have also omitted metrics which are complex conjugates or inverses of those shown, or are the same but with all indices primed, where relevant.) Also, not indicated in the table is the fact that $\eta_{\alpha\dot{\beta}}$ is positive definite for the Euclidean case and half-positive, half-negative otherwise. Finally, the dimension of the spinors is $2^{(D-1)/2}$ for D odd and $2^{(D-2)/2}$ (Weyl spinor) for D even. These $N \times N$ metrics define classical groups as subgroups of $GL(N,C)$:

$$\eta_{\alpha\beta} \to SO(N,C)$$

$$\Omega_{\alpha\beta} \to Sp(N,C)$$

$$\eta_\alpha{}^{\dot{\beta}} \to GL(N,R)$$

$$\Omega_\alpha{}^{\dot{\beta}} \to GL^*(N) \quad (\equiv U^*(N))$$

$$\eta_{\alpha\dot{\beta}} \to U(N) \quad (or\ U(\tfrac{N}{2},\tfrac{N}{2}))$$

$$\Omega_{\alpha\dot{\beta}} \to U(\tfrac{N}{2},\tfrac{N}{2})$$

$$\eta_{\alpha\beta}\ \eta_\alpha{}^{\dot{\beta}}\ \eta_{\alpha\dot{\beta}} \to SO(N) \quad (or\ SO(\tfrac{N}{2},\tfrac{N}{2}))$$

$$\eta_{\alpha\beta}\ \Omega_\alpha{}^{\dot{\beta}}\ \Omega_{\alpha\dot{\beta}} \to SO^*(N)$$

$$\Omega_{\alpha\beta}\ \eta_\alpha{}^{\dot{\beta}}\ \Omega_{\alpha\dot{\beta}} \to Sp(N)$$

$$\Omega_{\alpha\beta}\ \Omega_\alpha{}^{\dot{\beta}}\ \eta_{\alpha\dot{\beta}} \to USp(N) \quad (or\ USp(\tfrac{N}{2},\tfrac{N}{2}))$$

(When the matrix has a trace, the group can be factored into the corresponding "S"-group times an abelian factor $U(1)$ or $GL(1,R)$.)

The σ-matrices satisfy the obvious relation analogous to the γ-matrix anticommutation relations: Contract a pair of spinors on 2 σ-matrices and symmetrize in the vector indices and you get (twice) the metric for the vector representation (the SO(D_+,D_-) metric) times a Kronecker δ in the remaining spinor indices:

$$\sigma^{(a}{}_{\alpha\dot{\alpha}}\sigma^{b)\beta\dot{\alpha}} = \sigma^{(a}{}_{\alpha\alpha'}\sigma^{b)\beta\alpha'} = \sigma^{(a}{}_{[\alpha\gamma]}\sigma^{b)[\beta\gamma]} = \sigma^{(a}{}_{(\alpha\gamma)}\sigma^{b)(\beta\gamma)} = 2\eta^{ab}\delta_\alpha{}^\beta \quad , \qquad (5.3.11)$$

and similarly for expressions with dotted and undotted (or primed and unprimed) indices switched. (We have raised indices with spinor metrics when necessary.)

Although (irreducible) spinors thus have many differences in different dimensions, there are some properties which are dimension-independent, and it will prove useful to change notation to emphasize those similarities. We therefore define spinors which are real in all dimensions (or would be real after a complex similarity transformation, and therefore satisfy a generalized Majorana condition). For those kinds of spinors in the above table which are complex or pseudoreal, this means making a bigger spinor which contains the real and imaginary components of the previous one as independent components. If the original spinor was complex ($D_+ - D_-$ twice odd), the new spinor is reducible to an irreducible spinor and its inequivalent complex conjugate representation, which transform oppositely with respect to an internal U(1) generator ("γ_5"). If the original spinor was pseudoreal ($D_+ - D_- = 3,4,5$ *mod* 8), the new spinor reduces to 2 equivalent irreducible spinor representations, which transform as a doublet with respect to an internal SU(2).

The net result for these real spinors is that we have the following analog of the above table for those properties which hold for all values of D_+:

D_-	0 Euclidean	1 Lorentz	2 conformal	3
	$\psi_\alpha\ \psi_{\alpha'}$ $\eta_{\alpha\beta}$ $\gamma_{\alpha\beta'}$	$\psi_\alpha\ \psi^\alpha$ $\gamma_{(\alpha\beta)}\ \gamma^{(\alpha\beta)}$	$\psi_\alpha\ \psi_{\alpha'}$ $\Omega_{\alpha\beta}$ $\gamma_{\alpha\beta'}$	$\psi_\alpha\ \psi^\alpha$ $\gamma_{[\alpha\beta]}\ \gamma^{[\alpha\beta]}$

These γ-matrices satisfy the same relations as the σ-matrices in (5.3.11). (In fact, they are identical for $D_+ = D_-$ *mod* 8.) Their additional, D_+-dependent properties can be described by additional metrics: (1) the internal symmetry generators mentioned above; and (2) for D odd, a metric $M_{\alpha\beta}$ or $M_{\alpha\beta'}$ which relates the 2 types of spinors (since there are 2 independent irreducible spinor representations only for D even).

Similar methods of first-quantization will be applied in sect. 7.2 to the spinning string, which has spin-0 and spin-1/2 ground states. Classical mechanics actions for particles with other spins (or strings with ground states with other spins), i.e., gauge fields, are not known. (For the superstring, however, a nonmanifestly supersymmetric formalism can be obtained by a truncation of the spinning string, eliminating some of the ground states.) On the other hand, the BRST approach of chap. 3 allows the treatment of the quantum mechanics of arbitrary gauge fields. Furthermore, the superparticle, described in the following section, is described classical-mechanically by a spin-0 or spin-1/2 superfield, which includes component gauge fields, just as the string has component gauge fields in its excited modes.

5.4. Supersymmetric

The superparticle is obtained from the spinless particle by symmetrizing with respect to the supersymmetry of the higher-dimensional space in which the one-dimensional world line of the particle is imbedded. (For reviews of supersymmetry, see [1.17].) As for the spinless particle, a full understanding of this action consists of just understanding the algebras of the covariant derivatives and equations of motion. In order to describe arbitrary D, we work with the general real spinors of the previous section. The covariant derivatives are p_a (momentum) and d_α (anticommuting spinor), with

$$\{d_\alpha, d_\beta\} = 2\gamma^a{}_{\alpha\beta}p_a \tag{5.4.1}$$

(the other graded commutators vanish), where the γ matrices are symmetric in their spinor indices and satisfy

$$\gamma^{(a}{}_{\alpha\gamma}\gamma^{b)\beta\gamma} = 2\eta^{ab}\delta_\alpha{}^\beta \quad , \tag{5.4.2}$$

as described in the previous section. This algebra is represented in terms of coordinates x^a (spacetime) and θ^α (anticommuting), and their partial derivatives ∂_a and ∂_α, as

$$p_a = i\partial_a \quad , \quad d_\alpha = \partial_\alpha + i\gamma^a{}_{\alpha\beta}\theta^\beta\partial_a \quad . \tag{5.4.3}$$

These covariant derivatives are invariant under supersymmetry transformations generated by p_a and q_α, which form the algebra

$$\{q_\alpha, q_\beta\} = -2\gamma^a{}_{\alpha\beta}p_a \quad . \tag{5.4.4}$$

p_a is given above, and q_α is represented in terms of the same coordinates as

$$q_\alpha = \partial_\alpha - i\gamma^a{}_{\alpha\beta}\theta^\beta\partial_a \quad . \tag{5.4.5}$$

(See (5.3.1) for $D = 1$.) All these objects also transform covariantly under Lorentz transformations generated by

$$J_{ab} = -ix_{[a}p_{b]} + \tfrac{1}{4}\theta^\alpha\gamma_{[a\alpha\gamma}\gamma_{b]}{}^{\gamma\beta}\partial_\beta + M_{ab} \quad , \tag{5.4.6}$$

where we have included the (coordinate-independent) spin term M_{ab}. (In comparison with (2.2.4), the spin operator here gives just the spin of the superfield, which is a function of x and θ, whereas the spin operator there includes the $\theta\partial$ term, and thus gives the spin of the component fields resulting as the coefficients of a Taylor expansion of the superfield in powers of θ.)

As described in the previous section, even for "simple" supersymmetry (the smallest supersymmetry for that dimension), these spinors are reducible if the irreducible spinor representation isn't real, and reduce to the direct sum of an irreducible spinor and its complex conjugate. However, we can further generalize by letting the spinor represent more than one of such real spinors (and some of each of the 2 types that are independent when D is twice odd), and still use the same notation, with a single index representing all spinor components. (5.4.1-6) are then unchanged (except for the range of the spinor index). However, the nature of the supersymmetry representations will depend on D, and on the number of minimal supersymmetries. In the remainder of this section we'll stick to this notation to manifest those properties which are independent of dimension, and include such things as internal-symmetry generators when required for dimension-dependent properties.

For a massless, real, scalar field, $p^2 = 0$ is the only equation of motion, but for a massless, real, scalar superfield, the additional equation $\not{p}d = 0$ (where d is the spinor derivative) is necessary to impose that the superfield is a unitary representation of (on-shell) supersymmetry [5.12]: Since the hermitian supersymmetry generators q satisfy $\{q,q\} \sim p$, we have that $\{\not{p}q, \not{p}q\} \sim p^2\not{p} = 0$, but on unitary representations any hermitian operator whose square vanishes must also vanish, so $0 = \not{p}q = \not{p}d$ up to a term proportional to $p^2 = 0$. This means that only half the q's are nonvanishing. We can further divide these remaining q's in (complex) halves as creation and annihilation operators. A massless, irreducible representation of supersymmetry is then specified in this nonmanifestly Lorentz covariant approach by fixing the "Clifford" vacuum of these creation and annihilation operators to be an irreducible representation of the Poincaré group.

Unfortunately, the p^2 and $\not{p}d$ equations are not sufficient to determine an irreducible representation of supersymmetry, even for a scalar superfield (with certain exceptions in $D \leq 4$) since, although they kill the unphysical half of the q's, they don't restrict the Clifford vacuum. The latter restriction requires extra constraints in a manifestly Lorentz covariant formalism. There are several ways to find these additional constraints: One is to consider coupling to external fields. The simplest case is external super-Yang-Mills (which will be particularly relevant for strings). The generalization of the covariant derivatives is

$$d_\alpha \to \nabla_\alpha = d_\alpha + \Gamma_\alpha \quad ,$$

$$p_a \to \nabla_a = p_a + \Gamma_a \quad . \tag{5.4.7}$$

We thus have a graded covariant derivative ∇_A, $A = (a, \alpha)$. Without loss of generality, we consider cases where the only physical fields in the super-Yang-Mills multiplet are a vector and a spinor. The other cases (containing scalars) can be obtained easily by dimensional reduction. Then the commutation relations of the covariant derivatives become [5.13]

$$\begin{aligned} \{\nabla_\alpha \,, \nabla_\beta\} &= 2\gamma^a{}_{\alpha\beta}\nabla_a \quad , \\ [\nabla_\alpha \,, \nabla_a] &= 2\gamma_{a\alpha\beta}W^\beta \quad , \\ [\nabla_a \,, \nabla_b] &= F_{ab} \quad , \end{aligned} \tag{5.4.8}$$

where W^α is the super-Yang-Mills field strength (at $\theta = 0$, the physical spinor field), and consistency of the Jacobi (Bianchi) identities requires

$$\gamma_{a(\alpha\beta}\gamma^a{}_{\gamma)\delta} = 0 \quad . \tag{5.4.9}$$

This condition (when maximal Lorentz invariance is assumed, i.e., SO(D-1,1) for a taking D values) implies spacetime dimensions $D = 3, 4, 6, 10$, and "antispacetime" dimensions (the number of values of the index α) $D' = 2(D-2)$. (The latter identity follows from multiplying (5.4.9) by $\gamma^{b\alpha\beta}$ and using (5.4.2).) The generalization of the equations of motion is [5.14]

$$\begin{aligned} \not{p}d &\rightarrow \gamma^{a\alpha\beta}\nabla_a\nabla_\beta \quad , \\ \tfrac{1}{2}p^2 &\rightarrow \tfrac{1}{2}\nabla^a\nabla_a + W^\alpha\nabla_\alpha \quad , \end{aligned} \tag{5.4.10}$$

but closure of this algebra also requires new equations of motion which are certain Lorentz pieces of $\nabla_{[\alpha}\nabla_{\beta]}$. Specifically, in $D = 3$ there is only one Lorentz piece (a scalar), and it gives the usual field equations for a scalar multiplet [5.15]; in $D = 4$ the scalar piece again gives the usual equations for a chiral scalar multiplet, but the (axial) vector piece gives the chirality condition (after appropriate normal ordering); in $D = 6$ only the self-dual third-rank antisymmetric tensor piece appears in the algebra, and it gives equations satisfied by scalar multiplets (but not by tensor multiplets, which are also described by scalar superfield strengths, but can't couple minimally to Yang-Mills) [5.16]; but in $D = 10$ no multiplet is described because the only one possible would be Yang-Mills itself, but its field strength W^α carries a Lorentz index, and the equations described above (which apply only to scalars) need extra terms containing Lorentz generators.

Another way to derive the modifications is to use superconformal transformations. The superconformal groups [5.17] are actually easier to derive than the supersymmetry groups because they are just graded versions of classical groups. Specifically, the classical supergroups (see [5.18] for a review) have defining representations defined in terms of a metric $M_{A\dot{B}}$, which makes them unitary (or pseudounitary, if the metric isn't positive definite), and sometimes also a graded-symmetric metric M_{AB}, and thus $M_A{}^{\dot{B}}$ by combining them (and their inverses). The generators which have bosonic-bosonic or fermionic-fermionic indices are bosonic, and those with bosonic-fermionic are fermionic. (The choice of which parts of the A index are bosonic and which are fermionic can be reversed, but this doesn't affect the statistics of the group generators.) Since the bosonic subgroup of the supergroups with just the $M_{A\dot{B}}$ metric is the direct product of 2 unitary groups, those supergroups are called (S)SU(M|N) for M values of the index of one statistics and N of the other, where the S is because a (graded) trace condition is imposed, and there is a second S for M=N because then a second trace can be removed (so each of the 2 unitary subgroups becomes SU(N)). The supergroups which also have the graded-symmetric metric M_{AB} have a bosonic subgroup which is orthogonal in the sector where the metric is symmetric and symplectic in the sector where it is antisymmetric. In this case we choose the metric $M_{A\dot{B}}$ also to have graded symmetry, in such a way that the metric $M_A{}^{\dot{B}}$ obtained from their product is totally symmetric, so the defining representation is real, or totally antisymmetric, so the representation is pseudoreal. The former is generally called OSp(M|2N), and we call the latter OSp*(M|2N).

We next assume that the anticommuting generators of these supergroups are to be identified with the conformal generalization of the supersymmetry generators. Thus, one index is to be identified with an internal symmetry, and the other with a conformal spinor index. The conformal spinor reduces to 2 Lorentz spinors, one of which is the usual supersymmetry, the "square root" of translations, and the other of which is "S-supersymmetry," the square root of conformal boosts. The choice of supergroup then follows immediately from the graded generalization of the conformal spinor metrics appearing in the table of the previous section [5.19]:

D mod 8	superconformal	bosonic subgroup	dim-0/dilatations
0,4	(S)SU(N|ν,ν)	SU(ν,ν)$\otimes$SU(N)($\otimes$U(1))	SL(ν,C)$\otimes$(S)U(N)
1,3	OSp(N|2ν)	Sp(2ν)$\otimes$SO(N)	SL(ν,R)$\otimes$SO(N)
2	OSp(N|2ν) *or* (")$_+\otimes$(")$_-$	" (")$_+\otimes$(")$_-$	" (")$_+\otimes$(")$_-$
5,7	OSp*(2ν|2N)	SO*(2ν)$\otimes$USp(2N)	SL*(ν)$\otimes$USp(2N)
6	OSp*(2ν|2N) *or* (")$_+\otimes$(")$_-$	" (")$_+\otimes$(")$_-$	" (")$_+\otimes$(")$_-$

where "dim-0" are the generators which commute with dilatations (see sect. 2.2), so the last column gives Lorentz⊗internal symmetry (at least). ν is the dimension of the (irreducible) Lorentz spinor (1/2 that of the conformal spinor), N is the number of minimal supersymmetries, and $D(>2)$ is the dimension of (Minkowski) spacetime (with conformal group SO(D,2), so 2 less than the D in the previous table). The 2 choices for twice-odd D depend on whether we choose to represent the superconformal group on both primed and unprimed spinors. If so, there can be a separate N and N′. Again, we have used * to indicate groups which are Wick rotations such that the defining representation is pseudoreal instead of real. (SL* is sometimes denoted SU*.)

Unfortunately, as discussed in the previous section, the bosonic subgroup acting on the conformal spinor part of the defining representation, as defined by the spinor metrics (plus the trace condition, when relevant) gives a group bigger than the conformal group unless $D \leq 6$. However, we can still use $D \leq 6$, and perhaps some of the qualitative features of $D > 6$, for our analysis of massless field equations. We then generalize our analysis of sect. 2.2 from conformal to superconformal. It's sufficient to apply just the S-supersymmetry generators to just the Klein-Gordon operator. We then find [2.6,5.19]:

$$\tfrac{1}{2}p^2 \to \not{p}q = \not{p}d \to \begin{cases} \tfrac{1}{2}\{p^b\ ,\ J_{ab}\} + \tfrac{1}{2}\{p_a\ ,\ \Delta\} = p^b M_{ab} + p_a\left(d - \tfrac{D-2}{2}\right) \\ q_{[\alpha}q_{\beta]} + \cdots = d_{[\alpha}d_{\beta]} + \cdots \end{cases} ,\tag{5.4.11}$$

where the last expression means certain Lorentz pieces of dd plus certain terms containing Lorentz and internal symmetry generators. In particular, $\frac{1}{16}\gamma_{abc}{}^{\alpha\beta}d_\alpha d_\beta + \frac{1}{2}p_{[a}M_{bc]}$ is the supersymmetric analog of the Pauli-Lubansky vector [5.20]. The vector equation is (2.2.8) again, derived in essentially the same way.

For the constraints we therefore choose [5.21]

$$\begin{aligned} \mathcal{A} &= \tfrac{1}{2}p^2 \quad , \\ \mathcal{B}^\alpha &= \gamma^{a\alpha\beta}p_a d_\beta \quad , \\ \mathcal{C}_{abc} &= \tfrac{1}{16}\gamma_{abc}{}^{\alpha\beta}d_\alpha d_\beta + \tfrac{1}{2}p_{[a}M_{bc]} \quad , \\ \mathcal{D}_a &= M_a{}^b p_b + kp_a \quad ; \end{aligned}\tag{5.4.12a}$$

or, in matrix notation,

$$\begin{aligned} \mathcal{A} &= \tfrac{1}{2}p^2 \quad , \\ \mathcal{B} &= \not{p}d \quad , \end{aligned}$$

$$\begin{aligned} \mathcal{C}_{abc} &= \tfrac{1}{16} d\gamma_{abc} d + \tfrac{1}{2} p_{[a} M_{bc]} \quad , \\ \mathcal{D}_a &= M_a{}^b p_b + k p_a \quad ; \end{aligned} \tag{5.4.12b}$$

where out of (5.4.11) we have chosen $\mathcal{A}$ and $\mathcal{D}$ as for nonsupersymmetric theories (sect. 2.2), $\mathcal{B}$ for unitarity (as explained above), and just the Pauli-Lubansky part of the rest (which is all of it for D=10), the significance of which will be explained below.

These constraints satisfy the algebra

$$\begin{aligned} \{\mathcal{B}, \mathcal{B}\} &= 4\not{p}\mathcal{A} \quad , \\ [\mathcal{D}_a, \mathcal{D}_b] &= -2M_{ab}\mathcal{A} - p_{[a}\mathcal{D}_{b]} \quad , \\ [\mathcal{C}_{abc}, \mathcal{B}] &= -8\gamma_{abc} d\mathcal{A} \quad , \\ [\mathcal{C}_{abc}, \mathcal{D}_d] &= \eta_{d[a} p_b \mathcal{D}_{c]} \quad , \\ [\mathcal{C}^{abc}, \mathcal{C}_{def}] &= -\tfrac{1}{4}[\delta_{[d}{}^{[a} p_e \mathcal{C}^{bc]}{}_{f]} - (abc \leftrightarrow def)] - \tfrac{1}{128} d(4\gamma^{abc}{}_{def} - \delta_{[d}{}^{[a}\delta_e{}^b\gamma^{c]}{}_{f]})\mathcal{B} \quad , \\ rest &= 0 \quad , \end{aligned} \tag{5.4.13}$$

with some ambiguity in how the right-hand side is expressed due to the relations

$$\begin{aligned} p \cdot \mathcal{D} &= 2k\mathcal{A} \quad , \\ \not{p}\mathcal{B} &= 2d\mathcal{A} \quad , \\ d\mathcal{B} &= 2(tr\ I)\mathcal{A} \quad , \\ \tfrac{1}{6} p_{[a} \mathcal{C}_{bcd]} &= -\tfrac{1}{16} d\gamma_{abcd}\mathcal{B} \quad , \\ p^c \mathcal{C}_{cab} &= 2M_{ab}\mathcal{A} + p_{[a}\mathcal{D}_{b]} + \tfrac{1}{16} d\gamma_{ab}\mathcal{B} \quad . \end{aligned} \tag{5.4.14}$$

($\gamma_{ab} = \frac{1}{2}\gamma_{[a}\gamma_{b]}$, etc.)

In the case of supersymmetry with an internal symmetry group (extended supersymmetry, or even simple supersymmetry in D=5,6,7), there is an additional constraint analogous to $\mathcal{C}_{abc}$ for superisospin:

$$\mathcal{C}_{a,int} = \tfrac{1}{8} d\gamma_a \sigma_{int} d + p_a M_{int} \quad . \tag{5.4.15}$$

σ_{int} are the matrix generators of the internal symmetry group, in the representation to which d belongs, and M_{int} are those which act on the external indices of the superfield.

Unfortunately, there are few superspin-0 multiplets that are contained within spin-0, isospin-0 superfields (i.e., that themselves contain spin-0, isospin-0 component fields). In fact, the only such multiplets of physical interest in D>4 are N=1 Yang-Mills in D=9 and N=2 nonchiral supergravity in D=10. (For a convenient listing of multiplets, see [5.22].) However, by the method described in the previous section, spinor representations for the Lorentz group can be introduced. By including "γ-matrices" for internal symmetry, we can also introduce defining representations for the internal symmetry groups for which they are equivalent to the spinor representations of orthogonal groups (i.e., SU(2)=USp(2)=SO(3), USp(4)=SO(5), SU(4)=SO(6), SO(4)-vector=SO(3)-spinor⊗SO(3)′-spinor, SO(8)-vector=SO(8)′-spinor). Furthermore, arbitrary U(1) representations can be described by adding extra terms without introducing additional coordinates. This allows the description of most superspin-0 multiplets, but with some notable exceptions (e.g., 11D supergravity). However, these equations are not easily generalized to nonzero superspins, since, although the superspin operator is easy to identify in the light-cone formalism (see below), the corresponding operator would be nonlocal in a covariant description (or appear always with an additional factor of momentum).

We next consider the construction of mechanics actions. These equations describe only multiplets of superspin 0, i.e., the smallest representations of a given supersymmetry algebra, for reasons to be described below. (This is no restriction in D=3, where superspin doesn't exist, and in D=4 arbitrary superspin can be treated by a minor modification, since there superspin is abelian.) As described in the previous section, only spin-0 and spin-1/2 superfields can be described by classical mechanics, and we begin with spin-0, dropping spin terms in (5.4.12), and the generator $\mathcal{D}$. The action is then given by (3.1.10), where [5.23]

$$z^M = (x^m, \theta^\mu) \quad , \quad \pi_A = (p_a, id_\alpha) \quad ,$$

$$\dot{z}^M e_M{}^A(z) = (\dot{x} - i\dot{\theta}\gamma\theta, \dot{\theta}) \quad ,$$

$$i\mathcal{G}_i(\pi) = (\mathcal{A}, \mathcal{B}^\alpha, \mathcal{C}_{abc}) \quad . \tag{5.4.16}$$

Upon quantization, the covariant derivatives become

$$\pi_A = ie_A{}^M\partial_M = i(\partial_a, \partial_\alpha + i\gamma^a{}_{\alpha\beta}\theta^\beta\partial_a) \quad , \tag{5.4.17}$$

which are invariant under the supersymmetry transformations

$$\delta x = \xi - i\epsilon\gamma\theta \quad , \quad \delta\theta = \epsilon \quad . \tag{5.4.18}$$

The transformation laws then follow directly from (3.1.11), with the aid of (5.4.13) for the λ transformations.

The classical mechanics action can be quantized covariantly by BRST methods. In particular, the transformations generated by $\mathcal{B}$ [5.24] (with parameter κ) close on those generated by $\mathcal{A}$ (with parameter ξ):

$$\delta x = \xi p + i\kappa(\gamma d + \not{p}\gamma\theta) \quad , \quad \delta\theta = \not{p}\kappa \quad ,$$
$$\delta p = 0 \quad , \quad \delta d = 2p^2\kappa \quad ,$$
$$\delta g = \dot{\xi} + 4i\kappa\not{p}\psi \quad , \quad \delta\psi = \dot{\kappa} \quad . \qquad (5.4.19)$$

Because of the second line of (5.4.14), the ghosts have a gauge invariance similar to the original κ invariance, and then the ghosts of those ghosts again have such an invariance, etc., ad infinitum. This is a consequence of the fact that only half of θ can be gauged away, but there is generally no Lorentz representation with half the components of a spinor, so the spinor gauge parameter must itself be half gauge, etc. Although somewhat awkward, the infinite set of ghosts is straightforward to find. Furthermore, if derived from the light cone, the OSp(1,1|2) generators automatically contain this infinite number of spinors: There, θ is first-quantized in the same way as the Dirac spinor was second-quantized in sect. 3.5, and θ obtains an infinite number of components (as an infinite number of ordinary spinors) as a result of being a representation of a graded Clifford algebra (specifically, the Heisenberg algebras of γ^α and $\tilde{\gamma}^\alpha$). This analysis will be made in the next section.

On the other hand, the analysis of the constraints is simplest in the light-cone formalism. The $\mathcal{A}$, $\mathcal{B}$, and $\mathcal{D}$ equations can be solved directly, because they are all of the form $p \cdot f = p_+ f_- + \cdots$:

$$\mathcal{A} = 0 \quad \rightarrow \quad p_- = -\frac{1}{2p_+}{p_i}^2 \quad ,$$
$$\mathcal{B} = 0 \quad \rightarrow \quad \gamma_- d = -\frac{1}{2p_+}\gamma_i p_i \gamma_- \gamma_+ d \quad ,$$
$$\mathcal{D} = 0 \quad \rightarrow \quad M_{-i} = \frac{1}{p_+}({M_i}^j p_j + k p_i) \quad , \quad M_{-+} = k \quad , \qquad (5.4.20a)$$

where we have chosen the corresponding gauges

$$x_+ = 0 \quad ,$$
$$\gamma_+\theta = 0 \quad ,$$
$$M_{+i} = 0 \quad . \qquad (5.4.20b)$$

These solutions restrict the x's, θ's, and Lorentz indices, respectively, to those of the light cone. (Effectively, D is reduced by 2, except that p_+ remains.)

However, a superfield which is a function of a light-cone θ is not an irreducible representation of supersymmetry (except sometimes in D$\leq$4), although it is a unitary one. In fact, $\mathcal{C}$ is just the superspin operator which separates the representations: Due to the other constraints, all its components are linearly related to

$$\mathcal{C}_{+ij} = p_+ M_{ij} + \tfrac{1}{16} d\gamma_+\gamma_{ij} d \quad . \tag{5.4.21}$$

Up to a factor of p_+, this is the light-cone superspin: On an irreducible representation of supersymmetry, it acts as an irreducible representation of SO(D-2). In D=4 this can be seen easily by noting that the irreducible representations can be represented in terms of chiral superfields ($\bar{d} \doteq 0$) with different numbers of d's acting on them, and the $d\bar{d}$ in $\mathcal{C}$ just counts the numbers of d's. In general, if we note that the full light-cone Lorentz generator can be written as

$$\begin{aligned} J_{ij} &= -ix_{[i}p_{j]} + \tfrac{1}{2}\theta\gamma_{ij}\frac{\partial}{\partial\theta} + M_{ij} \\ &= -ix_{[i}p_{j]} - \frac{1}{16p_+}q\gamma_+\gamma_{ij}q + \frac{1}{16p_+}d\gamma_+\gamma_{ij}d + M_{ij} \\ &= \hat{J}_{ij} + \frac{1}{p_+}\mathcal{C}_{+ij} \quad , \end{aligned} \tag{5.4.22}$$

then, by expressing any state in terms of q's acting on the Clifford vacuum, we see that $\hat{J}$ gives the correct transformation for those q's and the x-dependence of the Clifford vacuum, so $\mathcal{C}/p_+$ gives the spin of the Clifford vacuum less the contribution of the qq term on it, i.e., the superspin. Unfortunately, the mechanics action can't handle spin operators for irreducible representations (either for M_{ij} or the superspin), so we must restrict ourselves not only to spin 0 (referring to the external indices on the superfield), but also superspin 0 (at least at the classical mechanics level). Thus, the remaining constraint $\mathcal{C} = 0$ is the only possible (first-class) constraint which can make the supersymmetry representation irreducible. The constraints (5.4.12) are therefore necessary and sufficient for deriving the mechanics action. However, if we allow the trivial kind of second-class constraints that can be solved in terms of matrices, we can generalize to spin-1/2. In principle, we could also do superspin-1/2, but this leads to covariant fields which are just those for superspin-0 with an extra spinor index tagged on, which differs by factors of momentum (with appropriate index contractions) from the desired expressions. (Thus, the superspin operator would be nonlocal on the latter.) Isospin-1/2 can be treated similarly (and superisospin-1/2, but again as a tagged-on index).

The simplest nontrivial example is D=4. (In D=3, C_{+ij} vanishes, since the transverse index i takes only one value.) There light-cone spinors have only 1 (complex) component, and so does C_{+ij}. For this case, we can (and must, for an odd number N of supersymmetries) modify C_{abc}:

$$C_{abc} = \tfrac{1}{16} d\gamma_{abc} d + \tfrac{1}{2} p_{[a} M_{bc]} + iH\epsilon_{abcd} p^d \quad , \tag{5.4.23}$$

where H is the "superhelicity." We then find

$$C_{+ij} = p_+ \epsilon_{ij} \left(\mathcal{M} + iH - i\frac{1}{4p_+}[d_{\mathbf{a}}, \bar{d}^{\mathbf{a}}] \right) , \tag{5.4.24}$$

where $M_{ij} = \epsilon_{ij}\mathcal{M}$ and $\{d_{\mathbf{a}}, \bar{d}^{\mathbf{b}}\} = p_+ \delta_{\mathbf{a}}{}^{\mathbf{b}}$, and $\mathbf{a}$ is an SU(N) index. The "helicity" h is given by $\mathcal{M} = -ih$, and we then find by expanding the field over chiral fields ϕ [5.25-27] ($\bar{d}\phi = 0$)

$$\psi \sim (d)^n \phi \quad \rightarrow \quad H = h + \tfrac{1}{4}(2n - N) \quad . \tag{5.4.25}$$

Specifying both the spin and superhelicity of the original superfield fixes both h and H, and thus determines n. Note that this requires H to be quarter-(odd-)integral for odd N. In general, the SU(N) representation of ϕ also needs to be specified, and the relevant part of (5.4.15) is

$$C_{+\mathbf{a}}{}^{\mathbf{b}} = p_+ \left[M_{\mathbf{a}}{}^{\mathbf{b}} - i\frac{1}{4p_+} \left([d_{\mathbf{a}}, \bar{d}^{\mathbf{b}}] - \tfrac{1}{4}\delta_{\mathbf{a}}{}^{\mathbf{b}} [d_{\mathbf{c}}, \bar{d}^{\mathbf{c}}] \right) \right] \quad , \tag{5.4.26}$$

and the vanishing of this quantity forces ϕ to be an SU(N)-singlet. (More general cases can be obtained simply by tacking extra indices onto the original superfield, and thus onto ϕ.)

We next consider 10D super Yang-Mills. The appropriate superfield is a Weyl or Majorana spinor, so we include terms as in the previous section in the mechanics action. To solve the remaining constraint, we first decompose SO(9,1) covariant spinors and γ-matrices to SO(8) light-cone ones as

$$d_\alpha = 2^{1/4} \begin{pmatrix} d_+ \\ d_- \end{pmatrix} \quad ,$$

$$\not{p}_{\alpha\beta} = \begin{pmatrix} \sqrt{2}p_+ & \not{p}_T \\ \not{p}_T{}^\dagger & -\sqrt{2}p_- \end{pmatrix} \begin{pmatrix} \Sigma & 0 \\ 0 & \Sigma \end{pmatrix} \quad , \quad \not{p}^{\alpha\beta} = \begin{pmatrix} \Sigma & 0 \\ 0 & \Sigma \end{pmatrix} \begin{pmatrix} \sqrt{2}p_- & \not{p}_T \\ \not{p}_T{}^\dagger & -\sqrt{2}p_+ \end{pmatrix} \quad ,$$

$$\not{a}_T \not{b}_T{}^\dagger + \not{b}_T \not{a}_T{}^\dagger = \not{a}_T{}^\dagger \not{b}_T + \not{b}_T{}^\dagger \not{a}_T = 2a_T \cdot b_T \quad ,$$

$$\not{p}_T{}^* = \Sigma \not{p}_T \Sigma \quad , \quad \Sigma = \Sigma^\dagger = \Sigma^* \quad , \quad \Sigma^2 = I \quad . \tag{5.4.27}$$

(We could choose the Majorana representation $\Sigma = I$, but other representations can be more convenient.) The independent supersymmetry-covariant derivatives are then

$$d_+ = \frac{\partial}{\partial\theta^+} + p_+\Sigma\theta^+ \quad , \quad p_T \quad , \quad p_+ \quad . \qquad (5.4.28)$$

In order to introduce chiral light-cone superfields, we further reduce SO(8) to SO(6)⊗SO(2)=U(4) notation:

$$d_+ = \sqrt{2}\begin{pmatrix} d_\mathrm{a} \\ \bar{d}^\mathrm{a} \end{pmatrix} \quad , \quad \not{p}_T = \begin{pmatrix} p_\mathrm{ab} & \delta_\mathrm{a}{}^\mathrm{b} p_L \\ \delta_\mathrm{b}{}^\mathrm{a} \bar{p}_L & \bar{p}^\mathrm{ab} \end{pmatrix} \quad , \quad \Sigma = \begin{pmatrix} 0 & I \\ I & 0 \end{pmatrix}$$

$$(\bar{p}^\mathrm{ab} = \tfrac{1}{2}\epsilon^\mathrm{abcd} p_\mathrm{cd}) \quad . \qquad (5.4.29)$$

In terms of this "euphoric" notation, the constraints C_{+ij} are written on the SO(6)-spinor superfield as

$$\frac{1}{p_+}C_{+\mathrm{a}}{}^\mathrm{b}\begin{pmatrix} \psi_\mathrm{c} \\ \bar{\psi}^\mathrm{c} \end{pmatrix} = i\tfrac{1}{2}\begin{pmatrix} \delta_\mathrm{c}{}^\mathrm{b}\psi_\mathrm{a} - \frac{1}{4}\delta_\mathrm{a}{}^\mathrm{b}\psi_\mathrm{c} \\ -\delta_\mathrm{a}{}^\mathrm{c}\bar{\psi}^\mathrm{b} + \frac{1}{4}\delta_\mathrm{a}{}^\mathrm{b}\bar{\psi}^\mathrm{c} \end{pmatrix} - i\frac{1}{4p_+}\left([d_\mathrm{a}, \bar{d}^\mathrm{b}] - \tfrac{1}{4}\delta_\mathrm{a}{}^\mathrm{b}[d_\mathrm{d}, \bar{d}^\mathrm{d}]\right)\begin{pmatrix} \psi_\mathrm{c} \\ \bar{\psi}^\mathrm{c} \end{pmatrix} ,$$

$$\frac{1}{p_+}C_+\begin{pmatrix} \psi_\mathrm{c} \\ \bar{\psi}^\mathrm{c} \end{pmatrix} = i\tfrac{1}{2}\begin{pmatrix} -\psi_\mathrm{c} \\ \bar{\psi}^\mathrm{c} \end{pmatrix} - i\frac{1}{4p_+}[d_\mathrm{a}, \bar{d}^\mathrm{a}]\begin{pmatrix} \psi_\mathrm{c} \\ \bar{\psi}^\mathrm{c} \end{pmatrix} ,$$

$$\frac{1}{p_+}C_{+\mathrm{ab}}\begin{pmatrix} \psi_\mathrm{c} \\ \bar{\psi}^\mathrm{c} \end{pmatrix} = i\tfrac{1}{2}\begin{pmatrix} \epsilon_\mathrm{abcd}\bar{\psi}^\mathrm{d} \\ 0 \end{pmatrix} - i\frac{1}{2p_+}d_\mathrm{a}d_\mathrm{b}\begin{pmatrix} \psi_\mathrm{c} \\ \bar{\psi}^\mathrm{c} \end{pmatrix} , \qquad (5.4.30)$$

and the complex conjugate equation for $C_+{}^\mathrm{ab}$. (Note that it is crucial that the original SO(10) superfield ψ^α was a spinor of chirality opposite to that of d_α in order to obtain soluble equations.) The solution to the first 2 equations gives ψ in terms of a chiral superfield ϕ,

$$\psi_\mathrm{a} = d_\mathrm{a}\phi \quad , \qquad (5.4.31a)$$

and that to the third equation imposes the self-duality condition [5.25-27]

$$\tfrac{1}{24}\epsilon^\mathrm{abcd} d_\mathrm{a}d_\mathrm{b}d_\mathrm{c}d_\mathrm{d}\phi = p_+{}^2\bar{\phi} \quad . \qquad (5.4.31b)$$

This can also be written as

$$\left\{\prod\left[(\tfrac{1}{2}p_+)^{-1/2}d\right]\right\}\phi = \bar{\phi} \quad \rightarrow \quad \int d\theta\, e^{\theta^\mathrm{a}\pi_\mathrm{a}p_+/2}\phi(\theta^\mathrm{a}) = [\phi(\bar{\pi}^\mathrm{a})]^* \quad . \qquad (5.4.32)$$

This corresponds to the fact that in the mechanics action τ reversal on θ^α includes multiplication by the charge conjugation matrix, which switches θ^a with $\bar{\theta}_\mathrm{a}$, which equals $-(2/p_+)\partial/\partial\theta^\mathrm{a}$ by the chirality condition $\bar{d}^\mathrm{a} = 0$.

These results are equivalent to those obtained from first-quantization of a mechanics action with $d_\alpha = 0$ as a second-class constraint [5.28]. (This is the analog of the constraint $\hat{\gamma} = 0$ of the previous section.) This is effectively the same as dropping the d terms from the action, which can then be written in second-order form by eliminating p by its equation of motion. This can be solved either by using a chiral superfield [5.27] as a solution to this constraint in a Gupta-Bleuler formalism,

$$\bar{d}^{\mathbf{a}}\phi = 0 \quad \rightarrow \quad d_{\mathbf{a}}\bar{\phi} = 0 \quad \rightarrow \quad \int \bar{\phi} d_+ \phi = 0 \quad , \tag{5.4.33}$$

or by using a superfield with a *real* 4-component θ [5.26] as a solution to this constraint before quantization (but after going to a light-cone gauge), determining half the remaining components of $\gamma_-\theta$ to be the canonical conjugates of the other half. However, whereas either of these methods with second-class constraints requires the breaking of manifest Lorentz covariance just for the formulation of the (field) theory, the method we have described above has constraints on the fields which are manifestly Lorentz covariant ((5.4.12)). Furthermore, this second-class approach requires that (5.4.32) be imposed in addition, whereas in the first-class approach it and the chirality condition automatically followed together from (5.4.21) (and the ordinary reality of the original SO(9,1) spinor superfield).

On the other hand, the formalism with second-class constraints can be derived from the first-class formalism *without* M_{ab} terms (and thus without $\mathcal{D}$ in (5.4.12)) [5.29]: Just as $\mathcal{A}$ and $\mathcal{B}$ were solved at the classical level to obtain (5.4.20), $\mathcal{C} = 0$ can also be solved classically. To be specific, we again consider $D = 10$. Then $\mathcal{C}_{+ij} = 0$ is equivalent to $d_{[\mu}d_{\nu]} = 0$ (where μ is an 8-valued light-cone spinor index). (They are just different linear combinations of the same 28 antisymmetric quadratics in d, which are the only nonvanishing d products classically.) This constraint implies the components of d are all proportional to the same anticommuting scalar, times different commuting factors:

$$d_\mu d_\nu = 0 \quad \rightarrow \quad d_\mu = c\zeta_\mu \quad , \tag{5.4.34a}$$

where c is anticommuting and ζ commuting. Furthermore, $\mathcal{C}_{+ij}$ are just SO(8) generators on d, and thus their gauge transformation can be used to rotate it in any direction, thus eliminating all but 1 component [5.30]. (This is clear from triality, since the spinor 8 representation is like the vector 8.) Specifically, we choose the gauge parameters of the $\mathcal{C}$ transformations to depend on ζ in such a way as to rotate ζ in any one direction, and then redefine c to absorb the remaining ζ factor:

$$\mathcal{C}\ gauge: \quad d_\mu = \delta_m{}^1 c \quad . \tag{5.4.34b}$$

In this gauge, the C constraint itself is trivial, since it is antisymmetric in d's. Finally, we quantize this one remaining component c of d to find

$$quantization: \quad c^2 = p_+ \quad \rightarrow \quad c = \pm\sqrt{p_+} \quad . \qquad (5.4.34c)$$

c has been determined only up to a sign, but there is a residual C gauge invariance, since the C rotation can also be used to rotate ζ in the opposite direction, changing its sign. After using the gauge invariance to make all but one component of ζ vanish, this sign change is the only part of the gauge transformation which survives. It can then be used to choose the sign in (5.4.34c). Thus, all the d's are determined (although 1 component is nonvanishing), and we obtain the same set of coordinates (x and q, no d) as in the second-class formalism. The C constraint can also be solved completely at the quantum mechanical level by Gupta-Bleuler methods [5.29]. The SO(D-2) generators represented by C are then divided up into the Cartan subalgebra, raising operators, and lowering operators. The raising operators are imposed as constraints (on the ket, and the lowering operators on the bra), implying only the highest-weight state survives, and the generators of the Cartan subalgebra are imposed only up to "normal-ordering" constants, which are just the weights of that state.

The components of this chiral superfield can be identified with the usual vector + spinor [5.26,27]:

$$\phi(x,\theta^{\mathbf{a}}) =$$
$$(p_+)^{-1}A_L(x) + \theta^{\mathbf{a}}(p_+)^{-1}\chi_{\mathbf{a}}(x) + \theta^{2\mathbf{ab}}A_{\mathbf{ab}}(x) + \theta^3{}_{\mathbf{a}}\bar{\chi}^{\mathbf{a}}(x) + \theta^4(p_+)\bar{A}_L(x) \quad , \quad (5.4.35)$$

and the $\theta = 0$ components of $\gamma_+\psi = (\psi_{\mathbf{a}}, \bar{\psi}^{\mathbf{a}})$ can be identified with the spinor. Alternatively, the vector + spinor content can be obtained directly from the vanishing of C_{+ij} of (5.4.21), without using euphoric notation: We first note that the Γ-matrices of M_{ij} are represented by 2 spinors, corresponding to the 2 different chiralities of spinors in SO(8). SO(8) has the property of "triality," which is the permutation symmetry of these 2 spinors with the vector representation. (All are 8-component representations.) Since the anticommutation relations of d are just a triality transformation of those of the Γ's (modulo p_+'s), they are represented by the other 2 representations: a spinor of the other chirality and a vector. The same holds for the representation of q. Thus, the direct product of the representations of Γ and d includes a singlet (superspin 0), picked out by $C_{+ij} = 0$, so the total SO(8) representation (generated by J_{ij} of (5.4.22)) is just that of q, a spinor (of opposite chirality) and a vector.

There is another on-shell method of analysis of (super)conformal theories that is manifestly covariant and makes essential use of spinors. This method expresses the fields in terms of the "spinor" representation of the superconformal group. (The ordinary conformal group is the case N=0.) The spinor is defined in terms of generalized γ-matrices ("twistors" [5.31] or "supertwistors" [5.32]):

$$\{\bar{\gamma}_A, \gamma^B] = \delta_A{}^B \quad , \tag{5.4.36}$$

where the index has been lowered by $M_{A\dot{B}}$, and the grading is such that the conformal spinor part has been chosen *commuting* and the internal part *anticommuting*, just as ordinary γ-matrices have bosonic (vector) indices but are anticommuting. The anticommuting γ's are then closely analogous to the light-cone supersymmetry generators $\gamma_- q$. The generators are then represented as

$$G^A{}_B \sim \gamma^A \bar{\gamma}_B \quad , \tag{5.4.37}$$

with graded (anti)symmetrization or traces subtracted, as appropriate. The case of OSp(1,1|2) has been treated in (3.5.1). A representation in terms of the usual superspace coordinates can then be generated by coset-space methods, as described in sect. 2.2. We begin by identifying the subgroup of the supergroup which corresponds to supersymmetry (by picking 1 of the 2 Lorentz spinors in the conformal spinor generator) and translations (by closure of supersymmetry). We then equate their representation in (5.4.37) (analogous to the $\hat{J}$'s of sect. 2.2) with their representation in (5.4.3,5) (analogous to the J's of sect. 2.2). (The constant γ-matrices of (5.4.3,5) should not be confused with the operators of (5.4.36).) This results in an expression analogous to (2.2.6), where $\Phi(0)$ is a function of half of (linear combinations of) the γ's of (5.4.36) (the other half being their canonical conjugates). (For example, the bosonic part of γ^A is a conformal spinor which is expressed as a Lorentz spinor ζ_α and its canonical conjugate. (5.4.37) then gives $p_a = \gamma_a{}^{\alpha\beta}\zeta_\alpha\zeta_\beta$, which implies $p^2 = 0$ in $D = 3,4,6,10$ due to (5.4.9).) We then integrate over these γ's to obtain a function of just the usual superspace coordinates x and θ. Due to the quadratic form of the momentum generator in terms of the γ's, it describes only positive energy. Negative energies, for antiparticles, can be introduced by adding to the field a term for the complex conjugate representation. At least in D=3,4,6,10 this superfield satisfies $p^2 = 0$ as a consequence of the explicit form of the generators (5.4.37), and as a consequence all the equations which follow from superconformal transformations. These equations form a superconformal tensor which can be written covariantly as an expression quadratic in $G^A{}_B$. In D=4 an additional U(1) acting on the twistor space can be identified as the (little group) helicity, and in D=6 a similar SU(2)(⊗SU(2)′ if the primed supergroup is also introduced) appears. ((5.4.34a) is also a supertwistor type of relation.)

The cases D=3,4,6,10 [5.33] are especially interesting not only for the above reason and (5.4.9) but also because their various spacetime groups form an interesting pattern if we consider these groups to be the same for these different dimensions except that they are over different generalized number systems **A** called "division algebras." These are generalizations of complex numbers which can be written as $z = z_0 + \sum_1^n z_i e_i$, $\{e_i, e_j\} = -2\delta_{ij}$, where n=0,1,3, or 7. Choosing for the different dimensions the division algebras

D	A
3	real
4	complex
6	quaternion
10	octonion

we have the correspondence

$$\left.\begin{array}{ll} SL_1(1,\mathbf{A}) & = SO(D-2) \\ SU(2,\mathbf{A}) & = SO(D-1) \\ SU(1,1,\mathbf{A}) & = SO(D-2,1) \\ SL(2,\mathbf{A}) & = SO(D-1,1) \\ SU'(4,\mathbf{A}) & = SO(D,2) \\ SU(N|4,\mathbf{A}) & = superconformal \end{array}\right\} /SO(D-3)$$

where SL_1 means only the real part (z_0) of the trace of the defining representation vanishes, by SU' we mean traceless and having the metric $\Omega_{\alpha\dot{\beta}}$ (vs. $\eta_{\alpha\dot{\beta}}$ for SU), and the graded SU has metric $M_{A\dot{B}} = (\eta_{a\dot{b}}, \Omega_{\alpha\dot{\beta}})$ (in that order). The $^{\bullet}$ refers to generalized conjugation $e_i \to -e_i$ (and the e_i are invariant under transposition, although their ordering inside the matrices changes). The "$/SO(D-3)$" refers to the fact that to get the desired groups we must include rotations of the $D-3$ e_i's among themselves. The only possible exception is for the $D = 10$ superconformal groups, which don't correspond to the $OSp(N|32)$ above, and haven't been shown to exist [5.34]. The light-cone form of the identity (5.4.9) ((7.3.17)) is equivalent to the division algebra identity $|xy| = |x||y|$ ($|x|^2 = xx^* = x_0{}^2 + \sum x_i{}^2$), where both the vector and spinor indices on the light-cone γ-matrices correspond to the index for (z_0, z_i) (all ranging over $D-2$ values) [5.35].

A similar first-quantization analysis will be made for the superstring in sect. 7.3.

5.5. SuperBRST

Instead of using the covariant quantization which would follow directly from the constraint analysis of (5.4.12), we will derive here the BRST algebra which follows from the light-cone by the method of sect. 3.6, which treats bosons and fermions symmetrically [3.16]. We begin with any (reducible) light-cone Poincaré representation which is also a supersymmetry representation, and extend also the light-cone supersymmetry generators to 4+4 extra dimensions. The resulting OSp(D+1,3|4) spinor does not commute with the BRST OSp(1,1|2) generators, and thus mixes physical and unphysical states. Fortunately, this extended supersymmetry operator q can easily be projected down to its OSp(1,1|2) singlet piece q_0. We begin with the fact that the light-cone supersymmetry generator is a tensor operator in a spinor representation of the Lorentz group:

$$[J_{ab}, q] = -\tfrac{1}{2}\gamma_{ab} q \quad , \tag{5.5.1}$$

where $\gamma_{ab} = \frac{1}{2}\gamma_{[a}\gamma_{b]}$. (All γ's are now Dirac γ-matrices, not the generalized γ's of (5.4.1,2).) As a result, its extension to 4+4 extra dimensions transforms with respect to the U(1)-type OSp(1,1|2) as

$$[J_{AB}, q\} = -\tfrac{1}{2}(\gamma_{AB} + \gamma_{A'B'})q \quad . \tag{5.5.2}$$

It will be useful to combine γ_A and $\gamma_{A'}$ into creation and annihilation operators as in (4.5.9):

$$\gamma_A = a_A + a^\dagger{}_A \quad , \quad \gamma_{A'} = i(a_A - a^\dagger{}_A) \quad ; \quad \{a_A, a^\dagger{}_B] = \eta_{AB} \tag{5.5.3}$$

$$\rightarrow \quad \tfrac{1}{2}(\gamma_{AB} + \gamma_{A'B'}) = a^\dagger{}_{[A}a_{B)} \quad . \tag{5.5.4}$$

($a^\dagger{}_A a_B$, without symmetrization, are a representation of $U(1,1|1,1)$.) We choose boundary conditions such that all "states" can be created by the creation operators $a^\dagger$ from a "vacuum" annihilated by the annihilation operators a. (This choice, eliminating states obtained from a second vacuum annihilated by $a^\dagger$, is a type of Weyl projection.) This vacuum is a fermionic spinor (acted on by γ_a) whose statistics are changed by $a^{\dagger\alpha}$ (but not by $a^\dagger{}_\pm$). If q is a real spinor, we can preserve this reality by choosing a representation where γ^A is real and $\gamma^{A'}$ is imaginary. (In the same way, for the ordinary harmonic oscillator the ground state can be chosen to be a real function of x, and the creation operator $\sim x - \partial/\partial x$ preserves the reality.) The corresponding charge-conjugation matrix is $C = i\gamma_{5'}$, where

$$\gamma_{5'} = \tfrac{1}{2}[\gamma_{+'}, \gamma_{-'}]e^{\pi \frac{1}{2}\{\gamma^{c'}, \gamma^{\bar{c}'}\}} \quad , \tag{5.5.5}$$

with $\gamma^{\alpha'} = (\gamma^{c'}, \gamma^{\bar{c}'})$. ($e^{i\gamma_{5'}\pi/2}$ converts to the representation where both γ^A and $\gamma^{A'}$ are real.)

We now project to the OSp(1,1|2) singlet

$$[J_{AB}, q_0\} = 0 \quad \rightarrow \quad q_0 = \delta(a^{\dagger A}a_A)q \quad , \tag{5.5.6}$$

where the Kronecker δ projects down to ground states with respect to these creation operators. It satisfies

$$\delta(a^\dagger a)a^\dagger = a\delta(a^\dagger a) = 0 \quad . \tag{5.5.7}$$

This projector can be rewritten in various forms:

$$\delta(a^{\dagger A}a_A) = \delta(a^{\dagger\alpha}a_\alpha)a_+a_-a^\dagger{}_+a^\dagger{}_- = \int_{-\pi}^{\pi} \frac{du}{2\pi}\, e^{iua^{\dagger A}a_A} \quad . \tag{5.5.8}$$

We next check that this symmetry of the physical states is the usual supersymmetry. We start with the light-cone commutation relations

$$\{q, \bar{q}\} = 2\mathcal{P}\not{p} \quad , \tag{5.5.9}$$

where $\mathcal{P}$ is a Weyl projector, when necessary, and, as usual, $\bar{q} = q^\dagger\eta$, with η the hermitian spinor metric satisfying $\gamma^\dagger\eta = \eta\gamma$. ($\eta$'s explicit form will change upon adding dimensions because of the change in signature of the Lorentz metric.) We then find

$$\{q_0, \bar{q}_0\} = \delta(a^\dagger a)2\mathcal{P}\not{p}\delta(a^\dagger a) = 2\mathcal{P}\delta(a^\dagger a)\gamma^a p_a \quad , \tag{5.5.10}$$

where the γ_A and $\gamma_{A'}$ terms have been killed by the $\delta(a^\dagger a)$'s on the left and right. The factors other than $2\gamma^a p_a$ project to the physical subspace (i.e., restrict the range of the extended spinor index to that of an ordinary Lorentz spinor). The analogous construction for the GL(1)-type OSp(1,1|2) fails, since in that case the corresponding projector $\delta(\eta^{BA}\gamma_A\gamma_{B'}) = \delta(\eta^{BA}\gamma_{A'}\gamma_B)^\dagger$ (whereas $\delta(a^\dagger a)$ is hermitian), and the 2 δ's then kill all terms in $\not{p}$ except $\eta^{BA}\gamma_{A'}p_B$.

As a special case, we consider arbitrary massless representations of supersymmetry. The light-cone representation of the supersymmetry generators is (cf. (5.4.20a))

$$q = q_+ - \frac{1}{2p_+}\gamma_+\gamma^i p_i q_+ \quad , \tag{5.5.11a}$$

where q_+ is a self-conjugate light-cone spinor:

$$\gamma_- q_+ = 0 \quad , \quad \{q_+, \bar{q}_+\} = 2\mathcal{P}\gamma_- p_+ \quad . \tag{5.5.11b}$$

Thus, q_+ has only half as many nonvanishing components as a Lorentz spinor, and only half of those are independent, the other half being their conjugates. The Poincaré algebra is then specified by

$$M_{ij} = \frac{1}{16p_+}\bar{q}\gamma_+\gamma_{ij}q + \breve{M}_{ij} \quad , \tag{5.5.12}$$

where $\breve{M}$ is an irreducible representation of SO(D-2), the superspin, specifying the spin of the Clifford vacuum of q_+. (Cf. (5.4.21,22). We have normalized the $\bar{q}q$ term for Majorana q.)

After adding 4+4 dimensions, q_+ can be Lorentz-covariantly further divided using $\gamma_{\pm'}$:

$$q_+ = \sqrt{p_+}\left(\frac{\partial}{\partial\bar{\theta}} + 2\gamma_-\theta\right) \quad ,$$

$$\gamma_-\frac{\partial}{\partial\bar{\theta}} = \gamma_+\theta = \gamma_{+'}\frac{\partial}{\partial\bar{\theta}} = \gamma_{-'}\theta = 0 \quad ,$$

$$\left\{\frac{\partial}{\partial\bar{\theta}}, \bar{\theta}\right\} = \mathcal{P}(\tfrac{1}{2}\gamma_-\gamma_+)(\tfrac{1}{2}\gamma_{+'}\gamma_{-'}) \quad . \tag{5.5.13}$$

After substitution of (5.5.11,12) into (5.5.6), we find

$$q_0 = \delta(a^\dagger a)\left[\sqrt{p_+}\frac{\partial}{\partial\bar{\theta}} + \frac{2}{\sqrt{p_+}}(\gamma^a p_a + \gamma^\alpha p_\alpha)\theta\right] . \tag{5.5.14}$$

From (5.5.12) we obtain the corresponding spin operators (for Majorana θ)

$$M_{ab} = \tfrac{1}{2}\bar{\theta}\gamma_{ab}\frac{\partial}{\partial\bar{\theta}} + \breve{M}_{ab} \quad , \quad M_{\alpha\beta} + M_{\alpha'\beta'} = \bar{\theta}a^\dagger{}_{(\alpha}a_{\beta)}\frac{\partial}{\partial\bar{\theta}} + \breve{M}_{\alpha\beta} + \breve{M}_{\alpha'\beta'} \quad ,$$

$$M_{-'+'} = -\tfrac{1}{2}\bar{\theta}\frac{\partial}{\partial\bar{\theta}} + \breve{M}_{-'+'} \quad , \quad M_{+'\alpha'} = \tfrac{1}{2}\bar{\theta}\gamma_-\gamma_{+'}\gamma_{\alpha'}\theta + \breve{M}_{+'\alpha'} \quad ,$$

$$M_{\alpha a} = \tfrac{1}{2}\bar{\theta}\gamma_\alpha\gamma_a\frac{\partial}{\partial\bar{\theta}} + \breve{M}_{\alpha a} \quad , \quad M_{-'\alpha'} = \tfrac{1}{16}\frac{\partial}{\partial\theta}\gamma_+\gamma_{-'}\gamma_{\alpha'}\frac{\partial}{\partial\bar{\theta}} + \breve{M}_{-'\alpha'} \quad . \tag{5.5.15}$$

Finally, we perform the unitary transformations (3.6.13) to find

$$q_0 = \delta(a^\dagger a)\left(\frac{\partial}{\partial\bar{\theta}} + 2\gamma^a p_a\theta\right) \quad . \tag{5.5.16}$$

The δ now projects out just the OSp(1,1|2)-singlet part of θ (i.e., the usual Lorentz spinor):

$$q_0 = \frac{\partial}{\partial\bar{\theta}_0} + \gamma^a p_a\theta_0 \quad ,$$

$$\frac{\partial}{\partial\bar{\theta}_0} = \delta(a^\dagger a)\frac{\partial}{\partial\bar{\theta}} \quad , \quad \theta_0 = 2\delta(a^\dagger a)\theta \quad , \quad \left\{\frac{\partial}{\partial\bar{\theta}_0}, \bar{\theta}_0\right\} = \mathcal{P}\delta(a^\dagger a) \quad . \qquad (5.5.17)$$

(5.5.15) can be substituted into (3.6.14). We then find the OSp(1,1|2) generators

$$J_{+\alpha} = ix_\alpha p_+ \quad , \quad J_{-+} = -ix_- p_+ \quad , \quad J_{\alpha\beta} = -ix_{(\alpha}p_{\beta)} + \widehat{M}_{\alpha\beta} \quad ,$$

$$J_{-\alpha} = -ix_- p_\alpha + \frac{1}{p_+}\left[-ix_\alpha \tfrac{1}{2}({p_a}^2 + p^\beta p_\beta) + \widehat{M}_\alpha{}^\beta p_\beta + \widehat{\mathcal{Q}}_\alpha\right] \quad ; \qquad (5.5.18a)$$

$$\widehat{M}_{\alpha\beta} = \bar{\theta}a^\dagger{}_{(\alpha}a_{\beta)}\frac{\partial}{\partial\bar{\theta}} + \check{M}_{\alpha\beta} + \check{M}_{\alpha'\beta'} \; ,$$

$$\widehat{\mathcal{Q}}_\alpha = -i\tfrac{1}{8}\overline{\left(\frac{\partial}{\partial\bar{\theta}} + i\gamma_-\gamma_{+'}\gamma^a p_a\theta\right)}\gamma_+\gamma_{-'}a^\dagger{}_\alpha\left(\frac{\partial}{\partial\bar{\theta}} - i\gamma_-\gamma_{+'}\gamma^a p_a\theta\right)$$
$$+ \left(\check{M}_{-'\alpha'} + \check{M}_\alpha{}^a p_a + \tfrac{1}{2}\check{M}_{+'\alpha'}{p_a}^2\right) \quad , \qquad (5.5.18b)$$

and $J_{ab} = -ix_{[a}p_{b]} + \frac{1}{2}\bar{\theta}\gamma_{ab}\partial/\partial\bar{\theta} + \check{M}_{ab}$ for the Lorentz generators. Finally, we can remove all dependence on $\gamma_\pm$ and $\gamma_{\pm'}$ by extracting the corresponding γ_0 factors contributing to the spinor metric η:

$$\gamma_0 = -i\tfrac{1}{2}(\gamma_+ - \gamma_-)(\gamma_{+'} - \gamma_{-'}) = {\gamma_0}^\dagger \quad , \quad \delta(a^\dagger a)\gamma_0 = \gamma_0\delta(a^\dagger a) = \delta(a^\dagger a) \quad ,$$

$$\frac{\partial}{\partial\bar{\theta}} \rightarrow \gamma_0\frac{\partial}{\partial\bar{\theta}} = i\tfrac{1}{2}\gamma_-\gamma_{+'}\frac{\partial}{\partial\bar{\theta}} \; , \quad \bar{\theta} \rightarrow \bar{\theta}\gamma_0 = i\tfrac{1}{2}\bar{\theta}\gamma_+\gamma_{-'} \quad ,$$

$$(\gamma_+, \gamma_{-'})\frac{\partial}{\partial\bar{\theta}} = (\gamma_+, \gamma_{-'})\theta = \bar{\theta}(\gamma_-, \gamma_{+'}) = \frac{\partial}{\partial\theta}(\gamma_-, \gamma_{+'}) = 0 \quad ,$$

$$\left\{\frac{\partial}{\partial\bar{\theta}}, \bar{\theta}\right] = \mathcal{P}(\tfrac{1}{2}\gamma_+\gamma_-)(\tfrac{1}{2}\gamma_{-'}\gamma_{+'}) \quad ; \qquad (5.5.19a)$$

and then convert to the harmonic oscillator basis with respect to these γ's:

$$\frac{\partial}{\partial\bar{\theta}} \rightarrow e^{a^\dagger{}_+ a^\dagger{}_-}\frac{\partial}{\partial\bar{\theta}} \quad , \quad \theta \rightarrow \tfrac{1}{2}e^{a^\dagger{}_+ a^\dagger{}_-}\theta \quad ,$$

$$\bar{\theta} \rightarrow \tfrac{1}{2}\bar{\theta}e^{a_+ a_-} \quad , \quad \frac{\partial}{\partial\theta} \rightarrow \frac{\partial}{\partial\theta}e^{a_+ a_-} \quad ;$$

$$a_\pm\frac{\partial}{\partial\bar{\theta}} = a_\pm\theta = \bar{\theta}a^\dagger{}_\pm = \frac{\partial}{\partial\theta}a^\dagger{}_\pm = 0 \; ;$$

$$\left\{\frac{\partial}{\partial\bar{\theta}}, \bar{\theta}\right] = \mathcal{P}(a_+ a_- a^\dagger{}_+ a^\dagger{}_-) \quad . \qquad (5.5.19b)$$

All $a_\pm$'s and $a^\dagger{}_\pm$'s can then be eliminated, and the corresponding projection operators (the factor multiplying $\mathcal{P}$ in (5.5.19b)) dropped. The only part of (5.5.18) (or the Lorentz generators) which gets modified is

$$\widehat{\mathcal{Q}}_\alpha = -\tfrac{1}{4}\bar{\mathcal{q}}a^\dagger{}_\alpha\mathcal{d} + \left(\check{M}_{-'\alpha'} + \check{M}_\alpha{}^a p_a + \tfrac{1}{2}\check{M}_{+'\alpha'}{p_a}^2\right) \quad ,$$

$$\not{q} = \frac{\partial}{\partial \theta} + \gamma^a p_a \theta \quad , \quad \not{d} = \frac{\partial}{\partial \theta} - \gamma^a p_a \theta \quad ,$$

$$q_0 = \delta(a^{\dagger\alpha} a_\alpha) \not{q} \quad , \quad d_0 = \delta(a^{\dagger\alpha} a_\alpha) \not{d} \quad . \tag{5.5.20}$$

If we expand the first term in $\widehat{Q}_\alpha$ level by level in $a^\dagger{}_\alpha$'s, we find a q at each level multiplying a d of the previous level. In particular, the first-level ghost $q_{1\alpha}$ multiplies the physical d_0. This means that $d_0 = 0$ is effectively imposed for only half of its spinor components, since the components of q are not all independent.

An interesting characteristic of this type of BRST (as well as more conventional BRST obtained by first-quantization) is that spinors obtain infinite towers of ghosts. In fact, this is necessary to allow the most general possible gauges. The simplest explicit example is BRST quantization of the action of sect. 4.5 for the Dirac spinor quantized in a gauge where the gauge-fixed kinetic operators are all p^2 instead of $\not{p}$. However, these ghosts are not all necessary for the gauge invariant theory, or for certain types of gauges. For example, for the type of gauge invariant actions for spinors described in sect. 4.5, the only parts of the infinite-dimensional OSp(D-1,1|2) spinors which are not pure gauge are the usual Lorentz spinors. (E.g., the OSp(D-1,1|2) Dirac spinor reduces to an ordinary SO(D-1,1) Dirac spinor.) For gauge-fixed, 4D N=1 supersymmetric theories, supergraphs use chains of ghost superfields which always terminate with chiral superfields. Chiral superfields can be irreducible off-shell representations of supersymmetry since they effectively depend on only half of the components of θ. (An analog also exists in 6D, with or without the use of harmonic superspace coordinates [4.12].) However, no chiral division of θ exists in 10D (θ is a real representation of SO(9,1)), so an infinite tower of ghost superfields is necessary for covariant background-field gauges. (For covariant non-background-field gauges, all but the usual finite Faddeev-Popov ghosts decouple.) Thus, the infinite tower is not just a property of the type of first-quantization used, but is an inherent property of the second-quantized theory. However, even in background-field gauges the infinite tower (except for the Faddeev-Popovs) contribute only at one loop to the effective action, so their evaluation is straightforward, and the only expected problem would be their summation.

The basic reason for the tower of θ's is the fact that only 1/4 (or, in the massive case, 1/2) of them appear in the gauge-invariant theory on-shell, but if θ is an irreducible Lorentz representation it's impossible to cancel 3/4 (or 1/2) of it covariantly. We thus effectively obtain the sums

$$1 - 1 + 1 - 1 + \cdots = \tfrac{1}{2} \quad , \tag{5.5.21a}$$

$$1 - 2 + 3 - 4 + \cdots = \tfrac{1}{4} \quad . \qquad (5.5.21b)$$

(The latter series is the "square" of the former.) The positive contributions represent the physical spinor (or θ) and fermionic ghosts at even levels, the negative contributions represent bosonic ghosts at odd levels (contributing in loops with the opposite sign), and the $\frac{1}{2}$ or $\frac{1}{4}$ represents the desired contribution (as obtained directly in light-cone gauges). Adding consecutive terms in the sum gives a nonconvergent (but nondivergent in case (5.5.21a)) result which oscillates about the desired result. However, there are unambiguous ways to regularize these sums. For example, if we represent the levels in terms of harmonic oscillators (one creation operator for (5.5.21a), and the 2 $a^\dagger{}_\alpha$'s for (5.5.21b)), these sums can be represented as integrals over coherent states (see (9.1.12)). For (5.5.21a), we have:

$$\begin{aligned} str(1) &= tr\left[(-1)^N\right] = \int \frac{d^2 z}{\pi}\, e^{-|z|^2} \left\langle z \middle| (-1)^{a^\dagger a} \middle| z \right\rangle \\ &= \int \frac{d^2 z}{\pi}\, e^{-|z|^2} \langle z | -z \rangle = \int \frac{d^2 z}{\pi}\, e^{-2|z|^2} = \tfrac{1}{2} \int \frac{d^2 z}{\pi}\, e^{-|z|^2} \\ &= \tfrac{1}{2} \quad , \end{aligned} \qquad (5.5.22)$$

where str is the supertrace; for (5.5.21b), the supertrace over the direct product corresponding to 2 sets of oscillators factors into the square of (5.5.22). (The corresponding partition function is $str(x^N) = 1/(1+x) = 1 - x + x^2 - \cdots$, and for 2 sets of oscillators $1/(1+x)^2 = 1 - 2x + 3x^2 - \cdots$.)

An interesting consequence of (5.5.21) is the preservation of the identity

$$D' = 2^{(D-k)/2} \quad ; \qquad D' = str_S(1) \quad , \quad D = str_V(1) \quad ; \qquad (5.5.23)$$

upon adding $(2,2|4)$ dimensions, where D' and D are the "superdimensions" of a spinor and vector, defined in terms of supertraces of the identity for that representation, and k is an integer which depends on whether the dimension is even or odd and whether the spinor is Weyl and/or Majorana (see sect. 5.3). k is unchanged by adding $(2,2|4)$ dimensions, D changes by addition of $4 - 4 = 0$, and, because of (5.5.21), D' changes by a factor of $2^2 \cdot (\frac{1}{2})^2 = 1$. This identity is important for super-Yang-Mills and superstrings.

Before considering the action for arbitrary supersymmetric theories, we'll first study the equations of motion, since the naive kinetic operators may require extra factors to write a suitable lagrangian. Within the OSp(1,1|2) formalism, the gauge-fixed field equations are (cf. (4.4.19))

$$(p^2 + M^2)\phi = 0 \qquad (5.5.24a)$$

when subject to the (Landau) gauge conditions [2.3]

$$\hat{\mathcal{Q}}_\alpha \phi = \widehat{M}_{\alpha\beta}\phi = 0 \quad . \tag{5.5.24b}$$

Applying these gauge conditions to the gauge transformations, we find the residual gauge invariance

$$\delta\phi = -i\tfrac{1}{2}\hat{\mathcal{Q}}_\alpha \Lambda^\alpha \quad , \quad \left[-\tfrac{3}{2}(p^2+M^2)+\hat{\mathcal{Q}}^2\right]\Lambda^\alpha = \widehat{M}_{\alpha\beta}\Lambda_\gamma + C_{\gamma(\alpha}\Lambda_{\beta)} = 0 \quad . \tag{5.5.24c}$$

(In the IGL(1) formalism, sect. 4.2, the corresponding equations involve just the $\hat{\mathcal{Q}}^+$ component of $\mathcal{Q}^\alpha$ and the $\widehat{M}^+$ and $\widehat{M}^3$ components of $\widehat{M}^{\alpha\beta}$, but are equivalent, since $\widehat{M}^+ = \widehat{M}^3 = 0 \rightarrow \widehat{M}^- = 0$, and $\hat{\mathcal{Q}}^+ = \widehat{M}^- = 0 \rightarrow \hat{\mathcal{Q}}^- = 0$.)

For simplicity, we consider the massless case, and $\breve{M}_{ij} = 0$. We can then choose the reference frame where $p_a = \delta_a{}^+ p_+$, and solve these equations in light-cone notation. (The +'s and −'s now refer to the usual Lorentz components; the unphysical x_-, p_+, and $\gamma_{\pm'}$ have already been eliminated.) The gauge conditions (5.5.24b) eliminate auxiliary degrees of freedom (as $\partial \cdot A = p_+ A_- = 0$ eliminates A_- in Yang-Mills), and (5.5.24c) eliminates remaining gauge degrees of freedom (as A_+ in light-cone-gauge Yang-Mills). We divide the spinors d, q, $\partial/\partial\theta$, and θ into halves using $\gamma_\pm$, and then further divide those into complex conjugate halves as creation and annihilation operators, as in (5.4.27,29):

$$d \rightarrow \gamma_+ d, \gamma_- d \rightarrow d_{\mathbf{a}}, \bar{d}^{\mathbf{a}}, \partial_{\mathbf{a}}, \bar{\partial}^{\mathbf{a}} \quad ,$$

$$q \rightarrow \gamma_+ q, \gamma_- q \rightarrow q_{\mathbf{a}}, \bar{q}^{\mathbf{a}}, \partial_{\mathbf{a}}, \bar{\partial}^{\mathbf{a}} \quad , \tag{5.5.25}$$

where the "−" parts of d and q are both just partial derivatives because the momentum dependence drops out in this frame, and $d, \bar{d}$ and $q, \bar{q}$ have graded harmonic oscillator commutators (up to factors of p_+). (5.5.20) then becomes

$$\hat{\mathcal{Q}}_\alpha \sim \bar{q}^{\mathbf{a}} a^\dagger{}_\alpha \partial_{\mathbf{a}} + q_{\mathbf{a}} a^\dagger{}_\alpha \bar{\partial}^{\mathbf{a}} + \bar{\partial}^{\mathbf{a}} a^\dagger{}_\alpha d_{\mathbf{a}} + \partial_{\mathbf{a}} a^\dagger{}_\alpha \bar{d}^{\mathbf{a}} \quad . \tag{5.5.26}$$

Since $\hat{\mathcal{Q}}_\alpha$ consists of terms of the form AB, either A or B can be chosen as the constraint in (5.5.24b), and the other will generate gauge transformations in (5.5.24c). We can thus choose either $\partial_{\mathbf{a}}$, or $\bar{d}^{\mathbf{a}}$ and $\bar{q}^{\mathbf{a}}$, and similarly for the complex conjugates, except for the Sp(2) singlets, where the choice is between $\partial_{\mathbf{a}}$ and just $\bar{d}^{\mathbf{a}}$ (and similarly for the complex conjugates). However, choosing both d and $\bar{d}$ (or both q and $\bar{q}$) for constraints causes the field to vanish, and choosing them both for gauge generators allows the field to be completely gauged away. As a result, the only consistent constraints and gauge transformations are

$$\bar{d}^{\mathbf{a}}\phi = (a_\alpha \bar{q}^{\mathbf{a}})\phi = \partial_{\mathbf{a}}\phi = 0 \quad , \quad \delta\phi = \bar{\partial}^{\mathbf{a}}\lambda_{\mathbf{a}} \quad , \tag{5.5.27a}$$

subject to the restriction that the residual gauge transformations preserve the gauge choice (explicitly, (5.5.24c), although it's more convenient to re-solve for the residual invariance in light-cone notation). (There is also a complex conjugate term in ϕ if it satisfies a reality condition. For each value of the index a, the choice of which oscillator is creation and which is annihilation is arbitrary, and corresponding components of d and $\bar{d}$ or q and $\bar{q}$ can be switched by changing gauges.) Choosing the gauge

$$\bar{\theta}^-{}_{\mathbf{a}}\phi = 0 \quad , \tag{5.5.27b}$$

for the residual gauge transformation generated by $\bar{\partial}^{\mathbf{a}} = \partial/\partial\bar{\theta}^-{}_{\mathbf{a}}$, the field becomes

$$\phi(\theta^+,\bar{\theta}^+,\theta^-,\bar{\theta}^-) = \delta(a_\alpha\theta^+)\delta(\bar{\theta}^-)\varphi(\theta_0{}^+,\bar{\theta}_0{}^+) \quad , \quad \bar{d}_0\varphi = 0 \quad . \tag{5.5.28}$$

φ is the usual chiral light-cone superfield (as in sect. 5.4), a function of only 1/4 of the usual Lorentz spinor θ_0. This agrees with the general result of equivalence to the light cone for U(1)-type 4+4-extended BRST given at the end of sect. 4.5.

Since the physical states again appear in the middle of the θ expansion (including ghost θ's), we can again use (4.1.1) as the action: In the light-cone gauge, from (5.5.28), integrating over the δ-functions,

$$S = \int dx\, d\theta_0{}^+\, d\bar{\theta}_0{}^+\, \bar{\varphi}\Box\varphi \quad , \tag{5.5.29}$$

which is the standard light-cone superspace action [5.25]. As usual for the expansion of superfields into light-cone superfields, the physical light-cone superfield appears in the middle of the non-light-cone-θ expansion of the gauge superfield, with auxiliary light-cone superfields appearing at higher orders and pure gauge ones at lower orders. Because some of the ghost θ's are commuting, we therefore expect an infinite number of auxiliary fields in the gauge-covariant action, as in the harmonic superspace formalism [4.12]. This may be necessary in general, because this treatment includes self-dual multiplets, such as 10D super-Yang-Mills. (This multiplet is superspin 0, and thus does not require the superspin $\check{M}_{ij}$ to be self-dual, so it can be treated in the OSp(1,1|2) formalism. However, an additional self-conjugacy condition on the light-cone superfield is required, (5.4.31b), and a covariant OSp(1,1|2) statement of this condition would be necessary.) However, in some cases (such as 4D N=1 supersymmetry) it should be possible to truncate out all but a finite number of these auxiliary superfields. This would require an (infinite) extension of the group OSp(1,1|2) (perhaps involving part of the unphysical supersymmetries $a_\alpha q$), in the same way that extending IGL(1) to OSp(1,1|2) eliminates Nakanishi-Lautrup auxiliary fields.

The unusual form of the $OSp(1,1|2)$ operators for supersymmetric particles may require new mechanics actions for them. It may be possible to derive these actions by inverting the quantization procedure, first using the BRST algebra to derive the hamiltonian and then finding the gauge-invariant classical mechanics lagrangian.

Exercises

(1) Derive (5.1.2) from (5.1.1) and (3.1.11).

(2) Show that, under the usual gauge transformation $A \to A+\partial\lambda$, $exp[-iq\int_{\tau_i}^{\tau_f} d\tau\, \dot{x}\cdot A(x)]$ transforms with a factor $exp\{-iq[\lambda(x(\tau_f)) - \lambda(x(\tau_i))]\}$. (In a Feynman path integral, this corresponds to a gauge transformation of the ends of the propagator.)

(3) Fourier transform (5.1.13), using (5.1.14). Explicitly evaluate the proper-time integral in the massless case to find the coordinate-space Green function satisfying $\Box G(x,x') = \delta^D(x-x')$ for arbitrary $D > 2$. Do $D = 2$ by differentiating with respect to x^2, then doing the proper-time integral, and finally integrating back with respect to x^2. (There is an infinite constant of integration which must be renormalized.) For comparison, do $D = 2$ by taking the limit from $D > 2$.

(4) Use the method described in (5.1.13,14) to evaluate the 1-loop propagator correction in ϕ^3 theory. Compare the corresponding calculation with the covariantized light-cone method of sect. 2.6. (See exercise (7) of that chapter.)

(5) Derive (5.2.1), and find J^3. Derive (5.2.2) and the rest of the OSp(1,1|2) algebra. Show these results agree with those of sect. 3.4.

(6) Quantize (5.3.10) in the 3 supersymmetric gauges described in that section, and find the corresponding IGL(1) (and OSp(1,1|2), when possible) algebras in each case, using the methods of sects. 3.2-3. Note that the methods of sect. 3.3 require some generalization, since commuting antighosts can be conjugate to the corresponding ghosts and still preserve Sp(2): $[C_\alpha, C_\beta] \sim C_{\alpha\beta}$. Show equivalence to the appropriate algebras of sects. 3.4-5.

(7) Derive the tables in sect. 5.3. (Review the group theory of SO(N) spinors, if necessary.) Use the tables to derive the groups, equivalent to SO(D_+,D_-) for $D \leq 6$, for which these spinors are the defining representation.

(8) Express the real-spinor γ-matrices of sect. 5.3 in terms of the σ-matrices there for arbitrary D_+ and D_-. Use the Majorana representation where the spinor

is not necessarily explicitly real, but equivalent to a real one, such that: (1) for complex representations, the bottom half of the spinor is the complex conjugate of the top half (each being irreducible); (2) for pseudoreal representations, the bottom half is the complex conjugate again but with the index converted with a metric to make it explicitly the same representation as the top; (3) for real representations, the spinor is just the real, irreducible one. Find the matrices representing the internal symmetry (U(1) for complex and SU(2) for pseudoreal).

(9) Check the Jacobi identities for the covariant derivatives whose algebra is given in (5.4.8). Check closure of the algebra (5.4.10) in $D = 10$, including the extra generator described in the text.

(10) Derive (5.4.13).

(11) Write the explicit action and transformation laws for (5.4.16).

(12) Write the explicit equations of motion (5.4.12), modified by (5.4.23), for a scalar superfield for N=1 supersymmetry in D=4. Show that this gives the usual covariant constraints and field equations (up to constants of integration) for the chiral scalar superfield (scalar multiplet). Do the same for a spinor superfield, and obtain the equations for the vector-multiplet field strength.

(13) Derive the explicit form of the twistor fields for D=3,4,6. Find an explicit expression for the supersymmetrized Pauli-Lubansky vector in D=4 in terms of supertwistors, and show that it automatically gives an explicit expression for the superhelicity H of (5.4.23) as an operator in supertwistor space. Show the supertwistor Pauli-Lubansky vector automatically vanishes in D=3, and derive an expression in D=6.

Similar methods can be applied to the one-handed modes of the heterotic string [1.13]. (Then in (6.2.6) only spacelike X's appear, so instead of $u_a u_b$ any positive-definite metric can be used, effectively summing over the one-handed X's.) Various properties of the actions (3.1.17, 6.2.6) have been discussed in the literature [6.3], particularly in relation to anomalies in the gauge symmetry of the lagrange multipliers upon naive lagrangian quantization. One simple way to avoid these anomalies while keeping a manifestly covariant 2D lagrangian is to add scalars ϕ with the squares of both $\partial_-\phi$ and $\partial_+\phi$ appearing in lagrange-multiplier terms [6.4]. Alternatively (or additionally), one can add Weyl-Majorana 2D spinors (i.e., real, 1-component, 1-handed spinors) whose nonvanishing energy-momentum tensor component couples to the appropriate lagrange multiplier. (E.g., a spinor with kinetic term $\psi\partial_+\psi$ appears also in the term $\lambda[(\partial_-\phi)^2 + \psi\partial_-\psi]$.) These nonpropagating fields appear together with scalars with only one or the other handedness or neither constrained, and unconstrained fermions which are Weyl and/or Majorana or neither. There are (at most) 2 lagrange multipliers, one for each handedness.

In the conformal gauge there is still a residue of the gauge invariance, which originally included not only 2D general coordinate transformations but also local rescalings of the 2D metric (since only its unit-determinant part appeared in the action). By definition, the subset of these transformations which leave (6.2.1) invariant is the conformal group. Unlike in higher dimensions, the 2D conformal group has an infinite number of generators. It can easily be shown that these transformations consist of the coordinate transformations (restricted by appropriate boundary conditions)

$$\sigma^{\pm\prime} = \zeta^{\pm}(\sigma^{\pm}) \quad , \tag{6.2.7a}$$

with $\pm$'s not mixing (corresponding to 2 1D general coordinate transformations), since these coordinate transformations have an effect on the metric which can be canceled by a local scaling:

$$d\sigma^{2\prime} = 2d\sigma^{+\prime}d\sigma^{-\prime} = \zeta^{+\prime}\zeta^{-\prime}2d\sigma^{+}d\sigma^{-} \quad . \tag{6.2.7b}$$

On shell, these transformations are sufficient to gauge away one Lorentz component of X, another being killed by the constraint (6.2.5). These 2 Lorentz components can be eliminated more directly by originally choosing stronger gauge conditions, as in the light-cone gauge.

The conformal gauge is a temporal gauge, since it is equivalent to setting the time components of the gauge field to constants: $g_{m0} = \eta_{m0}$. When generalized to

$D > 2$, it is the choice of Gaussian normal coordinates. We can instead choose a Lorentz gauge, $\partial_m g^{mn} = 0$. This is the De Donder gauge, or harmonic coordinates, which is standardly used in $D > 2$. We'll discuss this gauge in more detail in sect. 8.3.

6.3. Light cone

In a light-cone formalism [6.5] not only are more gauge degrees of freedom eliminated than in covariant gauges, but also more (Lorentz) auxiliary fields. We do the latter first by varying the action (6.1.1) with respect to all fields carrying a "−" Lorentz index (X_-, $P^m{}_-$):

$$\frac{\delta}{\delta X_-} \rightarrow \partial_m P^m{}_+ = 0 \quad ; \tag{6.3.1a}$$

$$\frac{\delta}{\delta P^m{}_-} \rightarrow g_{mn} = (A^r B_r)^{-1}(\epsilon_{pm}A^p\epsilon_{qn}A^q - B_m B_n) \quad ,$$

$$A^m = P^m{}_+ \quad , \quad B_m = \partial_m X_+ \quad . \tag{6.3.1b}$$

We next eliminate all fields with a "+" index by gauge conditions:

$$\tau : \quad X_+ = k\tau$$

$$\sigma : \quad P^0{}_+ = k \quad , \tag{6.3.2a}$$

where k is an arbitrary constant. (The same procedure is applied in light-cone Yang-Mills, where A_- is eliminated as an auxiliary field and A_+ as a gauge degree of freedom: see sect. 2.1.) The latter condition determines σ to be proportional to the amount of +-momentum between $\sigma = 0$ and the point at that value of σ (so the string length is proportional to $\int d\sigma P^0{}_+$, which is a constant, since $\partial_m P^m{}_+ = 0$). Thus, σ is determined up to a function of τ (corresponding to the choice of where $\sigma = 0$). However, $P^1{}_+$ is also determined up to a function of τ (since now $\partial_1 P^1{}_+ = 0$), so σ is completely determined, up to global translations $\sigma \rightarrow \sigma + constant$, by the further condition

$$P^1{}_+ = 0 \quad . \tag{6.3.2b}$$

(6.3.2) implies (6.2.1). For the open string, by the converse of the argument leading to the boundary condition (6.1.5), this determines the values of σ at the boundaries up to constants, so the remaining global invariance is used to choose $\sigma = 0$ at one boundary. For the closed string, the global invariance remains, and is customarily because the hamiltonian constraints appearing in (6.1.3b) (equivalent to (6.1.4)) can be expressed very simply in terms of them as

$$\widehat{P}^{(\pm)2} = 0 \quad , \tag{6.1.6b}$$

and because they have simple Poisson brackets with each other. For the open string, it's further useful to extend σ: If we choose coordinates such that $\sigma = 0$ for all τ at one end of the string, and such that (6.1.5) implies $X' = 0$ at that end, then we can define

$$X(\sigma) = X(-\sigma) \quad , \quad \widehat{P}(\sigma) = \frac{1}{\sqrt{2\alpha'}}(P^0 + X') = \widehat{P}^{(\pm)}(\pm\sigma) \ for \ \pm\sigma > 0 \quad , \tag{6.1.7a}$$

so the constraint (6.1.6b) simplifies to

$$\widehat{P}^2 = 0 \quad . \tag{6.1.7b}$$

6.2. Conformal gauge

The conformal gauge is given by the gauge conditions (on (6.1.1,3ac))

$$\mathcal{g}_{mn} = \eta_{mn} \quad , \tag{6.2.1}$$

where η is the 2D flat (Minkowski) space metric. (Since $\mathcal{g}$ is unit-determinant, it has only 2 independent components, so the 2 gauge parameters of (6.1.2) are sufficient to determine it completely.) As for the particle, this gauge can't be obtained everywhere, so it's imposed everywhere except the boundary in τ. Then variation of $\mathcal{g}$ at initial or final τ implies (6.1.4) there, and the remaining field equations then imply it everywhere. In this gauge those equations are

$$P_m = -\partial_m X \quad , \quad \partial_m P^m = 0 \quad \rightarrow \quad \partial^2 X = 0 \quad . \tag{6.2.2}$$

It is now easy to see that the endpoints of the string travel at the speed of light. (With slight generalization, this can be shown in arbitrary gauges.) From (6.1.4,5) and (6.2.1,2), we find that $dX = d\tau \dot{X}$, and thus $dX \cdot dX = d\tau^2 \dot{X}^2 = d\tau^2(\dot{X}^2 + X'^2) = 0$. (6.2.2) is most easily solved by the use of 2D light-cone coordinates

$$\sigma^{\pm} = \frac{1}{\sqrt{2}}(\sigma^1 \mp \sigma^0) \rightarrow \eta_{mn} = \begin{pmatrix} 0 & 1 \\ 1 & 0 \end{pmatrix} = \eta^{mn} \quad , \tag{6.2.3}$$

where 2D indices now take the values $\pm$. We then have

$$\partial_+ \partial_- X = 0 \rightarrow X = \tfrac{1}{2}[\hat{X}^{(+)}(\tau + \sigma) + \hat{X}^{(-)}(\tau - \sigma)] \quad . \tag{6.2.4a}$$

For the open string, the boundary condition at one boundary, chosen to be $\sigma = 0$, is

$$(\partial_+ + \partial_-)X = 0 \to \dot{\hat{X}}^{(+)}(\tau) = \dot{\hat{X}}^{(-)}(\tau) \to \hat{X}^{(+)}(\tau) = \hat{X}^{(-)}(\tau) \quad , \qquad (6.2.4b)$$

without loss of generality, since the constant parts of $\hat{X}^{(\pm)}$ appear in X only as their sum. Thus, the modes of the open string correspond to the modes of one handedness of the closed string. The boundary condition at the other boundary of the open string, taken as $\sigma = \pi$, and the "boundary" condition of the closed string, which is simply that the "ends" at $\sigma = \pm\pi$ are the same point (and thus the closed string X is periodic in σ with period 2π, or equivalently X and X' have the same values at $\sigma = \pi$ as at $\sigma = -\pi$) both take the form

$$\dot{\hat{X}}^{(\pm)}(\tau + 2\pi) = \dot{\hat{X}}^{(\pm)}(\tau) \quad \to \quad \hat{X}^{(\pm)}(\tau + 2\pi) = \hat{X}^{(\pm)}(\tau) + 4\pi\alpha' p^{(\pm)} \quad , \qquad (6.2.4c)$$

$$p^{(+)} = p^{(-)} \quad . \qquad (6.2.4d)$$

The constraints (6.1.4) also simplify:

$$P^{\pm 2}(\tau, \sigma) = \tfrac{1}{2}\hat{X}^{(\pm)\prime 2}(\tau \pm \sigma) = 0 \quad . \qquad (6.2.5)$$

These constraints will be used to build the BRST algebra in chapt. 8.

The fact that the modes of the open string correspond to half the modes of the closed string (except that both have 1 zero-mode) means that the open string can be formulated as a closed string with modes of one handedness (clockwise or counterclockwise). This is accomplished by adding to the action (6.1.1) for the closed string the term

$$S_1 = \int \frac{d^2\sigma}{2\pi} \tfrac{1}{2} u_a u_b \lambda_{mn} \tfrac{1}{2}(g^{mp} - \epsilon^{mp}) \tfrac{1}{2}(g^{nq} - \epsilon^{nq})(\partial_p X^a)(\partial_q X^b) \quad , \qquad (6.2.6)$$

where u is a constant, timelike or lightlike (but not spacelike) vector ($u^2 \leq 0$), and $\epsilon^{+-} = -1$. λ is a lagrange multiplier which constrains $(u \cdot \partial_- X)^2 = 0$, and thus $u \cdot \partial_- X = 0$, in the gauge (6.2.1). Together with (6.2.5), this implies the Lorentz covariant constraint $\partial_- X^a = 0$, so X depends only on $\tau - \sigma$, as in (3.1.14-17). Thus, the formulation using (6.2.6) is Lorentz covariant even though S_1 is not manifestly so (because of the constant vector u). We can then identify the new X with the $\hat{X}$ of (6.2.4). Since λ itself appears multiplied by $\partial_- X$ in the equations of motion, it thus drops out, implying that it's a gauge degree of freedom which, like g, can be gauged away except at infinity.

6. CLASSICAL MECHANICS

6.1. Gauge covariant

In this chapter we'll consider the mechanics action for the string and its gauge fixing, as a direct generalization of the treatment of the particle in the previous chapter.

The first-order action for string mechanics is obtained by generalizing the 1-dimensional particle mechanics world-line of (5.1.1) to a 2-dimensional world sheet [6.1]:

$$S = \frac{1}{\alpha'}\int \frac{d^2\sigma}{2\pi}\ \left[(\partial_{\mathbf{m}}X)\cdot P^{\mathbf{m}} + \mathscr{g}_{\mathbf{mn}}\tfrac{1}{2}P^{\mathbf{m}}\cdot P^{\mathbf{n}}\right] \quad , \tag{6.1.1}$$

where $X(\sigma^{\mathbf{m}})$ is the position in the higher-dimensional space in which the world sheet is imbedded of the point whose location in the world sheet itself is given by $\sigma^{\mathbf{m}} = (\sigma^0,\sigma^1) = (\tau,\sigma)$, $d^2\sigma = d\sigma^0 d\sigma^1 = d\tau d\sigma$, $\partial_{\mathbf{m}} = (\partial_0,\partial_1) = (\partial/\partial\tau,\partial/\partial\sigma)$, and $\mathscr{g}_{\mathbf{mn}} = (-g)^{-1/2}g_{\mathbf{mn}}$ is the unit-determinant part of the 2D metric. (Actually, it has determinant -1.) $1/2\pi\alpha'$ is both the string tension and the rest-mass per unit length. (Their ratio, the square of the velocity of wave propagation in the string, is unity in units of the speed of light: The string is relativistic.) This action is invariant under 2D general coordinate transformations (generalizing (5.1.4); see sect. 4.1):

$$\begin{aligned}
\delta X &= \epsilon^{\mathbf{m}}\partial_{\mathbf{m}}X \quad , \\
\delta P^{\mathbf{m}} &= \partial_{\mathbf{n}}(\epsilon^{\mathbf{n}}P^{\mathbf{m}}) - P^{\mathbf{n}}\partial_{\mathbf{n}}\epsilon^{\mathbf{m}} \quad , \\
\delta\mathscr{g}^{\mathbf{mn}} &= \partial_{\mathbf{p}}(\epsilon^{\mathbf{p}}\mathscr{g}^{\mathbf{mn}}) - \mathscr{g}^{\mathbf{p}(\mathbf{m}}\partial_{\mathbf{p}}\epsilon^{\mathbf{n})} \quad .
\end{aligned} \tag{6.1.2}$$

Other forms of this action which result from eliminating various combinations of the auxiliary fields $P^{\mathbf{m}}$ and $\mathscr{g}_{\mathbf{mn}}$ are

$$S = \frac{1}{\alpha'}\int \frac{d^2\sigma}{2\pi}\left[\dot{X}\cdot P^0 - \frac{1}{\mathscr{g}_{11}}\tfrac{1}{2}(P^{02}+X'^2) - \frac{\mathscr{g}_{01}}{\mathscr{g}_{11}}P^0\cdot X'\right] \tag{6.1.3a}$$

$$= \frac{1}{\alpha'}\int \frac{d^2\sigma}{2\pi}\left[\dot{X}\cdot P^0 - \lambda_+\tfrac{1}{4}(P^0 + X')^2 - \lambda_-\tfrac{1}{4}(P^0 - X')^2\right] \qquad (6.1.3b)$$

$$= -\frac{1}{\alpha'}\int \frac{d^2\sigma}{2\pi}\, g^{mn}\tfrac{1}{2}(\partial_m X)\cdot(\partial_n X) \qquad (6.1.3c)$$

$$= \frac{1}{\alpha'}\int \frac{d^2\sigma}{2\pi}\left[(\partial_m X)\cdot P^m + \sqrt{-det\ P^m\cdot P^n}\right] \qquad (6.1.3d)$$

$$= -\frac{1}{\alpha'}\int \frac{d^2\sigma}{2\pi}\sqrt{-det\,(\partial_m X)\cdot(\partial_n X)} = -\frac{1}{\alpha'}\int \sqrt{-(dX^a \wedge dX^b)^2} \quad , \qquad (6.1.3e)$$

where $\frac{1}{\alpha'}P^0$ is the momentum (σ-)density (the momentum $p = \frac{1}{\alpha'}\int \frac{d\sigma}{2\pi}\, P^0$), $' = \partial/\partial\sigma$, and to obtain (6.1.3d) we have used the determinant of the g equations of motion

$$P^m\cdot P^n - \tfrac{1}{2}g^{mn} g_{pq} P^p\cdot P^q = 0 \quad . \qquad (6.1.4)$$

As a consequence of this equation and the equation of motion for P, the 2D metric is proportional to the "induced" metric $\partial_m X\cdot\partial_n X$ (as appears in (6.1.3e)), which results from measuring distances in the usual Minkowski way in the D-dimensional space in which the 2D surface is imbedded (using $dX = d\sigma^m\partial_m X$). (The equations of motion don't determine the proportionality factor, since only the unit-determinant part of the metric appears in the action.) In analogy to the particle, (6.1.4) also represents the generators of 2D general-coordinate transformations. (6.1.3a) is the hamiltonian form, (6.1.3b) is a rewriting of the hamiltonian form to resemble the example (3.1.14) (but with the *indefinite-metric sum* of squares of both left- *and* right-handed modes constrained), (6.1.3c) is the second-order form, and (6.1.3e) is the area swept out by the world sheet [6.2]. If the theory is derived from the form (6.1.3b), there are 2 sets of transformation laws of the form (3.1.15) (with appropriate sign differences), and (6.1.2,3c) can then be obtained as (3.1.16,17).

For the open string, the X equations of motion also imply certain boundary conditions in σ. (By definition, the closed string has no boundary in σ.) Varying the $(\partial X)\cdot P$ term and integrating by parts to pull out the δX factor, besides the equation of motion term ∂P we also get a surface term $n_m P^m$, where n_m is the normal to the boundary. If we assume 2D coordinates such that the position of the σ boundaries are constant in τ, then we have the boundary condition

$$P^1 = 0 \quad . \qquad (6.1.5)$$

It's convenient to define the quantities

$$\widehat{P}^{(\pm)} = \frac{1}{\sqrt{2\alpha'}}(P^0 \pm X') \qquad (6.1.6a)$$

dealt with in the quantum theory by imposing a constraint of invariance under this transformation on the field or first-quantized wave function.

The length of the string is then given by integrating (6.3.2a): $p_+ = \frac{1}{\alpha'}\int \frac{d\sigma}{2\pi} P^0{}_+ = (1/2\pi\alpha')k \cdot length$. The two most convenient choices are

$$k = 1 \quad \rightarrow \quad length = 2\pi\alpha' p_+$$
$$length = \pi\ (2\pi) \quad \rightarrow \quad k = 2\alpha' p_+\ (\alpha' p_+)\ for\ open\ (closed) \quad . \tag{6.3.3}$$

For the free string the latter choice is more convenient for the purpose of mode expansions. In the case of interactions k must be constant even through interactions, and therefore can't be identified with the value of p_+ of each string, so the former choice is made. Note that for $p_+ < 0$ the string then has negative length. It is then interpreted as an *anti*string (or outgoing string, as opposed to incoming string). The use of negative lengths is particularly useful for interactions, since then the vertices are (cyclically) symmetric in all strings: e.g., a string of length 1 breaking into 2 strings of length $\frac{1}{2}$ is equivalent to a string of length $\frac{1}{2}$ breaking into strings of length 1 and $-\frac{1}{2}$.

The action now becomes

$$S = \int d\tau \left\{ \dot{x}_- p_+ + \frac{1}{\alpha'}\int \frac{d\sigma}{2\pi} \left[(\partial_{\mathbf{m}} X_i) P^{\mathbf{m}}{}_i + \eta_{\mathbf{mn}} \tfrac{1}{2} P^{\mathbf{m}}{}_i P^{\mathbf{n}}{}_i \right] \right\} \quad , \tag{6.3.4a}$$

or, in hamiltonian form,

$$S = \int d\tau \left\{ \dot{x}_- p_+ + \frac{1}{\alpha'}\int \frac{d\sigma}{2\pi} \left[\dot{X}_i P^0{}_i - \tfrac{1}{2}(P^0{}_i{}^2 + X'_i{}^2) \right] \right\} \quad . \tag{6.3.4b}$$

X_i is found as in (6.2.4), but X_+ is given by (6.3.2), and X_- is given by varying the original action with respect to the auxiliary fields $g_{\mathbf{mn}}$ and $P^{\mathbf{m}}{}_+$, conjugate to those varied in (6.3.1):

$$\frac{\delta}{\delta g_{\mathbf{mn}}} \rightarrow P^{\mathbf{m}}{}_- = -P^{\mathbf{m}}{}_i P^0{}_i - \tfrac{1}{2}\delta_0{}^{\mathbf{m}} \eta_{\mathbf{np}} P^{\mathbf{n}}{}_i P^{\mathbf{p}}{}_i$$
$$\frac{\delta}{\delta P^{\mathbf{m}}{}_+} \rightarrow X_- = x_- - \int \frac{d\sigma}{2\pi} P^1{}_- + constant \quad , \tag{6.3.5}$$

where the constant is chosen to cancel the integral when integrated over σ, so that $\frac{1}{\alpha'}\int \frac{d\sigma}{2\pi} \dot{X}_- P^0{}_+ = \dot{x}_- p_+$ in (6.3.4). (This term has been restored in (6.3.4) in order to avoid using equations of motion to eliminate any coordinates whose equations of motion involve time derivatives, and are thus not auxiliary. In (6.3.1a) all but the zero-mode part of the X_- equation can be used to solve for all but a σ-independent part of $P^1{}_+$, without inverting time derivatives.)

After the continuous 2D symmetries have been eliminated by coordinate choices, certain discrete symmetries remain: σ and τ reversal. As in the particle case, τ reversal corresponds to a form of charge conjugation. However, the open string has a group theory factor associated with each end which is the complex-conjugate representation of that at the other end (so for unitary representations they can cancel for splitting or joining strings), so for charge conjugation the 2 ends should switch, which requires σ reversal. Furthermore, the closed string has clockwise and counterclockwise modes which are distinguishable (especially for the heterotic string), so again we require σ reversal to accompany τ reversal to keep $\sigma^{\pm}$ from mixing. We therefore define charge conjugation to be the simultaneous reversal of τ and σ (or $\sigma \rightarrow 2\pi\alpha' p_{+} - \sigma$ to preserve the positions of the boundaries of the open string). On the other hand, some strings are *nonoriented* (as opposed to the *oriented* ones above) in that solutions with σ reversed are not distinguished (corresponding to open strings with real representations for the group theory factors, or closed strings with clockwise modes not separated from counterclockwise). For such strings we also need to define a σ reversal which, because of its action on the 2D surface, is called a "twist". In the quantum theory these invariances are imposed as constraints on the fields or first-quantized wave functions (see chapt. 10).

Exercises

(1) Derive (6.1.3) from (6.1.1). Derive (6.1.5).

(2) Derive (6.3.1).

string, this means that the "leading Regge trajectory" $N(M^2)$ (see sec. 9.1) has half the slope and twice the intercept for the closed string as for the open string. If we apply the additional constraint of symmetry of the state under interchange of the 2 sets of string operators, the state is symmetric under interchange of $\sigma \leftrightarrow -\sigma$, and is therefore "nonoriented"; otherwise, the string is "oriented", the clockwise and counterclockwise modes being distinguishable (so the string can carry an arrow to distinguish it from a string that's flipped over).

From now on we choose units

$$\alpha' = \tfrac{1}{2} \quad . \tag{7.1.13}$$

In the case of the open bosonic string, the free light-cone Poincaré generators can be obtained from the covariant expressions (the obvious generalization of the particle expressions, because of (7.1.2), since X_a and $P^0{}_a$ are defined to Lorentz transform as vectors and X to translate by a constant),

$$J_{ab} = -i\int_{-\pi}^{\pi} \frac{d\sigma}{2\pi} X_{[a}(\sigma)P^0{}_{b]}(\sigma) \quad , \quad p_a = \int \frac{d\sigma}{2\pi} P^0{}_a(\sigma) \quad , \tag{7.1.14a}$$

by substituting the gauge condition (6.3.2) and free field equations (6.1.7) (using (7.1.7))

$$\hat{X}_+(\sigma) = p_+\sigma \quad , \quad \hat{P}^2 - 2 = 0 \quad \rightarrow \quad \hat{P}_- = -\frac{\hat{P}_i{}^2 - 2}{2p_+}$$

$$\rightarrow \quad J_{ij} = -ix_{[i}p_{j]} + \sum a^\dagger{}_{n[i}a_{nj]} \quad ,$$

$$J_{+i} = ix_i p_+ \quad , \quad J_{-+} = -ix_- p_+ \quad ,$$

$$J_{-i} = -i(x_- p_i - x_i p_-) + \sum a^\dagger{}_{n[-}a_{ni]} \quad ,$$

$$p_- = -\frac{1}{2p_+}\left[p_i{}^2 + 2\left(\sum n a^\dagger{}_n{}^i a_{ni} - 1\right)\right] \quad , \quad a_{n-} = -\frac{1}{p_+}(p^i a_{ni} + \Delta_n) \quad ,$$

$$\Delta_n = i\frac{1}{\sqrt{n}}\left(\tfrac{1}{2}\sum_{m=1}^{n-1}\sqrt{m(n-m)}\,a_m{}^i a_{n-m,i} - \sum_{m=1}^{\infty}\sqrt{m(n+m)}\,a^\dagger{}_m{}^i a_{n+m,i}\right) \quad . \tag{7.1.14b}$$

(We have added the normal-ordering constant of (7.1.8) to the constraint (6.1.7).) As for arbitrary light-cone representations, the Poincaré generators can be expressed completely in terms of the independent coordinates x_i and x_-, the corresponding momenta p_i and p_+, the mass operator M, and the generators of a spin SO(D−1), M_{ij} and M_{im}. For the open bosonic string, we have the usual oscillator representation of the SO(D−2) spin generators

$$M_{ij} = -i\int' \hat{X}_i \hat{P}_j = \sum_n a^\dagger{}_{n[i}a_{nj]} \quad , \tag{7.1.15a}$$

of the interactions will still look different from a true closed string.) Either way, the new boundary conditions (periodicity) allow (Fourier) expansion of all operators in terms of exponentials instead of cosines or sines. Furthermore, $\widehat{P}$ contains all of X and $\delta/\delta X$ (except x, which is conjugate to $p \sim \int d\sigma \widehat{P}$; i.e., all the translationally invariant part). In particular,

$$H = \int_{-\pi}^{\pi} \frac{d\sigma}{2\pi} \tfrac{1}{2}\widehat{P}_i{}^2 \quad . \tag{7.1.4}$$

The commutation relations are

$$[\widehat{P}_i(\sigma_1), \widehat{P}_j(\sigma_2)] = 2\pi i \delta'(\sigma_2 - \sigma_1)\delta_{ij} \quad . \tag{7.1.5}$$

For the closed string, we define 2 $\widehat{P}$'s by

$$\widehat{P}^{(\pm)}(\sigma) = \frac{1}{\sqrt{2\alpha'}}\left[\alpha' i \frac{\delta}{\delta X(\pm\sigma)} \pm X'(\pm\sigma)\right] = \frac{1}{\sqrt{2\alpha'}}\hat{X}^{(\pm)\prime}(\sigma) \quad . \tag{7.1.6}$$

Then operators which are expressed in terms of integrals over σ also include sums over $\pm$: e.g., $H = H^{(+)} + H^{(-)}$, with $H^{(\pm)}$ given in terms of $\widehat{P}^{(\pm)}$ by (7.1.4).

In order to compare with particles, we'll need to expand all operators in Fourier modes. In practical calculations, functional techniques are easier, and mode expansions should be used only as the final step: external line factors in graphs, or expansion of the effective action. (Similar remarks apply in general field theories to expansion of superfields in components and expansion of fields about vacuum expectation values.) For $\widehat{P}$ for the open string (suppressing Lorentz indices),

$$\widehat{P}(\sigma) = \sum_{n=-\infty}^{\infty} \alpha_n e^{-in\sigma} \quad ,$$

$$\alpha_0 = \sqrt{2\alpha' }p \quad , \quad \alpha_n = (\alpha_{-n})^\dagger = -i\sqrt{n}a_n{}^\dagger \quad ,$$

$$[p, x] = i \quad , \quad [\alpha_m, \alpha_n] = n\delta_{m+n,0} \quad , [a_m, a_n{}^\dagger] = \delta_{mn} \quad . \tag{7.1.7a}$$

We also have

$$\hat{X}(\sigma) = \hat{X}^\dagger(\sigma) = \hat{X}^*(-\sigma) = (x + 2\alpha' p\sigma) + \sqrt{2\alpha'}\sum_1^\infty \frac{1}{\sqrt{n}}(a_n{}^\dagger e^{-in\sigma} + a_n e^{in\sigma}) \quad ,$$

$$X(\sigma) = \tfrac{1}{2}\left[\hat{X}(\sigma) + \hat{X}(-\sigma)\right] \quad , \quad P^0(\sigma) = \alpha' i \frac{\delta}{\delta X(\sigma)} = \tfrac{1}{2}\left[\hat{X}'(\sigma) + \hat{X}'(-\sigma)\right] . \tag{7.1.7b}$$

(The $\dagger$ and $*$ have the usual matrix interpretation if the operators are considered as matrices acting on the Hilbert space: The $\dagger$ is the usual operatorial hermitian eliminates independent zero-modes, while

$$0 = \Delta p_- = -\frac{1}{p_+}(M^{2(+)} - M^{2(-)}) = -\frac{2}{p_+}\Delta N \qquad (7.1.19b)$$

is then the usual remaining light-cone constraint equating the numbers of left-handed and right-handed modes.

The generators of the Lorentz subgroup again take the form (2.3.5), as for the open string, and the operators appearing in J_{ab} are expressed in terms of the open-string ones appearing in (7.1.15) as

$$M_{ij} = \sum M^{(\pm)}{}_{ij} \quad ,$$

$$M^2 = 2\sum M^{2(\pm)} = 4(N-2) \quad , \quad N = \sum N^{(\pm)} \quad , \quad \Delta N = N^{(+)} - N^{(-)} \quad ,$$

$$M_{im}M = 2\sum (M_{im}M)^{(\pm)} \ . \qquad (7.1.20)$$

(Since J_{ab} and p_a are expressed as sums, so are M_{ab} and M. This causes objects quadratic in these operators to be expressed as twice the sums in the presence of the constraint $\Delta p_a = 0 \to p^{(\pm)}{}_a = \frac{1}{2}p_a$.)

These Poincaré algebras will be used to derive the OSp(1,1|2) algebras used in finding gauge-invariant actions in sects. 8.2 and 11.2.

7.2. Spinning

In this section we'll describe a string model with fermions obtained by introducing a 2D supersymmetry into the world sheet [7.1,5.1], in analogy to sect. 5.3, and derive the corresponding Poincaré algebra. The description of the superstring obtained by this method isn't manifestly spacetime supersymmetric, so we'll only give a brief discussion of this formalism before giving the present status of the manifestly supersymmetric formulation. In both cases, (free) quantum consistency requires $D = 10$.

For simplicity, we go directly to the (first-)quantized formalism, since as far as the field theory is concerned the only relevant information from the free theory is how to construct the covariant derivatives from the coordinates, and then the equations of motion from the covariant derivatives. After quantization, when (in the Schrödinger picture) the coordinates depend only on σ, there still remains a 1D supersymmetry in σ-space [7.2]. The covariant derivatives are simply the 1D the usual (anti)symmetry of the adjoint representation (as a matrix acting on the vector representation) is imposed not just by switching the indices, but by flipping the whole string (switching the indices and $\sigma \leftrightarrow \pi - \sigma$). As a result, N odd gives matrices with the symmetry of the adjoint representation (by including an extra "−" sign in the condition to put the massless vector in that representation), while N even gives matrices with the opposite symmetry. (A particular curiosity is the "SO(1)" open string, which has *only* N even, and no massless particles.) Such strings, because of their symmetry, are thus "nonoriented". On the other hand, if the group is just unitary, there is no symmetry condition, and the string is "oriented". (The ends can be labeled with arrows pointing in opposite directions, since the hermitian conjugate state can be thought of as the corresponding "antistring".)

The closed string is treated analogously. The mode expansions of the coordinates and operators are given in terms of 2 sets of hatted operators $\hat{X}^{(\pm)}$ or $\hat{P}^{(\pm)}$ expanded over ${\alpha^{(\pm)}}_n$:

$$X(\sigma) = \tfrac{1}{2}\left[\hat{X}^{(+)}(\sigma) + \hat{X}^{(-)}(-\sigma)\right] \quad , \quad P^0(\sigma) = \tfrac{1}{2}\left[\hat{X}^{(+)\prime}(\sigma) + \hat{X}^{(-)\prime}(-\sigma)\right] \quad ;$$

$$\hat{X}^{(\pm)}(\sigma) = \hat{X}^{(\pm)\dagger}(\sigma) = \hat{X}^{(\pm)*}(-\sigma) \quad \rightarrow \quad {\alpha^{(\pm)}}_n = {{\alpha^{(\pm)}}_{-n}}^\dagger = -{{\alpha^{(\pm)}}_n}^* \quad . \tag{7.1.9}$$

However, because of the periodicity condition (6.2.4c), and since the zero-modes (from (7.1.7)) appear only as their sum, they are not independent:

$$x^{(\pm)} = x \quad , \quad p^{(\pm)} = \tfrac{1}{2}p \quad . \tag{7.1.10}$$

Operators which have been integrated over σ (as in (7.1.4)) can be expressed as sums over two sets ($\pm$) of open-string operators: e.g.,

$$H = H^{(+)} + H^{(-)} = \tfrac{1}{4}\alpha' {p_i}^2 + N - 2 \quad , \quad N = N^{(+)} + N^{(-)} \quad . \tag{7.1.11}$$

Remembering the constraint under global σ translations

$$i\frac{d}{d\sigma} \equiv \int \frac{d\sigma}{2\pi} X' \cdot i\frac{\delta}{\delta X} = \Delta N \equiv N^{(+)} - N^{(-)} = 0 \quad , \tag{7.1.12}$$

the spectrum is then given by the direct product of 2 open strings, but with the states constrained so that the 2 factors (from the 2 open strings) give equal contributions to the mass. The masses squared are thus −4, 0, 4, 8, ... times $1/\alpha'$, and the highest spin at any mass level is $N = \frac{1}{2}\alpha' M^2 + 2$ (with the corresponding state given by the symmetric traceless product of N ${a_1}^\dagger$'s, half of which are from one open-string set and half from the other). Compared with $\alpha' M^2 + 1$ for the open

7. LIGHT-CONE QUANTUM MECHANICS

7.1. Bosonic

In this section we will quantize the light-cone gauge bosonic string described in sect. 6.3 and derive the Poincaré algebra, which is a special case of that described in sect. 2.3.

The quantum mechanics of the free bosonic string is described in the light-cone formalism [6.5] in terms of the independent coordinates $X_i(\sigma)$ and x_-, and their canonical conjugates $P^0{}_i(\sigma)$ and p_+, with "time" ($x_+ = 2\alpha' p_+ \tau$) dependence given by the hamiltonian (see (6.3.4b))

$$H = \frac{1}{\alpha'}\int \frac{d\sigma}{2\pi}\, \tfrac{1}{2}(P^0{}_i{}^2 + X_i'^2) \quad . \tag{7.1.1}$$

Functionally,

$$\frac{1}{\alpha'}P^0{}_i(\sigma) = i\frac{\delta}{\delta X_i(\sigma)} \quad , \quad \left[\frac{\delta}{\delta X_i(\sigma_1)}, X_j(\sigma_2)\right] = \delta_{ij}2\pi\delta(\sigma_2 - \sigma_1) \quad . \tag{7.1.2}$$

(Note our unconventional normalization for the functional derivative.)

For the open string, it's convenient to extend σ from $[0,\pi]$ to $[-\pi,\pi]$ as in (6.1.7) by defining

$$X(-\sigma) = X(\sigma) \quad , \quad \widehat{P}(\sigma) = \frac{1}{\sqrt{2\alpha'}}\left(\alpha' i\frac{\delta}{\delta X} + X'\right) = \frac{1}{\sqrt{\alpha'}}P^{\pm}(\pm\sigma)\; for\; \pm\sigma > 0 \quad , \tag{7.1.3a}$$

so that $\widehat{P}$ is periodic (and $\sim P^+$ or P^-, as in (6.2.5), for $\sigma >$ or < 0), *or* to express the open string as a closed string with modes which propagate only clockwise (or only counterclockwise) in terms of $\hat{X}$ of sect. 6.2:

$$\widehat{P}(\sigma) = \frac{1}{\sqrt{2\alpha'}}\hat{X}'(\sigma) \quad , \tag{7.1.3b}$$

which results in the same $\widehat{P}$ (same boundary conditions and commutation relations). The latter interpretation will prove useful for graph calculations. (However, the form where $\int'$ means the zero-modes are dropped. The mass operator M and the remaining SO(D−1) operators M_{im} are given by

$$M^2 = -2p_+ \int' \widehat{P}_- = \int' (\widehat{P}_i{}^2 - 2) = 2\left(\sum n a^\dagger{}_n{}^i a_{ni} - 1\right) = 2(N-1) \quad ,$$

$$M_{im} M = ip_+ \int' \hat{X}_i \widehat{P}_- = \sum (a^\dagger{}_{ni}\Delta_n - \Delta^\dagger{}_n a_{ni}) \quad . \qquad (7.1.15b)$$

The usual light-cone formalism for the closed string is not a true light-cone formalism, in the sense that not all constraints have been solved explicitly by eliminating variables: The one constraint that remains is that the contribution to the "energy" p_- from the clockwise modes is equal to that from the counterclockwise ones. As a result, the naive Poincaré algebra does not close [4.10]: Using the expressions

$$J_{-i} = -i(x_- p_i - x_i p_-) + \sum_{n,\pm} a^{\dagger(\pm)}{}_{n[-} a^{(\pm)}{}_{ni]} \quad ,$$

$$p_- = -\frac{1}{2p_+}\left[p_i{}^2 + 4\left(N^{(+)} + N^{(-)} - 2\right)\right] \quad , \qquad (7.1.16a)$$

we find

$$[J_{-i}, J_{-j}] = -\frac{4}{p_+{}^2}\Delta N \Delta J_{ij} \quad , \qquad (7.1.16b)$$

where ΔJ_{ij} is the difference between the (+) and (−) parts of J_{ij}.

We instead define 2 sets of open-string light-cone Poincaré generators $J^{(\pm)}{}_{ab}$ and $p^{(\pm)}{}_a$, built out of independent zero- and nonzero-modes. The closed-string Poincaré generators are then [4.10]

$$J_{ab} = J^{(+)}{}_{ab} + J^{(-)}{}_{ab} \quad , \quad p_a = p^{(+)}{}_a + p^{(-)}{}_a \quad . \qquad (7.1.17a)$$

Since the operators

$$\Delta p_a = p^{(+)}{}_a - p^{(-)}{}_a \qquad (7.1.17b)$$

commute with themselves and transform as a vector under the Lorentz algebra, we can construct a Poincaré algebra from just J_{ab} and Δp_a. This is the Poincaré algebra whose extension is relevant for string field theory; it closes off shell. (This holds in either light-cone or covariant quantization.) However, as described above, this results in an unphysical doubling of zero-modes. This can be fixed by applying the constraints (see (7.1.10))

$$\Delta p_a = 0 \quad . \qquad (7.1.18)$$

In the light-cone formalism,

$$\Delta p_+ = \Delta p_i = 0 \qquad (7.1.19a)$$

conjugate, whereas the * is the usual complex conjugate as for functions, which changes the sign of momenta. Combined they give the operatorial transpose, which corresponds to integration by parts, and thus also changes the sign of momenta, which are derivatives.) H is now defined with normal ordering:

$$H = \tfrac{1}{2} : \sum_{-\infty}^{\infty} \alpha_n \cdot \alpha_{-n} : + \, constant$$

$$= \alpha' p_i{}^2 + \sum_{1}^{\infty} n a_n{}^\dagger \cdot a_n + constant = \alpha' p_i{}^2 + N + constant \quad . \qquad (7.1.8a)$$

In analogy to ordinary field theory, $i\partial/\partial\tau + H = \alpha' p^2 + N + constant.$

The constant in H has been introduced as a finite renormalization after the infinite renormalization done by the normal ordering: As in ordinary field theory, wherever infinite renormalization is necessary to remove infinities, finite renormalization should also be considered to allow for the ambiguities in renormalization prescriptions. However, also as in ordinary field theories, the renormalization is required to respect all symmetries of the classical theory possible (otherwise the symmetry is anomalous, i.e., not a symmetry of the quantum theory). In the case of any light-cone theory, the one symmetry which is never manifest (i.e., an automatic consequence of the notation) is Lorentz invariance. (This sacrifice was made in order that unitarity would be manifest by choosing a ghost-free gauge with only physical, propagating degrees of freedom.) Thus, in order to prove Lorentz invariance isn't violated, the commutators of the Lorentz generators $J_{ab} = \frac{1}{\alpha'} \int \frac{d\sigma}{2\pi} X_{[a} P^0{}_{b]}$ must be checked. All are trivial except $[J_{-i}, J_{-j}] = 0$ because X_- and $P^0{}_-$ are quadratic in X_i and $P^0{}_i$ by (6.3.5). The proof is left as an (important) exercise for the reader that the desired result is obtained only if $D = 26$ and the renormalization constant in H and $P^0{}_-$ ($H = -\frac{1}{\alpha'} \int \frac{d\sigma}{2\pi} P^0{}_-$) is given by

$$H = \alpha' p_i{}^2 + N - 1 \equiv \alpha'(p_i{}^2 + M^2) \quad . \qquad (7.1.8b)$$

The constant in H also follows from noting that the first excited level ($a_{1i}{}^\dagger \, |0\rangle$) is just a transverse vector, which must be massless. The mass spectrum of the open string, given by the operator M^2, is then a harmonic oscillator spectrum: The possible values of the mass squared are -1, 0, 1, 2, ... times $1/\alpha'$, with spins at each mass level running as high as $N = \alpha' M^2 + 1$. (The highest-spin state is created by the symmetric traceless part of N $a_1{}^\dagger$'s.)

Group theory indices are associated with the ends of the open string, so the string acts like a matrix in that space. If the group is orthogonal or symplectic, supersymmetrization of the $\hat{P}$ operators of sect. 7.1 (or those of sect. 6.1 at the classical level, using Poisson brackets):

$$\{\hat{D}_a(\sigma_1,\theta_1),\hat{D}_b(\sigma_2,\theta_2)\} = \eta_{ab}\mathrm{d}_2 2\pi\delta(\sigma_2-\sigma_1)\delta(\theta_2-\theta_1) \quad , \tag{7.2.1a}$$

$$\hat{D}(\sigma,\theta) = \hat{\Psi}(\sigma) + \theta\hat{P}(\sigma)$$

$$\to [\hat{P},\hat{P}] = as\ before \quad , \quad \{\hat{\Psi}_a(\sigma_1),\hat{\Psi}_b(\sigma_2)\} = \eta_{ab}2\pi\delta(\sigma_2-\sigma_1) \quad . \tag{7.2.1b}$$

The 2D superconformal generators (the supersymmetrization of the generators (6.1.7b) of 2D general coordinate transformations, or of just the residual conformal transformations after the lagrange multipliers have been gauged away) are then

$$\tfrac{1}{2}\hat{D}\mathrm{d}\hat{D} = (\tfrac{1}{2}\hat{\Psi}\cdot\hat{P}) + \theta(\tfrac{1}{2}\hat{P}^2 + \tfrac{1}{2}i\hat{\Psi}'\cdot\hat{\Psi}) \quad . \tag{7.2.2}$$

There are 2 choices of boundary conditions:

$$\hat{D}(\sigma,\theta) = \pm\hat{D}(\sigma+2\pi,\pm\theta) \to \hat{P}(\sigma) = \hat{P}(\sigma+2\pi) \quad , \quad \hat{\Psi}(\sigma) = \pm\hat{\Psi}(\sigma+2\pi) \quad . \tag{7.2.3}$$

The + choice gives fermions (the Ramond model), while the − gives bosons (Neveu-Schwarz).

Expanding in modes, we now have, in addition to (7.1.7),

$$\hat{\Psi}(\sigma) = \sum_{-\infty}^{\infty}\gamma_n e^{-in\sigma} \quad \to \quad \{\gamma^a{}_m,\gamma^b{}_n\} = \eta^{ab}\delta_{m+n,0} \quad , \quad \gamma_{-n} = \gamma_n{}^\dagger \quad , \tag{7.2.4}$$

where m,n are integral indices for the fermion case and half-(odd)integral for the bosonic. The assignment of statistics follows from the fact that, while the γ_n's are creation operators $\gamma_n = d_n{}^\dagger$ for $n>0$, they are γ matrices $\gamma_0 = \gamma/\sqrt{2}$ for $n=0$ (as in the particle case, but in relation to the usual γ matrices now have Klein transformation factors for both d_n's and, in the BRST case, the ghost $\hat{C}$, or else the d_n's and $\hat{C}$ are related to the usual by factors of γ_{11}). However, as in the particle case, a functional analysis shows that this assignment can only be maintained if the number of anticommuting modes is even; in other words,

$$\gamma_{11}(-1)^{\sum d^\dagger d} = 1 \quad . \tag{7.2.5}$$

(In terms of the usual γ-matrices, the γ-matrices here also contain the Klein transformation factor $(-1)^{\sum d^\dagger d}$. However, in practice it's more convenient to use the γ-matrices of (7.2.4-5), which anticommute with all fermionic operators, and equate them directly to the usual matrices after all fermionic oscillators have been eliminated.) As usual, if γ_0 is represented explicitly as matrices, the hermiticity in

$$\begin{aligned}
\frac{1}{2\pi}[\mathcal{C}_{\alpha\beta}(1),\mathcal{C}_{\gamma\delta}(2)] &= -2\delta(2-1)P_a\gamma^a{}_{[\gamma|[\alpha}\mathcal{C}_{\beta]|\delta]} \quad , \\
\frac{1}{2\pi}[\mathcal{C}_{\alpha\beta}(1),\mathcal{D}_a(2)] &= 2i\delta'(2-1)\gamma_a{}^{\gamma\delta}\gamma_{b\gamma[\alpha}P^b\mathcal{C}_{\beta]\delta}(1) \\
&\quad -4i\delta(2-1)(\gamma_{a\gamma[\alpha}D'_{\beta]}\mathcal{B}^\gamma + P^b\gamma_{ba[\alpha}{}^\delta\mathcal{C}'_{\beta]\delta}) \quad , \\
\frac{1}{2\pi}[\mathcal{D}_a(1),\mathcal{D}_b(2)] &= -\delta''(2-1)\gamma_{ab\alpha}{}^\beta D_\beta\mathcal{B}^\alpha(1) + (2) \\
&\quad +\delta'(2-1)\left[-4iP_{(a}\mathcal{D}_{b)} + \eta_{ab}(3D'_\alpha\mathcal{B}^\alpha - D_\alpha\mathcal{B}^{\alpha\prime})\right](1) - (2) \\
&\quad +\delta(2-1)\left[-2i(3P'_{[a}\mathcal{D}_{b]} - P_{[a}\mathcal{D}'_{b]})\right. \\
&\quad \left. +2\gamma_{ab\alpha}{}^\beta(3D'_\beta\mathcal{B}'^\alpha - D''_\beta\mathcal{B}^\alpha) + \gamma_{abc}{}^{\alpha\beta}(3P'^c\mathcal{C}'_{\alpha\beta} - P''^c\mathcal{C}_{\alpha\beta})\right] \quad .
\end{aligned} \tag{7.3.8}$$

(Due to identities like $\not{P}DD \sim \mathcal{B}D \sim \not{P}\mathcal{C}$, there are other forms of some of these relations.)

BRST quantization can again be performed, and there are an infinite number of ghosts, as in the particle case. However, a remaining problem is to find the appropriate ground state (and corresponding string field). Considering the results of sect. 5.4, this may require modification of the generators (7.3.7) and BRST operator, perhaps to include Lorentz generators (acting on the ends of the string?) or separate contributions from the BRST transformations of Yang-Mills field theory (the ground state of the open superstring, or of a set of modes of one handedness of the corresponding closed strings). On the other hand, the ground state, rather than being Yang-Mills, might be purely gauge degrees of freedom, with Yang-Mills appearing at some excited level, so modification would be unnecessary. The condition $Q^2 = 0$ should reproduce the conditions $D = 10$, $\alpha_0 = 1$.

The covariant derivatives and constraints can also be derived from a 2D lagrangian of the general form (3.1.10), as for the superparticle [7.5]. This classical mechanics Lagrangian imposes weaker constraints than the Green-Schwarz one [7.6] (which sets $D_\alpha = 0$ via Gupta-Bleuler), and thus should not impose stronger conditions.

On the other hand, quantization in the light-cone formalism is understood. Spinors are separated into halves, with the corresponding separation of the γ matrices giving the splitting of vectors into transverse and longitudinal parts, as in (5.4.27). The light-cone gauge is then chosen as

$$P_+(\sigma) = p_+ \quad , \quad \Theta^- = 0 \quad . \tag{7.3.9}$$

the case of the Ramond string, comparing to (2.3.5), we find in place of (7.1.15)

$$M^2 = 2\sum n\left(a^\dagger{}_n{}^i a_{ni} + d^\dagger{}_n{}^i d_{ni}\right) \quad ,$$

$$M_{ij} = \tfrac{1}{4}\gamma_{[i}\gamma_{j)} + \sum\left(a^\dagger{}_{n[i}a_{nj)} + d^\dagger{}_{n[i}d_{nj)}\right) \quad ,$$

$$M_{im}M = \tfrac{1}{2}\left(\gamma_i\widetilde{M} + \tfrac{1}{2}\gamma^j\Sigma_{ji} + \Sigma_i{}^j\gamma_j\right) + \left[\left(a^\dagger{}_{ni}\Delta_n - \Delta^\dagger{}_n a_{ni}\right) + \left(d^\dagger{}_{ni}\Xi_n - \Xi^\dagger{}_n d_{ni}\right)\right] \quad ,$$

$$\widetilde{M} = \Sigma^i{}_i \quad , \quad \Sigma_{ij} = i\sum\sqrt{2n}\left(d^\dagger{}_{ni}a_{nj} - d_{ni}a^\dagger{}_{nj}\right) \quad ,$$

$$p_- = -\frac{1}{2p_+}(p_i{}^2 + M^2) \quad , \quad \gamma_- = -\frac{1}{p_+}(\gamma^i p_i + \widetilde{M}) \quad ,$$

$$a_{n-} = -\frac{1}{p_+}\left(p^i a_{ni} - i\tfrac{1}{2}\sqrt{\tfrac{n}{2}}\gamma^i d_{ni} + \Delta_n\right) \quad , \quad d_{n-} = -\frac{1}{p_+}\left(p^i d_{ni} + i\sqrt{\tfrac{n}{2}}\gamma^i a_{ni} + \Xi_n\right) \quad ,$$

$$\Delta_n = i\frac{1}{\sqrt{n}}\left\{\tfrac{1}{2}\sum_{m=1}^{n-1}\left[\sqrt{m(n-m)}\;a_m{}^i a_{n-m,i} + (m - \tfrac{n}{2})d_m{}^i d_{m-n,i}\right]\right.$$
$$\left. - \sum_{m=1}^{\infty}\left[\sqrt{m(n+m)}\;a^\dagger{}_m{}^i a_{n+m,i} + (m + \tfrac{n}{2})d^\dagger{}_m{}^i d_{n+m,i}\right]\right\} \quad ,$$

$$\Xi_n = i\left[\sum_{m=1}^{n-1}\sqrt{m}\;a_m{}^i d_{n-m,i} + \sum_{m=1}^{\infty}\left(\sqrt{n+m}\;d^\dagger{}_m{}^i a_{n+m,i} - \sqrt{m}\;a^\dagger{}_m{}^i d_{n+m,i}\right)\right] \quad . \tag{7.2.8}$$

This algebra can be applied directly to obtain gauge-invariant actions, as was described in sect. 4.5.

7.3. Supersymmetric

We now obtain the superstring [7.3] as a combined generalization of the bosonic string and the superparticle, which was described in sect. 5.4.

Although the superstring can be formulated as a truncation of the spinning string, a manifestly supersymmetric formulation is expected to have the usual advantages that superfields have over components in ordinary field theories: simpler constructions of actions, use of supersymmetric gauges, easier quantum calculations, no-renormalization theorems which follow directly from analyzing counterterms, etc. As usual, the free theory can be obtained completely from the covariant derivatives and equations of motion [7.5]. The covariant derivatives are defined by their affine Lie, or Kač-Moody (or, as applied to strings, "Kač-Kradle"), algebra of the form

$$\frac{1}{2\pi}[\mathcal{G}_i(\sigma_1), \mathcal{G}_j(\sigma_2)\} = \delta(\sigma_2 - \sigma_1)f_{ij}{}^k\mathcal{G}_k(\sigma_1) + i\delta'(\sigma_2 - \sigma_1)g_{ij} \quad , \tag{7.3.1}$$

light-cone formalism, or BRST invariance in the yet-to-be-constructed covariant formalism).

Similarly, light-cone expressions for q_α can be obtained from the covariant ones:

$$\begin{aligned} q_+ &= \int \frac{d\sigma}{2\pi} \left(\frac{\delta}{\delta\Theta^+} - p_+ \Sigma \Theta^+ \right) \equiv \int \frac{d\sigma}{2\pi} Q_+ \quad , \\ q_- &= \int \frac{d\sigma}{2\pi} \frac{1}{\sqrt{2}p_+} \widehat{\not{P}}_T{}^\dagger Q_+ \quad . \end{aligned} \tag{7.3.14}$$

If the superstring is formulated directly in the light cone, (7.3.14) can be used as the starting point. Q_+ and D_+ can be considered as independent variables (instead of Θ^+ and $\delta/\delta\Theta^+$), defined by their self-conjugate commutation relations (analogous to those of $\widehat{P}$):

$$\begin{aligned} \frac{1}{2\pi}\{Q_+(\sigma_1), Q_+(\sigma_2)\} &= -\delta(\sigma_2 - \sigma_1) 2 p_+ \Sigma \quad , \\ \frac{1}{2\pi}\{D_+(\sigma_1), D_+(\sigma_2)\} &= \delta(\sigma_2 - \sigma_1) 2 p_+ \Sigma \quad . \end{aligned} \tag{7.3.15}$$

However, as described above and for the particle, D_+ is unnecessary for describing physical polarizations, so we need not introduce it. In order to more closely study the closure of the algebra (7.3.14), we introduce more light-cone spinor notation (see sect. 5.3): Working in the Majorana representation $\Sigma = I$, we introduce $(D-2)$-dimensional Euclidean γ-matrices as

$$\not{p}_T \rightarrow \gamma^i{}_{\mu\nu'} p_i \quad ,$$

$$\gamma^{(i}{}_{\mu\mu'} \gamma^{j)}{}_{\nu\mu'} = 2\delta^{ij} \delta_{\mu\nu} \quad , \quad \gamma^{(i}{}_{\mu\mu'} \gamma^{j)}{}_{\mu\nu'} = 2\delta^{ij} \delta_{\mu'\nu'} \quad , \tag{7.3.16}$$

where not only vector indices i but also spinor indices μ and μ' can be raised and lowered by Kronecker δ's, and primed and unprimed spinor indices are not necessarily related. (However, as for the covariant indices, there may be additional relations satisfied by the spinors, irrelevant for the present considerations, that differ in different dimensions.) Closure of the supersymmetry algebra (on the momentum in the usual way (5.4.4), but in light-cone notation, and with the light-cone expressions for p_i, p_+, and p_-) then requires the identity (related to (7.3.16) by "triality")

$$\gamma_{i(\mu|\mu'} \gamma^i{}_{|\nu)\nu'} = 2\delta_{\mu\nu} \delta_{\mu'\nu'} \quad . \tag{7.3.17}$$

This identity is actually (7.3.3b) in light-cone notation, and the equality of the dimensions of the spinor and vector can be derived by tracing (7.3.17) with $\delta_{\mu\nu}$.

$$P_a = \widehat{P}_a + i\gamma_{a\alpha\beta}\Theta^\alpha\Theta'^\beta ,$$
$$\Omega^\alpha = i\Theta'^\alpha \quad . \tag{7.3.5}$$

These are invariant under supersymmetry generated by

$$q_\alpha = \int \frac{d\sigma}{2\pi}\left(\frac{\delta}{\delta\Theta^\alpha} - \gamma^a{}_{\alpha\beta}\widehat{P}_a\Theta^\beta - \tfrac{1}{6}i\gamma^a{}_{\alpha\beta}\gamma_{a\gamma\delta}\Theta^\beta\Theta^\gamma\Theta'^\delta\right) \quad ,$$
$$p_a = \int \frac{d\sigma}{2\pi}\widehat{P}_a \quad , \tag{7.3.6}$$

where $\{q_\alpha, q_\beta\} = -2\gamma^a{}_{\alpha\beta}p_a$.

The smallest (generalized Virasoro) algebra which includes generalizations of the operators $\frac{1}{2}\widehat{P}^2$ of the bosonic string and $\frac{1}{2}p^2$ and $\not{p}d$ of the superparticle is generated by

$$\mathcal{A} = \tfrac{1}{2}P^2 + \Omega^\alpha D_\alpha = \tfrac{1}{2}\widehat{P}^2 + i\Theta'^\alpha\frac{\delta}{\delta\Theta^\alpha} \quad ,$$
$$\mathcal{B}^\alpha = \gamma^{a\alpha\beta}P_a D_\beta \quad ,$$
$$\mathcal{C}_{\alpha\beta} = \tfrac{1}{2}D_{[\alpha}D_{\beta]} \quad ,$$
$$\mathcal{D}_a = i\gamma_a{}^{\alpha\beta}D_\alpha D'_\beta \quad . \tag{7.3.7}$$

Note the similarity of $\mathcal{A}$ to (5.4.10), (7.2.6a), and (8.1.10,12). The algebra generated by these operators is (classically)

$$\frac{1}{2\pi}[\mathcal{A}(1),\mathcal{A}(2)] = i\delta'(2-1)[\mathcal{A}(1)+\mathcal{A}(2)] \quad ,$$
$$\frac{1}{2\pi}[\mathcal{A}(1),\mathcal{B}^\alpha(2)] = i\delta'(2-1)[\mathcal{B}^\alpha(1)+\mathcal{B}^\alpha(2)] \quad ,$$
$$\frac{1}{2\pi}[\mathcal{A}(1),\mathcal{C}_{\alpha\beta}(2)] = i\delta'(2-1)[\mathcal{C}_{\alpha\beta}(1)+\mathcal{C}_{\alpha\beta}(2)] \quad ,$$
$$\frac{1}{2\pi}[\mathcal{A}(1),\mathcal{D}_a(2)] = i\delta'(2-1)[\mathcal{D}_a(1)+2\mathcal{D}_a(2)] \quad ,$$
$$\frac{1}{2\pi}\{\mathcal{B}^\alpha(1),\mathcal{B}^\beta(2)\} = i\delta'(2-1)\tfrac{1}{2}\gamma^{a\alpha\gamma}\gamma_a{}^{\beta\delta}[\mathcal{C}_{\gamma\delta}(1)+\mathcal{C}_{\gamma\delta}(2)] + 4\delta(2-1)\cdot$$
$$\cdot[\gamma^{a\alpha\beta}(P_a\mathcal{A}+\tfrac{1}{8}\mathcal{D}_a) + (\delta_\gamma{}^{(\alpha}\delta_\delta{}^{\beta)} - \tfrac{1}{2}\gamma_a{}^{\alpha\beta}\gamma^a{}_{\gamma\delta})\Omega^\gamma\mathcal{B}^\delta] \quad ,$$
$$\frac{1}{2\pi}[\mathcal{B}^\alpha(1),\mathcal{C}_{\beta\gamma}(2)] = 4\delta(2-1)[\delta_{[\beta}{}^\alpha D_{\gamma]}\mathcal{A} + (\delta_\delta{}^\epsilon\delta_{[\beta}{}^\alpha - \tfrac{1}{2}\gamma_a{}^{\alpha\epsilon}\gamma^a{}_{\delta[\beta})\Omega^\delta\mathcal{C}_{\gamma]\epsilon}] \quad ,$$
$$\frac{1}{2\pi}[\mathcal{B}^\alpha(1),\mathcal{D}_a(2)] = -2i\delta'(2-1)[2\gamma_a{}^{\alpha\beta}D_\beta\mathcal{A} + (3\delta_\gamma{}^\epsilon\gamma_a{}^{\alpha\delta} - \gamma_{ab\gamma}{}^\epsilon\gamma^{b\alpha\delta})\cdot$$
$$\cdot\Omega^\gamma\mathcal{C}_{\delta\epsilon}](1) + 2i\delta(2-1)[4\gamma_a{}^{\alpha\beta}D'_\beta\mathcal{A}$$
$$+ (3\delta_\gamma{}^\epsilon\gamma_a{}^{\alpha\delta} - \gamma_{ab\gamma}{}^\epsilon\gamma^{b\alpha\delta})\Omega^\gamma\mathcal{C}'_{\delta\epsilon} - i\gamma_{a\beta\gamma}\gamma^{b\alpha\gamma}\Omega^\beta\mathcal{D}_b] \quad ,$$

(7.2.4) means pseudohermiticity with respect to the time component $\gamma_0{}^0$ (the indefinite metric of the Hilbert space of a Dirac spinor). However, if γ_0 is instead represented as operators (as, e.g., creation and annihilation operators, as for the usual operator representation of SU(N)⊂SO(2N)), no explicit metric is necessary (being automatically included in the definition of hermitian conjugation for the operators).

The rest is similar to the bosonic formalism, and is straightforward in the 1D superfield formalism. For example,

$$H = \int \frac{d\sigma}{2\pi}\, d\theta\, \tfrac{1}{2}\hat{D}\mathrm{d}\hat{D} = \int \frac{d\sigma}{2\pi}\,(\tfrac{1}{2}\hat{P}^2 + \tfrac{1}{2}i\hat{\Psi}'\hat{\Psi}) = \tfrac{1}{2}(p^2 + M^2) \quad ,$$

$$M^2 = 2\sum_{n>0} n(a_n{}^\dagger \cdot a_n + d_n{}^\dagger \cdot d_n) \quad . \tag{7.2.6a}$$

For the fermionic sector we also have (from (7.2.2))

$$\int \frac{d\sigma}{2\pi}\, \hat{\Psi}\cdot\hat{P} = \frac{1}{\sqrt{2}}(\not{p} + \widetilde{M}) \quad , \quad \widetilde{M}^2 = M^2 \quad . \tag{7.2.6b}$$

(As described above, the d's anticommute with γ, and thus effectively include an implicit factor of γ_{11}. $\widetilde{M}$ is thus analogous to the $\not{M}$ of (4.5.12).) The fermionic ground state is massless (especially due to the above chirality condition), but the bosonic ground state is a tachyon. (The latter can most easily be seen, as for the bosonic string, by noting that the first excited level consists of only a massless vector.) However, consistent quantum interactions require truncation to the spectrum of the superstring described in the next section. This means, in addition to the chirality condition (7.2.5) in the fermionic sector, the restriction in the bosonic sector to even M^2 [7.3]. (Unlike the fermionic sector, odd M^2 is possible because of the half-integral mode numbers.) As for the bosonic string, besides determining the ground-state masses Lorentz invariance also fixes the dimension, now $D = 10$.

In the light-cone formulation of the spinning string we have instead of (7.1.14) [7.4]

$$J_{ab} = \int \frac{d\sigma}{2\pi}\left(-iX_{[a}P^0{}_{b]} + \tfrac{1}{2}\hat{\Psi}_{[a}\hat{\Psi}_{b]}\right) \quad ,$$

$$\hat{X}_+ = p_+\sigma \quad , \quad \hat{\Psi}_+ = 0 \quad ; \quad \hat{P}^2 + i\hat{\Psi}'\cdot\hat{\Psi} = \hat{\Psi}\cdot\hat{P} = 0$$

$$\to \quad \hat{P}_- = -\frac{1}{2p_+}\left(\hat{P}_i{}^2 + i\hat{\Psi}'^i\hat{\Psi}_i\right) \quad , \quad \hat{\Psi}_- = -\frac{1}{p_+}\hat{\Psi}^i\hat{P}_i \quad . \tag{7.2.7}$$

The resulting component expansion for the Neveu-Schwarz string is similar to the non-spinning bosonic string, with extra contributions from the new oscillators. For

Other operators are then eliminated by auxiliary field equations:

$$\mathcal{A} = 0 \quad \rightarrow \quad P_- = -\frac{1}{p_+}\left(\tfrac{1}{2}\widehat{P}_i{}^2 + i\Theta^{+\prime}D_+\right) \quad ,$$

$$\mathcal{B} = 0 \quad \rightarrow \quad D_- = \frac{1}{\sqrt{2}p_+}\widehat{\not{P}}_T{}^{\dagger}D_+ \quad . \tag{7.3.10}$$

The remaining coordinates are $x_\pm$, $X_i(\sigma)$, and $\Theta^+(\sigma)$, and the remaining operators are

$$D_+ = \frac{\delta}{\delta\Theta^+} + p_+\Sigma\Theta^+ \quad , \quad P_i = \widehat{P}_i \quad , \quad \Omega^+ = i\Theta^{+\prime} \quad . \tag{7.3.11}$$

However, instead of imposing $\mathcal{C} = \mathcal{D} = 0$ quantum mechanically, we can solve them classically, in analogy to the particle case. The $\mathcal{C}_{+ij}(\sigma)$ are now *local* (in σ) SO(8) generators, and can be used to gauge away all but 1 Lorentz component of D_+, by the same method as (5.4.34ab) [5.30,29]. After this, $\mathcal{D}_+$ is just the product of this one component times its σ-derivative. Furthermore, $\mathcal{D}_+$ is a Virasoro algebra for D_+, and can thus be used to gauge away all but the zero-mode [5.30,29] (as the usual one $\mathcal{A}$ did for P_+ in (7.3.9)), after this constraint implies D_+ factors in a way analogous to (5.4.34a):

$$\mathcal{D}_+ = 0 \quad \rightarrow \quad D_+ = c\xi(\sigma) \quad . \tag{7.3.12}$$

(The proof is identical, since $\mathcal{D}_+ = 0$ is equivalent to $D_+(\sigma_1)D_+(\sigma_2) = 0$.) We are thus back to the particle case for D, with a single mode remaining, satisfying the commutation relation $c^2 = p_+ \rightarrow c = \pm\sqrt{p_+}$, so D is completely determined. Alternatively, as for the particle, we could consider $D = 0$ as a second-class constraint [7.6], or impose the condition $D_+ = 0$ (which eliminates all auxiliary string fields), as a Gupta-Bleuler constraint. This requires a further splitting of the spinors, as in (5.4.29), and the Gupta-Bleuler constraint is again a chirality condition, as in (5.4.33). $\mathcal{C} = \mathcal{D} = 0$ are then also satisfied ala Gupta-Bleuler (with appropriate "normal ordering"). Thus, in a "chiral" representation (as in ordinary supersymmetry) we have a chiral, "on-shell" string superfield, or wave function, $\Phi[x_\pm, X_i(\sigma), \Theta^a(\sigma)]$, which satisfies a light-cone field equation

$$\left(i\frac{\partial}{\partial x_+} + H\right)\Phi = 0 \quad ,$$

$$H = -p_- = -\frac{1}{\alpha'}\int\frac{d\sigma}{2\pi}\,\widehat{P}_- = \int\frac{d\sigma}{2\pi}\left(\tfrac{1}{2}\widehat{P}_i{}^2 + i\Theta^{a\prime}\frac{\delta}{\delta\Theta^a}\right) \tag{7.3.13}$$

The dimension of spacetime $D = 10$ and the constant in H (zero) are determined by considerations similar to those of the bosonic case (Lorentz invariance in the where f are the algebra's structure constants and g its (not necessarily Cartan) metric (both constants). The zero-modes of these generators give an ordinary (graded) Lie algebra with structure constants f. The Jacobi identities are satisfied if and only if

$$f_{[ij|}{}^l f_{l|k)}{}^m = 0 \quad , \tag{7.3.2a}$$

$$f_{i(j|}{}^l g_{l|k]} = 0 \quad , \tag{7.3.2b}$$

where the first equation is the usual Jacobi identity of a Lie algebra and the second states the total (graded) antisymmetry of the structure constants with index lowered by the metric g. In this case, we wish to generalize $\{d_\alpha, d_\beta\} = 2\gamma^a{}_{\alpha\beta}p_a$ for the superparticle and $[\widehat{P}_a(\sigma_1), \widehat{P}_b(\sigma_2)] = 2\pi i\delta'(\sigma_2 - \sigma_1)\eta_{ab}$ for the bosonic string. The simplest generalization consistent with the Jacobi identities is:

$$\begin{aligned}
\{D_\alpha(\sigma_1), D_\beta(\sigma_2)\} &= 2\pi\delta(\sigma_2 - \sigma_1)2\gamma^a{}_{\alpha\beta}P_a(\sigma_1) \quad , \\
[D_\alpha(\sigma_1), P_a(\sigma_2)] &= 2\pi\delta(\sigma_2 - \sigma_1)2\gamma_{a\alpha\beta}\Omega^\beta(\sigma_1) \quad , \\
\{D_\alpha(\sigma_1), \Omega^\beta(\sigma_2)\} &= 2\pi i\delta'(\sigma_2 - \sigma_1)\delta_\alpha{}^\beta \quad , \\
[P_a(\sigma_1), P_b(\sigma_2)] &= 2\pi i\delta'(\sigma_2 - \sigma_1)\eta_{ab} \quad , \\
[P,\Omega] = \{\Omega,\Omega\} &= 0 \quad ,
\end{aligned} \tag{7.3.3a}$$

$$\gamma_{a(\alpha\beta}\gamma^a{}_{\gamma)\delta} = 0 \quad . \tag{7.3.3b}$$

(7.3.2b) requires the introduction of the operator Ω, and (7.3.2a) then implies (7.3.3b). This supersymmetric set of modes (as $\widehat{P}$ for the bosonic string) describes a complete open string or half a closed string, so two such sets are needed for the closed superstring, while the heterotic string needs one of these plus a purely bosonic set.

Note the analogy with the super-Yang-Mills algebra (5.4.8):

$$(D_\alpha, P_a, \Omega^\alpha) \leftrightarrow (\nabla_\alpha, \nabla_a, W^\alpha) \quad , \tag{7.3.4}$$

and also that the constraint (7.3.3b) occurs on the γ-matrices, which implies $D = 3, 4, 6$, or 10 [7.6] when the maximal Lorentz invariance is assumed (i.e., all of SO(D$-$1,1) for the D-vector P_a).

This algebra can be solved in terms of $\widehat{P}_a$, a spinor coordinate $\Theta^\alpha(\sigma)$, and its derivative $\delta/\delta\Theta^\alpha$:

$$D_\alpha = \frac{\delta}{\delta\Theta^\alpha} + \gamma^a{}_{\alpha\beta}\widehat{P}_a\Theta^\beta + \tfrac{1}{2}i\gamma^a{}_{\alpha\beta}\gamma_{a\gamma\delta}\Theta^\beta\Theta^\gamma\Theta'^\delta \quad ,$$

Returning to deriving the light-cone formalism from the covariant one, we can also obtain the light-cone expressions for the Poincaré generators, which should prove important for covariant quantization, via the OSp(1,1|2) method. As in general, they are completely specified by M_{ij}, $M_{im}M$, and M^2:

$$M_{ij} = \int^{\prime} -i\hat{X}_i\hat{P}_j + \mathcal{M}_{ij} \quad ,$$

$$M_{im}M = \int^{\prime} i\hat{X}_i p_+ \hat{P}_- + \mathcal{M}_i{}^j \hat{P}_j \quad ,$$

$$M^2 = \int^{\prime} -2p_+\hat{P}_- \quad , \qquad (7.3.18)$$

where

$$-p_+\hat{P}_- = \tfrac{1}{2}\hat{P}_i{}^2 + i\frac{1}{8p_+}Q\gamma_+Q' - i\frac{1}{8p_+}\mathcal{D}_+ \quad ,$$

$$\mathcal{M}_{ij} = -\frac{1}{16p_+}Q\gamma_+\gamma_{ij}Q + \frac{1}{16p_+}\mathcal{C}_{+ij} \quad , \qquad (7.3.19)$$

contain all D dependence (as opposed to X and Q dependence) only in the form of $\mathcal{C}$ and $\mathcal{D}$, which can therefore be dropped. (Cf. (5.4.22). γ_+ picks out Q_+ from Q, as in (5.4.27).)

We can now consider deriving the BRST algebra by the method of adding 4+4 dimensions to the light-cone (sects. 3.6, 5.5). Unfortunately, adding 4+4 dimensions doesn't preserve (7.3.17). In fact, from the analysis of sect. 5.3, we see that to preserve the symmetries of the σ-matrices requires increasing the number of commuting dimensions by a multiple of 8, and the number of time dimensions by a multiple of 4. This suggests that this formalism may need to be generalized to adding 8+8 dimensions to the light-cone (4 space, 4 time, 8 fermionic). Coincidentally, the light-cone superstring has 8+8 physical (σ-dependent) coordinates, so this would just double the number of oscillators. Performing the reduction from OSp(4,4|8) to OSp(2,2|4) to OSp(1,1|2), if one step is chosen to be U(1)-type and the other GL(1)-type, it may be possible to obtain an algebra which has the benefits of both formalisms.

As for the bosonic string, the closed superstring is constructed as the direct product of 2 open strings (1 for the clockwise modes and 1 for the counterclockwise): The hamiltonian is the sum of 2 open-string ones, and the closed-string ground state is the product of 2 open-string ones. In the case of type I or IIB closed strings, the 2 sets of modes are the same kind, and the former (the bound-state of type I open strings) is nonoriented (in order to be consistent with the $N = 1$

solved in a Gupta-Bleuler fashion in light-cone notation. The difference between that and actual light-cone quantization is that in the light-cone quantization $\hat{P}_-$ is totally eliminated at the classical level, whereas in the light-cone notation for the covariant-gauge quantization the constraint is used to determine the dependence on the $\pm$ oscillators in terms of the transverse oscillators. One way to do this would be to start with a state constructed from just transverse oscillators (as in light-cone quantization) and add in terms involving longitudinal oscillators until the constraints are satisfied (or actually half of them, ala Gupta-Bleuler). A simpler way is to start at the classical level in an arbitrary conformal gauge with transverse oscillators, and then conformally transform them to the light-cone gauge to see what a transverse oscillator (in the physical sense, not the light-cone-index sense) looks like. We thus wish to consider

$$d\sigma' \hat{P}_i{}'(\sigma') = d\sigma \hat{P}_i(\sigma) \quad ,$$

$$\sigma' = \frac{1}{p_+}\hat{X}_+(\sigma) = \sigma + \textit{oscillator-terms} \quad . \tag{8.1.7}$$

(Without loss of generality, we can work at $x_+ = 0$. Equivalently, we can explicitly subtract x_+ from $\hat{X}_+$ everywhere the latter appears in this derivation.) If we consider the same transformation on $\hat{P}_+$ (using $\partial\hat{X}/\partial\sigma \equiv \hat{P}$), we find $\hat{P}_+{}' = p_+$, the light-cone gauge. (8.1.7) can be rewritten as

$$\hat{P}'(\sigma_1) = \int d\sigma_2\, \delta\left(\sigma_1 - \frac{1}{p_+}\hat{X}_+(\sigma_2)\right)\hat{P}(\sigma_2) \quad . \tag{8.1.8}$$

(8.1.7) follows upon replacing σ_1 with $\sigma_1{}'$ and integrating out the δ-function (with the Jacobian giving the conformal weight factor). A more convenient form for quantization comes from the mode expansion: Multiplying by $e^{in\sigma_1}$ and integrating,

$$\alpha_n{}' = \int \frac{d\sigma}{2\pi}\, e^{in\hat{X}_+(\sigma)/p_+}\,\hat{P}(\sigma) \quad . \tag{8.1.9}$$

These ("DDF") operators [8.2] (with normal ordering, as usual, upon quantization, and with transverse Lorentz index i, and $n > 0$) can be used to create all physical states. Due to their definition in terms of a conformal transformation from an arbitrary conformal gauge to a completely fixed (light-cone) gauge, they are automatically conformally invariant: i.e., they commute with $\mathcal{G}$. (This can be verified to remain true after quantization.) Consequently, states constructed from them satisfy the Gupta-Bleuler constraints, since the conformal generators push past these operators to hit the vacuum. Thus, these operators allow the construction of the

Find $X^a(\sigma, \tau)$. In the center-of-mass frame, find the energy and spin, and relate them.

(2) Do (1) in the conformal gauge by using (7.1.7) for $\hat{X}^a$ for $a = 0, 1$ (same n), and applying the constraint (6.2.5). Compare results.

(3) Prove the light-cone Poincaré algebra closes only for $D = 26$, and determines the constant in (7.1.8a).

(4) Find explicit expressions for all the states at the 4 lowest mass levels of the open bosonic string. For the massive levels, combine SO(D−2) representations into SO(D−1) ones. Do the same for the 4 lowest (nontrivial) mass levels of the closed string.

(5) Derive (7.1.15), including the expressions in terms of σ-integrals. What happens to the part of this integral symmetric in ij for M_{ij}?

(6) Derive (7.2.8).

(7) Derive (7.3.17), both from (7.3.3b) and closure of (7.3.14). Show that it implies $D - 2 = 1, 2, 4, 8$.

(8) Show that (7.3.5) satisfies (7.3.3). Show that (7.3.6) gives a supersymmetry algebra, and that the operators of (7.3.5) are invariant. Check (7.3.8) till you drop.

(9) Find all the states in the spinning string at the tachyonic, massless, and first massive levels. Show that, using the truncation of sect. 7.2, there are equal numbers of bosons and fermions at each level. Construct the same states using the X and Q oscillators of sect. 7.3.

(The $\propto_0 - 1$ just replaces the 1 in (8.1.1) with $\propto_0$.) Corresponding to (8.1.2b), we now have the exact quantum mechanical commutation relations (after normal ordering)

$$[\hat{\mathbf{L}}_m, \hat{\mathbf{L}}_n] = (n-m)\hat{\mathbf{L}}_{m+n} + \left[\frac{D-26}{12}(m^3 - m) + 2(\propto_0 - 1)m\right]\delta_{m,-n} \quad . \qquad (8.1.15)$$

The terms linear in m in the anomalous terms (those not appearing in the classical result (8.1.2)) are trivial, and can be arbitrarily modified by adding a constant to $\hat{\mathbf{L}}_0$. That the remaining term is $\sim m^3$ follows from (1D) dimensional analysis: $[\hat{\mathcal{G}},\hat{\mathcal{G}}] \sim \delta'\hat{\mathcal{G}} + \delta'''$, since the first term implies $\hat{\mathcal{G}} \sim 1/\sigma^2$ dimensionally, so only $\delta''' \sim 1/\sigma^4$ can be used. The values of the coefficients in these terms can also be determined by evaluating just the vacuum matrix elements $\langle 0|\,[\hat{\mathbf{L}}_{-n}, \hat{\mathbf{L}}_n]\,|0\rangle$ for $n = 1, 2$. Further examining these terms, we see that the ghost contributions are necessary to cancel those from the physical coordinates (which have coefficient D), and do so only for $D = 26$. The remaining anomaly cancels for $\propto_0 = 1$. Under the same conditions one can show that $Q^2 = 0$. Thus, in the covariant formalism, where Lorentz covariance is manifest and not unitarity (the opposite of the light-cone formalism), $Q^2 = 0$ is the analog of the light cone's $[J_{-i}, J_{-j}] = 0$ (and the calculation is almost identical, so the reader will have little trouble modifying his previous calculation). $\widehat{C}$ and $\delta/\delta\widehat{C}$ can be expanded in zero-modes and creation and annihilation operators, as $\widehat{P}$ ((7.1.7a)), but the creation operators in $\widehat{C}$ are canonically conjugate to the annihilation operators in $\delta/\delta\widehat{C}$, and vice versa:

$$\widehat{C} = c + \sum_1^\infty \frac{1}{\sqrt{n}}(c_n{}^\dagger e^{-in\sigma} + c_n e^{in\sigma}) \quad ,$$

$$\frac{\delta}{\delta\widehat{C}} = \frac{\partial}{\partial c} + \sum_1^\infty \sqrt{n}(-i\tilde{c}_n{}^\dagger e^{-in\sigma} + i\tilde{c}_n e^{in\sigma}) \quad ;$$

$$\{c_m, \tilde{c}_n{}^\dagger\} = i\delta_{mn} \quad , \quad \{\tilde{c}_m, c_n{}^\dagger\} = -i\delta_{mn} \quad . \qquad (8.1.16)$$

(Since the IGL(1) formalism is directly related to the OSp(1,1|2), as in sect. 4.2, we have normalized the oscillators in a way that will make the Sp(2) symmetry manifest in the next section.) The physical states are obtained by hitting $|0\rangle$ with $a^\dagger$'s but also requiring $Q\,|\psi\rangle = 0$; states $|\psi\rangle = Q\,|\chi\rangle$ are null states (pure gauge). The condition of being annihilated by Q is equivalent to being annihilated by $\mathbf{L}_n$ for $n \leq 0$ (i.e., the "nonpositive energy" part of $\mathcal{G}(\sigma)$, which is now normal ordered and includes the $-\propto_0$ term of (8.1.10,14)), which is just the constraint in Gupta-Bleuler quantization. $\mathbf{L}_0$ is simply the Lorentz-covariantization of H of (7.1.8) (i.e., all transverse indices replaced with Lorentz indices).

correspond to the 1D general coordinate transformations (in "zeroth-quantized" notation)

$$\left[\lambda_2(\sigma)\frac{\partial}{\partial\sigma}, \lambda_1(\sigma)\frac{\partial}{\partial\sigma}\right] = \lambda_{[2}\lambda_{1]}{}'\frac{\partial}{\partial\sigma} \quad , \tag{8.1.3b}$$

giving the structure constants

$$f(\sigma_1, \sigma_2; \sigma_3) = 2\pi\delta'(\sigma_2 - \sigma_1)\left[\delta(\sigma_1 - \sigma_3) + \delta(\sigma_2 - \sigma_3)\right] \quad , \tag{8.1.4a}$$

or

$$\lambda_1{}^j\lambda_2{}^i f_{ij}{}^k \quad \leftrightarrow \quad \lambda_1\frac{\overleftrightarrow{\partial}}{\partial\sigma}\lambda_2 \quad . \tag{8.1.4b}$$

More generally, we have operators whose commutation relations

$$[\mathcal{G}(\sigma_1), \mathcal{O}(\sigma_2)] = 2\pi\left[\delta(\sigma_2 - \sigma_1)\mathcal{O}'(\sigma_2) + w\delta'(\sigma_2 - \sigma_1)\mathcal{O}(\sigma_2)\right] \tag{8.1.5a}$$

represent the transformation properties of a 1D tensor of (covariant) rank w, or a scalar density of weight w:

$$\left[\int \lambda\mathcal{G}, \mathcal{O}\right] = \lambda\mathcal{O}' + w\lambda'\mathcal{O} \quad . \tag{8.1.5b}$$

Equivalently, in terms of 2D conformal transformations, it has scale weight w. (Remember that conformal transformations in $D = 2$ are equivalent to 1D general coordinate transformations on $\sigma^{\pm}$: See (6.2.7).) In particular, we see from (8.1.2a) that $\mathcal{G}$ itself is a 2nd-rank-covariant (as opposed to contravariant) tensor: It is the energy-momentum tensor of the mechanics action. (It was derived by varying that action with respect to the metric.) The finite form of these transformations follows from exponentiating the Lie algebra represented in (8.1.3): (8.1.5) can then also be rewritten as the usual coordinate transformations

$$\left(\frac{\partial\sigma'}{\partial\sigma}\right)^w \mathcal{O}'(\sigma') = \mathcal{O}(\sigma) \quad , \tag{8.1.6a}$$

where the primes here stand for the transformed quantities (not σ-derivatives) or as

$$(d\sigma')^w\mathcal{O}'(\sigma') = (d\sigma)^w\mathcal{O}(\sigma) \quad , \tag{8.1.6b}$$

indicating their tensor structure. In particular, a covariant vector ($w = 1$) can be integrated to give an invariant. $\widehat{P}$ is such a vector (and the momentum p the corresponding conformal invariant), which is why $\mathcal{G}$ has twice its weight (by (8.1.1)).

Before performing the BRST quantization of this algebra, we relate it to the light-cone quantization of the previous chapter. The constraints (8.1.1) can be supersymmetry of the open string, rather than the $N = 2$ supersymmetry generated by the 2 sets of modes of oriented, type II strings). Type IIA closed strings have Θ's with the opposite chirality between the two sets of modes (i.e., one set has a Θ^α while the other has a Θ_α). The ground states of these closed strings are supergravity ($N = 1$ supergravity for type I and $N = 2$ for type II). The heterotic string is a closed string for which one set of modes is bosonic (with the usual tachyonic scalar ground state) while the other is supersymmetric (with the usual supersymmetric Yang-Mills ground state). The lowest-mass physical states, due to the $\Delta N = 0$ restriction, are the product of the massless sector of each set (since now $\Delta N = H^{(+)} - H^{(-)} = (N^{(+)} - 1) - N^{(-)}$). The dimension of spacetime for the 2 sets of modes is made consistent by compactification of some of the 26 dimensions of one and some (or none) of the 10 of the other onto a torus, leaving the same number of noncompactified dimensions (at least the physical 4) for both sets of modes. These compactified bosonic modes can also be fermionized (see the next section), giving an equivalent formulation in which the extra dimensions don't explicitly appear: For example, fermionization of 16 of the dimensions produces 32 (real) fermionic coordinates, giving an SO(32) internal symmetry (when the fermions are given the same boundary conditions, all periodic or all antiperiodic). The resulting spectrum for the massless sector of heterotic strings consists of supergravity coupled to supersymmetric Yang-Mills with $N = 1$ (in 10D counting) supersymmetry. The vectors gauging the Cartan subalgebra of the full Yang-Mills group are the obvious ones coming from the toroidal compactification (i.e., those that would be obtained from the noncompactified theory by just dropping dependence on the compactified coordinates), while the rest correspond to "soliton" modes of the compactified coordinates for which the string winds around the torus. As for the dimension and ground-state mass, quantum consistency restricts the allowed compactifications, and in particular the toroidal compactifications are restricted to those which, in the case of compactification to $D = 10$, give Yang-Mills group SO(32) or $E_8 \otimes E_8$. (These groups give anomaly-free 10D theories in their massless sectors. There is also an SO(16)×SO(16) 10D-compactification which can be considered to have broken N=1 supersymmetry. There are other 10D-compactifications which have tachyons.)

Some aspects of the interacting theory will be described in chapts. 9 and 10.

Exercises

(1) Use (7.1.7) as the *classical* solution for $\hat{X}^i$, and set $a_{ni} = 0$ for $n \neq 1$ and $i \neq 1$.

physical Hilbert space within the formalism of covariant-gauge quantization, and allow a direct comparison with light-cone quantization.

On the other hand, for most purposes it is more convenient to solve the constraints as covariantly as possible (which is we why we are working with covariant-gauge quantization in the first place). The next step is the IGL(1) algebra [3.4]

$$Q = \int \frac{d\sigma}{2\pi}\, \widehat{C}\left(-i\tfrac{1}{2}\widehat{P}^2 + \widehat{C}'\frac{\delta}{\delta\widehat{C}} + i\propto_0\right) \equiv -i\int \frac{d\sigma}{2\pi}\, \widehat{C}\mathcal{A} \ ,$$

$$J^3 = \int \frac{d\sigma}{2\pi}\, \widehat{C}\frac{\delta}{\delta\widehat{C}} \ . \tag{8.1.10}$$

Expanding in the ghost zero-mode

$$c = \int \frac{d\sigma}{2\pi}\, \widehat{C} \tag{8.1.11}$$

we also find (see (3.4.3b))

$$p^2 + M^2 = 2\int \mathcal{A} \ , \quad M^+ = -i\int \widehat{C}\widehat{C}' \ . \tag{8.1.12}$$

$\propto_0$ (the intercept of the leading Regge trajectory) is a constant introduced, as in the light-cone formalism, because of implicit normal ordering. The only ambiguous constant in J^3 is an overall one, which we choose to absorb into the zero-mode term so that it appears as $c\partial/\partial c$, so that physical fields have vanishing ghost number. (This also makes $J^{3\dagger} = 1 - J^3$.) In analogy to the particle, $\widehat{C}$ is a momentum (defined to be periodic on $\sigma \in [-\pi, \pi]$), as follows from consideration of τ reversal in the classical mechanics action, but here τ reversal is accompanied by σ reversal in order to avoid switching + and − modes. (In the classical action the ghost is odd under such a transformation, since it carries a 2D vector index, as does the gauge parameter, while the antighost is even, carrying 2 indices, as does the gauge-fixing function g_{mn}.) $\widehat{C}$ can also be separated into odd and even parts, which is useful when similarly separating $\widehat{P}$ as in (7.1.3a):

$$\widehat{C} = C + \widetilde{C} \quad , \quad C(-\sigma) = C(\sigma) \quad , \quad \widetilde{C}(-\sigma) = -\widetilde{C}(\sigma) \ . \tag{8.1.13}$$

We now pay attention to the quantum effects. Rather than examining the BRST algebra, we look at the IGL(1)-invariant Virasoro operators (from (3.2.13))

$$\hat{\mathcal{G}} = \mathcal{G} + \widehat{C}'\frac{\delta}{\delta\widehat{C}} + \left(\widehat{C}\frac{\delta}{\delta\widehat{C}}\right)' + i(\propto_0 - 1) \ . \tag{8.1.14}$$

8. BRST QUANTUM MECHANICS

8.1. IGL(1)

We first describe the form of the BRST algebra obtained by first-quantization of the bosonic string by the method of sect. 3.2, using the constraints found in the conformal (temporal) gauge in sect. 6.2.

The residual gauge invariance in the covariant gauge is conformal transformations (modified by the constraint that they preserve the position of the boundaries). After quantization in the Schrödinger picture (where the coordinates have no τ dependence), the Virasoro operators [8.1]

$$\mathcal{G}(\sigma) = -i\left[\tfrac{1}{2}\widehat{P}^2(\sigma) - 1\right] \quad , \tag{8.1.1}$$

with $\widehat{P}$ as in (7.1.3) but for all Lorentz components, generate only these transformations (instead of the complete set of 2D general coordinate transformations they generated when left as arbitrary off-shell functions of σ and τ in the classical mechanics). Using the hamiltonian form of BRST quantization, we first find the classical commutation relations (Poisson brackets, neglecting the normal-ordering constant in (8.1.1))

$$\begin{aligned}[\mathcal{G}(\sigma_1), \mathcal{G}(\sigma_2)] &= 2\pi\delta'(\sigma_2 - \sigma_1)[\mathcal{G}(\sigma_1) + \mathcal{G}(\sigma_2)] \\ &= 2\pi\left[\delta(\sigma_2 - \sigma_1)\mathcal{G}'(\sigma_2) + 2\delta'(\sigma_2 - \sigma_1)\mathcal{G}(\sigma_2)\right] \quad , \end{aligned} \tag{8.1.2a}$$

or in mode form

$$i\mathcal{G}(\sigma) = \sum \mathbf{L}_n e^{-in\sigma} \quad \rightarrow$$

$$[\mathbf{L}_m, \mathbf{L}_n] = (n - m)\mathbf{L}_{m+n} \quad . \tag{8.1.2b}$$

These commutation relations, rewritten as

$$\left[\int \frac{d\sigma_1}{2\pi}\,\lambda_1(\sigma_1)\mathcal{G}(\sigma_1), \int \frac{d\sigma_2}{2\pi}\,\lambda_2(\sigma_2)\mathcal{G}(\sigma_2)\right] = \int \frac{d\sigma}{2\pi}\,\lambda_{[2}(\sigma)\lambda_{1]}{}'(\sigma)\mathcal{G}(\sigma) \quad , \tag{8.1.3a}$$

An interesting fact about the Virasoro algebra (8.1.15) (and its generalizations, see below) is that, after an appropriate shift in $\mathbf{L}_0$ (namely, the choice of $\alpha_0 = 1$ in this case), the anomaly does not appear in the Sp(2) (=SL(2)=SU(1,1)=SO(2,1)= projective group) subalgebra given by $n = 0, \pm 1$ [8.3], independent of the representation (in this case, D). Furthermore, unlike the whole Virasoro algebra (even when the anomaly cancels), we can define a state which is left invariant by this Sp(2). Expanding (8.1.14) in modes (as in (7.1.7a, 8.1.2b)), the only term in $\hat{\mathbf{L}}_1$ containing no annihilation operators is $\sim p \cdot a_1{}^\dagger$, so we choose $p = 0$. Then $\hat{\mathbf{L}}_0 = 0$ requires the state be on-shell, which means it's in the usual massless sector (Yang-Mills). Further examination then shows that this state is uniquely determined to be the state corresponding to a constant ($p = 0$) Yang-Mills ghost field C. It can also be shown that this is the only gauge-invariant, BRST-invariant state (i.e., in the "cohomology" of Q) of that ghost-number J^3 [8.4]. Since it has the same ghost number as the gauge parameter Λ (see (4.2.1)), this means that it can be identified as the only gauge invariance of the theory which has no inhomogeneous term: Any gauge parameter of the form $\Lambda = Q\epsilon$ not only leaves the free action invariant, but also the interacting one, since upon gauge fixing it's a gauge invariance of the ghosts (which means the ghosts themselves require ghosts), which must be maintained at the interacting level for consistent quantization. However, any parameter satisfying $Q\Lambda = 0$ won't contribute to the free gauge transformation of the physical fields, but may contribute at the interacting level. In fact, gauge transformations in the cohomology of Q are just the global invariances of the theory, or at least those which preserve the second-quantized vacuum about which the decomposition into free and interacting has been defined. Since the BRST transformation $\delta\Phi = Q\Phi$ is just the gauge transformation with the gauge parameter replaced by the ghost, this transformation parameter appears in the field in the same position as would the corresponding ghost. For the bosonic string, the only massless physical field is Yang-Mills, and thus the only global invariance is the usual global nonabelian symmetry. Thus, the state invariant under this Sp(2) directly corresponds to the global invariance of the string theory, and to its ghost. This Sp(2) symmetry can be maintained at the interacting level in tree graph calculations (see sect. 9.2), especially for vertices, basically due to the fact that tree graphs have the same global topology as free strings. In such calculations it's therefore somewhat more convenient to expand states about this Sp(2)-invariant "vacuum" instead of the usual one. (We now refer to the first-quantized vacuum with respect to which free fields are defined. It's redefinition is unrelated to the usual vacuum redefinitions of field theory, which are inhomogeneous in the fields.) This effectively switches the role of the corresponding pair of ghost oscillators (just the $n = 1$ mode) between creation and annihilation operators.

The closed string [4.5] is quantized similarly, but with 2 sets of modes ($\pm$; except that there are still just one x and p), and we can separate

$$\widehat{C}_{\pm}(\sigma) = (C \pm \widetilde{C})(\pm\sigma) \tag{8.1.17}$$

corresponding to (7.1.6).

Since $\mathcal{A}$ commutes with both M^3 and M^+, it is Sp(2)-invariant. Thus, the modified Virasoro operators $\mathbf{\mathring{L}}_n$ it gives (in analogy to (8.1.2b), or, more specifically, the nonnegative-energy ones), and in particular their fermionic parts, can be used to generate (BRST) Sp(2)-invariant states, with the exception of the zeroth and first fermionic Virasoro operators (the projective subgroup), which vanish on the vacuum. We will now show that these operators, together with the bosonic oscillators, are sufficient to generate *all* such states, i.e., the complete set of physical fields [4.1]. (By physical fields we mean all fields appearing in the gauge-invariant action, including Stueckelberg fields and unphysical Lorentz components.) This is seen by bosonizing the two fermionic coordinates into a single additional bosonic coordinate, whose contribution to the Virasoro operators includes a term linear in the new oscillators, but lacking the first mode. This corresponds to the fact that M^+ contains a term linear in the annihilation operator of the first mode. Thus, the Virasoro operators generate excitations in all but the first mode of the new coordinate, and the condition $M^+ = 0$ kills only excitations in the first mode. J^3 is just the zero mode of the new coordinate, so its vanishing (which then implies $T = 0$) completes the derivation.

The bosonization is essentially the same as the standard procedure [8.5], except for differences due to the indefinite metric of the Hilbert space of the ghosts. The fermionic coordinates can be expressed in terms of a bosonic coordinate $\hat{\chi}$ (analogous to $\hat{X}$) as

$$\widehat{C} = e^{\hat{\chi}} \ , \quad \frac{\delta}{\delta\widehat{C}} = e^{-\hat{\chi}} \ , \tag{8.1.18}$$

with our usual implicit normal ordering (with both terms $\hat{q} + \hat{p}\sigma$ of the zero mode appearing in the same exponential factor). Note the hermiticity of these fermionic coordinates, due to the lack of i's in the exponents. (For physical bosons and fermions, we would use $\hat{\psi} = e^{i\hat{\varphi}}$, $\hat{\psi}^\dagger = e^{-i\hat{\varphi}}$, with $\hat{\psi}$ canonically conjugate to $\hat{\psi}^\dagger$.) $\hat{\chi}$

has the mode expansion

$$\hat{\chi} = (\hat{q} + \hat{p}\sigma) + \sum_{n=1}^{\infty} \frac{1}{\sqrt{n}}(\hat{a}_n e^{in\sigma} + \hat{a}^\dagger{}_n e^{-in\sigma}) \quad ;$$

$$[\hat{p}, \hat{q}] = -i \quad , \quad [\hat{a}_m, \hat{a}^\dagger{}_n] = -\delta_{mn} \quad . \qquad (8.1.19)$$

By comparison with (7.1.7), we see that this coordinate has a timelike metric (i.e., it's a ghost). Using

$$: e^{a\hat{\chi}(\sigma)} :: e^{b\hat{\chi}(\sigma')} : \; = \; : e^{a\hat{\chi}(\sigma)+b\hat{\chi}(\sigma')} : \left[2i \; sin\tfrac{\sigma'-\sigma}{2} + \epsilon\right]^{ab} \quad , \qquad (8.1.20)$$

we can verify the fermionic anticommutation relations, as well as

$$J^3 = i\hat{p} + \tfrac{1}{2} \quad , \quad M^+ = \int \frac{d\sigma}{2\pi} e^{2\hat{\chi}} \quad ; \qquad (8.1.21a)$$

$$\mathcal{A} = \tfrac{1}{2}\left(\widehat{P}^2 - \hat{\chi}'^2 - \hat{\chi}'' - \tfrac{9}{4}\right) \quad . \qquad (8.1.21b)$$

Since J^3 is quantized in integral values, χ is defined to exist on a circle of imaginary length with anticyclic boundary conditions. (The imaginary eigenvalues of this hermitian operator are due to the indefiniteness induced by the ghosts into the Hilbert-space metric.) Conversely, choosing such values for $i\hat{p}$ makes $\widehat{C}$ periodic in σ. The SU(2) which follows from J^3 and M^+ is not the usual one constructed in bosonization [8.6] because of the extra factors and inverses of $\partial/\partial\sigma$ involved (see the next section).

Since we project onto $\hat{p} = i\frac{1}{2}$ when acting on Φ, we find for the parts of M^+ and $\mathcal{A}$ linear in χ oscillators when acting on Φ

$$M^+ = e^{2\hat{q}} 2\hat{a}_1 + \cdots \quad , \quad \check{\mathbf{L}}_n = \tfrac{1}{2}\sqrt{n}(n-1)\hat{a}^\dagger{}_n + \cdots \quad . \qquad (8.1.22)$$

This shows how the constraint $T = 0$ essentially just eliminates the zeroth and first oscillators of χ.

We have seen some examples above of Virasoro operators defined as expressions quadratic in functions of σ (and their functional derivatives). More generally, we can consider a bosonic (periodic) function $\hat{f}(\sigma)$ with arbitrary weight w. In order to obtain the transformation law (8.1.5), we must have

$$\mathcal{G}(\sigma) = \hat{f}'\frac{\delta}{\delta\hat{f}} - w\left(\hat{f}\frac{\delta}{\delta\hat{f}}\right)' \quad , \qquad (8.1.23a)$$

up to an overall normal-ordering constant (which we drop). By manipulations like those above, we find

$$[\mathbf{L}_m, \mathbf{L}_n] = (n-m)\mathbf{L}_{m+n} + \left\{\left[(w-\tfrac{1}{2})^2 - \tfrac{1}{12}\right] m^3 - \tfrac{1}{6}m\right\} \delta_{m,-n} \quad . \tag{8.1.23b}$$

Since $\hat{f}'$ and $\delta/\delta\hat{f}$ (or $\hat{f}$ and $-\delta/\delta\hat{f}'$) have the same commutation relations as two $\widehat{P}$'s, but with off-diagonal metric $\eta_{ab} = \binom{0\,1}{1\,0}$, for $w = 0$ (or $w = 1$) the algebra (8.1.23) must give just twice the contribution to the anomaly as a single $\widehat{P}$. This agrees exactly with (8.1.15) (the D term). For fermionic $\hat{f}$, the anomalous terms in (8.1.23) have the opposite overall sign. In that case, $\hat{f}$ and $\delta/\delta\hat{f}$ have the anticommutation relations of 2 physical fermions (see sect. 7.2), again with the off-diagonal metric, and $w = \frac{1}{2}$ gives the Virasoro operators for 2 physical fermions (i.e., as in the above bosonic case, $\mathcal{G}$ can be rewritten as the sum of 2 independent $\mathcal{G}$'s). The anomaly for a single physical fermion in thus given by half of that in (8.1.23b), with opposite sign. Another interesting case is $w = -1$ (or 2), which, for fermions, gives the ghost contribution of (8.1.15) (the non-D terms, for $\propto_0 = 0$; comparing (8.1.14) with (8.1.23a), we see $\widehat{C}$ has $w = -1$ and thus $\delta/\delta\widehat{C}$ has $w = 2$). Thus, (8.1.23) is sufficient to give all the Virasoro algebras which are homogeneous of second order in 1D functions. By the method of bosonization (8.1.18), the fermionic case of (8.1.23a) can be rewritten as

$$\mathcal{G} = i\left[\tfrac{1}{2}\widehat{\mathcal{P}}^2 + (\tfrac{1}{2} - w)\widehat{\mathcal{P}}' + \tfrac{1}{8}\right] \quad , \tag{8.1.23c}$$

where $\hat{f} = exp\,\hat{\chi}$ in terms of a timelike coordinate χ ($\widehat{\mathcal{P}} = \hat{\chi}'$). For $w = \frac{1}{2}$, this gives an independent demonstration that 2 physical fermions give the same anomaly as 1 physical boson (modulo the normal-ordering constant), since they are physically equivalent (up to the boundary conditions on the zero-modes). (There are also factors of i that need to be inserted in various places to distinguish physical bosons and fermions from ghost ones, but these don't affect the value of the anomaly.)

As before, these Virasoro operators correspond to 2D energy-momentum tensors obtained by varying an action with respect to the 2D metric. Using the vielbein formalism of sect. 4.1, we first note that the Lorentz group has only one generator, which acts very simply on the light-cone components of a covariant tensor:

$$M_{\mathbf{ab}} = \epsilon_{\mathbf{ab}}\mathcal{M} \quad (\epsilon_{+-} = 1) \quad \rightarrow \quad [\mathcal{M}, \psi_{(s)}] = s\psi_{(s)} \quad , \tag{8.1.24}$$

where for tensors s is the number of "+" indices minus "−" indices. However, since the 2D Lorentz group is abelian, this generalizes to arbitrary "spin," half-integral as well as irrational. The covariant derivative can then be written as

$$\nabla_a = e_a + \omega_a \mathcal{M} \quad ,$$

$$\omega_a = \tfrac{1}{2}\epsilon^{cb}\omega_{abc} = \tfrac{1}{2}\epsilon^{cb}c_{bca} = -\epsilon_{ab}e\partial_m e^{-1}e^{am} \quad . \tag{8.1.25}$$

We also have the only nonvanishing component of the curvature R_{abcd} given by (4.1.31):

$$e^{-1}R = \epsilon^{mn}\partial_n\omega_m = \partial_m\left[e^{am}\partial_n(e^{-1}e_a{}^n)\right] \quad . \tag{8.1.26}$$

The covariant action corresponding to (8.1.23a) is

$$S \sim \int d^2x\; e^{-1}(\psi^{(+)}{}_{1-w}\nabla_-\psi^{(+)}{}_w + \psi^{(-)}{}_{1-w}\nabla_+\psi^{(-)}{}_w) \quad ,$$

$$[\mathcal{M}, \psi^{(\pm)}{}_w] = \pm w\psi^{(\pm)}{}_w \quad , \tag{8.1.27}$$

where $\psi^{(\pm)}{}_w$ corresponds to $\hat{f}$ and $\psi^{(\pm)}{}_{1-w}$ to $\delta/\delta\hat{f}$. For open-string boundary conditions, $\psi^{(\pm)}{}_w$ are combined to form $\hat{f}$ (as, for $w=1$, $P^\pm$ combined to form $\hat{P}$ in (7.1.3a)); for closed strings, the 2 functions can be used independently (as the usual ($\pm$) modes for closed strings). We thus see that the spin is related to the weight (at least for these free, classical fields) as $s = \pm w$ for $\psi^{(\pm)}{}_w$. The action corresponding to (8.1.23c) (neglecting the normal-ordering constant) is [8.7]

$$S \sim \int d^2x\; e^{-1}\;\left[\tfrac{1}{2}\chi\Box\chi + (w - \tfrac{1}{2})R\chi\right] \quad . \tag{8.1.28a}$$

(We have dropped some surface terms, as in (4.1.36).) The fact that (8.1.28a) represents a particle with spin can be seen in (at least) 2 ways. One way is to perform a duality transformation [8.8]: (8.1.28a) can be written in first-order form as

$$S \sim \int d^2x\; e^{-1}\;\left[\tfrac{1}{2}(F_a)^2 + F^a\nabla_a\chi + (w - \tfrac{1}{2})R\chi\right] \quad . \tag{8.1.28b}$$

(Note that $\nabla\chi$ is the field strength for χ under the global invariance $\chi \to \chi + constant$. In that respect, the second term in (8.1.28b) is like a "Chern-Simons" term, since it can only be written as the product of 1 field with 1 field strength, in terms of the fields ω_a and χ and their field strengths R and $\nabla_a\chi$.) Eliminating F by its equation of motion gives back (8.1.28a), while eliminating χ gives

$$S \sim \int d^2x\; e^{-1}\,\tfrac{1}{2}(G_a)^2 \quad ,$$

$$G_a = -\epsilon_{ab}\nabla^b\phi \quad , \quad [\mathcal{M}, \phi] = w - \tfrac{1}{2} \quad . \tag{8.1.28c}$$

(Actually, since (8.1.28a) and (8.1.28c) are equivalent on shell, we could equally well have started with (8.1.28c) and avoided this discussion of duality transformations.

However, (8.1.28a) is a little more conventional-looking, and the one that more commonly appears in the literature.) The unusual Lorentz transformation law of ϕ follows from the fact that it's the logarithm of a tensor:

$$[\mathcal{M}, e^{\phi}] = (w - \tfrac{1}{2})e^{\phi} \quad , \quad [\mathcal{M}, e^{-\phi}] = (\tfrac{1}{2} - w)e^{-\phi} \quad . \tag{8.1.29}$$

This is analogous to (8.1.18), but the weights there are increased by $\frac{1}{2}$ by quantum effects. (More examples of this effect will be discussed in sect. 9.1.)

Another way to see that χ has an effective Lorentz weight w is to look at the relationship between Lorentz weights and weights under 2D general coordinate transformations (or 1D, or 2D conformal transformations), as in (8.1.5). This follows from the fact that conformally invariant theories, when coupled to gravity, become locally scale invariant theories (even without introducing the scale compensator ϕ of (4.1.34)). (Conversely, conformal transformations can be defined as the subgroup of general coordinate + local scale transformations which leaves the vacuum invariant.) This means that we can gauge-transform e to 1, or, equivalently, redefine the nongravitational fields to cancel all dependence on e. Then $e_a{}^m$ appears only in the unit-determinant combination $\mathfrak{e}_a{}^m = e^{-1/2}e_a{}^m$. The weights w then appear in the scale transformation which leaves (8.1.27) invariant:

$$\psi_w{}' = e^{w\zeta}\psi_w \quad . \tag{8.1.30}$$

(This has the same form as a local Lorentz transformation, but with different relative signs for the fields in the $(\pm)$ terms of (8.1.27). This is related to the fact that, upon applying the equations of motion, and in the conformal gauge, the 2 sets of fields depend respectively on $\tau \mp \sigma$, and therefore have independent conformal transformations on these $\pm$ coordinates, except as related by boundary conditions for the open string.) Choosing $e^{\zeta} = e^{-1/2}$ then replaces the ψ's with fields which are scale-invariant, but transform under general coordinate transformations as densities of weight w (i.e., as tensors times $e^{-w/2}$). In the conformal gauge, these densities satisfy the usual free (from gravity) field equations, since the vielbein has been eliminated (the determinant by redefinition, the rest by choice of general coordinate gauge). Similar remarks apply to (8.1.28), but it's not scale invariant. To isolate the scale noninvariance of that action, rather than make the above scale transformation, we make a nonlocal redefinition of χ in (8.1.28a) which reduces to the above type of scale transformation in the conformal gauge $\mathfrak{e}_a{}^m = \delta_a{}^m$:

$$\chi \to \chi - (w - \tfrac{1}{2})\frac{1}{\Box}R \quad . \tag{8.1.31}$$

In the conformal gauge, $R = -\frac{1}{2}\Box\, ln\, e$. (Remember: χ is like the logarithm of a tensor.) Under this redefinition, the action becomes

$$S \rightarrow \int d^2x\, e^{-1} \left[\tfrac{1}{2}\chi\Box\chi - \tfrac{1}{2}(w-\tfrac{1}{2})^2 R\frac{1}{\Box}R\right] \qquad (8.1.32)$$

(Note that χ now satisfies the usual scalar field equation.) The redefined field is now scale invariant, and the scale noninvariance can now be attributed to the second term, which is the same kind of term responsible for the conformal (Virasoro) anomalies at the quantum level (i.e., the 1-loop contribution to the 2D field theory in a background gravitational field). In fact, the conformally invariant action (4.1.39), with a factor proportional to $1/(D-2)$, is the dimensionally regularized expression responsible for the anomaly: Although (4.1.39) is conformally invariant in arbitrary D, subtraction of the divergent (i.e., with a coefficient $1/(D-2)$) R term, which is conformally invariant only in $D = 2$ (as follows from considering the $D \rightarrow 2$ limit of (4.1.39) without multiplying by $1/(D-2)$), leaves a renormalized (finite) action which, in the limit $D \rightarrow 2$, is just the second term of (8.1.32). Thus, the second term in (8.1.32) contributes classically to the anomaly of (8.1.23b), the remaining contribution being the usual quantum contribution of the scalar. (On the other hand, in the fermionic theory from which (8.1.28) can be derived by quantum mechanical bosonization, all of the anomaly is quantum mechanical.)

$D < 26$ can also be quantized (at least at the tree level), but there is an anomaly in 2D local scale invariance which causes $det(g_{mn})$ to reappear at the quantum level [8.9] (or, in the light-cone formalism, an extra "longitudinal" Lorentz component of X [1.3,4]); however, there are complications at the one-loop level which have not yet been resolved.

Presently the covariant formulation of string interactions is understood only within the IGL(1) formalism (although in principle it's straightforward to obtain the OSp(1,1|2) formalism by eliminating the auxiliary fields, as in sect. 4.2). These interactions will be discussed in sect. 12.2.

8.2. OSp(1,1|2)

We next use the light-cone Poincaré algebra of the string, obtained in sect. 7.1, to derive the OSp(1,1|2) formulation as in sect. 3.4, which can be used to find the gauge-invariant action. We then relate this to the first-quantized IGL(1) of the previous section by the methods of sect. 4.2. This OSp(1,1|2) formalism can also be derived from first quantization simply by treating the zero-mode of g as g for the particle (sect. 5.1), and the other modes of the metric as in the conformal gauge (sect. 6.2).

All operators come directly from the light-cone expressions of sect. 7.1, using the dimensional extension of sect. 2.6, as described in the general case in sect. 3.4. In particular, to evaluate the action and its gauge invariance in the form (4.1.6), we'll need the expressions for $M_{\alpha\beta}$, M^2, and $\mathcal{Q}_\alpha = M_\alpha{}^a p_a + M_{\alpha m}M$ given by using $i = (a,\alpha)$ in (7.1.14,15). Thus,

$$M_{\alpha\beta} = -i\int' \hat{X}_\alpha \hat{P}_\beta = \sum_n a^\dagger{}_{n(\alpha}a_{n\beta)} \quad ,$$

$$M_{\alpha a} = -i\int' \hat{X}_\alpha \hat{P}_a = \sum_n a^\dagger{}_{n[\alpha}a_{na]} \,,$$

$$M^2 = \int' (\hat{P}_i{}^2 - 2) = 2\left(\sum n a^\dagger{}_n{}^i a_{ni} - 1\right) = 2(N-1) \,,$$

$$M_{\alpha m}M = -i\tfrac{1}{2}\int' \hat{X}_\alpha \hat{P}_j{}^2 = \sum \left(a^\dagger{}_{n\alpha}\Delta_n - \Delta^\dagger{}_n a_{n\alpha}\right) \quad ,$$

$$\Delta_n = i\frac{1}{\sqrt{n}}\left(\tfrac{1}{2}\sum_{m=1}^{n-1}\sqrt{m(n-m)}a_m{}^i a_{n-m,i} - \sum_{m=1}^{\infty}\sqrt{m(n+m)}a^\dagger{}_m{}^i a_{n+m,i}\right) \quad , \tag{8.2.1}$$

where again the i summation is over both a and α, representing modes coming from both the physical $X^a(\sigma)$ (with x^a identified as the usual spacetime coordinate) and the ghost modes $X^\alpha(\sigma)$ (with x^α the ghost coordinates of sect. 3.4), as in the mode expansion (7.1.7).

To understand the relation of the first-quantized BRST quantization [4.4,5] to that derived from the light cone (and from the OSp(1,1|2)), we show the Sp(2) symmetry of the ghost coordinates. We first combine all the ghost oscillators into an "isospinor" [4.1]:

$$C^\alpha = \frac{1}{\partial/\partial\sigma}\left(\hat{C}', \frac{\delta}{\delta\hat{C}}\right) \quad . \tag{8.2.2}$$

This isospinor directly corresponds (except for lack of zero modes) to $\hat{X}^\alpha$ of the OSp(1,1|2) formalism from the light cone: We identify

$$\hat{X}^\alpha = (x^\alpha + p^\alpha\sigma) + C^\alpha \quad , \tag{8.2.3}$$

and we can thus directly construct objects which are manifestly covariant under the Sp(2) of $M_{\alpha\beta}$. The periodic inverse derivative in (8.2.2) is defined in terms of the saw-tooth function

$$\frac{1}{\partial/\partial\sigma}f(\sigma) = \int d\sigma' \, \tfrac{1}{2}\left[\epsilon(\sigma-\sigma') - \frac{1}{\pi}(\sigma-\sigma')\right] f(\sigma') \quad , \tag{8.2.4}$$

where $\epsilon(\sigma) = \pm 1$ for $\pm\sigma > 0$. The product of the derivative with this inverse derivative, in either order, is simply the projection operator which subtracts out the zero mode. (For example, $\mathcal{C}^+$ is just $\widehat{\mathcal{C}}$ minus its zero mode.) Along with $\widehat{P}_a$, this completes the identification of the nonzero-modes of the two formalisms. We can then rewrite the other relevant operators (8.1.10,12) in terms of $\mathcal{C}^\alpha$:

$$p^2 + M^2 = \int \frac{d\sigma}{2\pi}\,(\widehat{P}_a{}^2 + \mathcal{C}^{\alpha\prime}\mathcal{C}_\alpha{}' - 2) \quad ,$$

$$\mathcal{Q}_\alpha = -i\tfrac{1}{2}\int \frac{d\sigma}{2\pi}\,\mathcal{C}_\alpha(\widehat{P}_a{}^2 + \mathcal{C}^{\alpha\prime}\mathcal{C}_\alpha{}' - 2) \quad ,$$

$$M_{\alpha\beta} = -i\tfrac{1}{2}\int \frac{d\sigma}{2\pi}\,\mathcal{C}_{(\alpha}\mathcal{C}_{\beta)}{}' \quad . \tag{8.2.5}$$

Again, all definitions include normal ordering. This first-quantized IGL(1) can then be seen to agree with that derived from OSp(1,1|2) in (8.2.1) by expanding in zero-modes.

For the closed string, the OSp(1,1|2) algebra is extended to an IOSp(1,1|2) algebra following the construction of (7.1.17): As for the open-string case, the D-dimensional indices of the light-cone formalism are extended to (D+4)-dimensional indices, but just the values $A = (\pm, \alpha)$ are kept for the BRST algebra. To obtain the analog of the IGL(1) formalism, we perform the transformation (3.4.3a) for both left and right-handed modes. The extension of IGL(1) to GL(1|1) analogous to that of OSp(1,1|2) to IOSp(1,1|2) uses the subalgebras $(Q, J^3, p_-, p^{\tilde c} = \partial/\partial c)$ of the IOSp(1,1|2)'s of each set of open-string operators. After dropping the terms containing $\partial/\partial\tilde c$, x_- drops out, and we can set $p_+ = 1$ to obtain:

$$\begin{aligned}
Q &\rightarrow -ic\tfrac{1}{2}(p_a{}^2 + M^2) + M^+ i\frac{\partial}{\partial c} + \mathcal{Q}^+ \quad , \\
J^3 &\rightarrow c\frac{\partial}{\partial c} + M^3 \quad , \\
p_- &\rightarrow -\tfrac{1}{2}(p_a{}^2 + M^2) \quad , \\
p^{\tilde c} &\rightarrow \frac{\partial}{\partial c} \quad .
\end{aligned} \tag{8.2.6}$$

These generators have the same algebra as N=2 supersymmetry in one dimension, with Q and $p^{\tilde c}$ corresponding to the two supersymmetry generators (actually the complex combination and its complex conjugate), J^3 to the O(2) generator which scales them, and p_- the 1D momentum. The closed-string algebra GL(1|1) is then constructed in analogy to the IOSp(1,1|2), taking sums for the J's and differences for the p's.

The application of this algebra to obtaining the gauge-invariant action will be discussed in chapt. 11.

8.3. Lorentz gauge

We will next consider the OSp(1,1|2) algebra which corresponds to first-quantization in the Lorentz gauge [3.7], as obtained by the methods of sect. 3.3 when applied to the Virasoro algebra of constraints.

From (3.3.2), for the OSp(1,1|2) algebra we have

$$Q^\alpha = \int \frac{d\sigma}{2\pi}\Big[\, C^\alpha(\tfrac{1}{2}\widehat{P}^2 - 1) + \tfrac{1}{2}C^{(\alpha}C^{\beta)\prime} i\frac{\delta}{\delta C^\beta} - Bi\frac{\delta}{\delta C_\alpha} + \tfrac{1}{2}(C^\alpha B' - C^{\alpha\prime}B)i\frac{\delta}{\delta B} + \tfrac{1}{4}C^\beta(C_\beta{}'C^\alpha)' i\frac{\delta}{\delta B}\Big] \quad . \tag{8.3.1}$$

B is conjugate to the time-components of the gauge field, which in this case means the components g^{00} and g^{01} of the unit-determinant part of the world-sheet metric (see chapt. 6). This expression can be simplified by the unitary transformation $Q^\alpha \to UQ^\alpha U^{-1}$ with

$$\ln U = -\tfrac{1}{2}\int (\tfrac{1}{2}C^\alpha C_\alpha)\frac{\delta}{\delta B}{}' \quad . \tag{8.3.2}$$

We then have the OSp(1,1|2) (from (3.3.7)):

$$J_{-\alpha} = \int C_\alpha\left(-i\tfrac{1}{2}\widehat{P}^2 + i + C^{\beta\prime}\frac{\delta}{\delta C^\beta} + B'\frac{\delta}{\delta B}\right) - B\frac{\delta}{\delta C^\alpha} \quad ,$$

$$J_{\alpha\beta} = \int C_{(\alpha}\frac{\delta}{\delta C^{\beta)}} \quad , \quad J_{+\alpha} = 2\int C_\alpha\frac{\delta}{\delta B} \quad ,$$

$$J_{-+} = \int 2B\frac{\delta}{\delta B} + C^\alpha\frac{\delta}{\delta C^\alpha} \quad . \tag{8.3.3}$$

A gauge-fixed kinetic operator for string field theory which is invariant under the full OSp(1,1|2) can be derived,

$$K = \tfrac{1}{2}\left\{J_-{}^\alpha, \left[J_{-\alpha}, \int i\frac{\delta}{\delta B}\right]\right\} = \int \tfrac{1}{2}\widehat{P}^2 - 1 + C^{\alpha\prime}i\frac{\delta}{\delta C^\alpha} + B'i\frac{\delta}{\delta B} = \tfrac{1}{2}(p^2 + M^2) \quad ; \tag{8.3.4}$$

as the zero-mode of the generators $\widehat{\mathcal{G}}(\sigma)$ from (3.3.10):

$$\widehat{\mathcal{G}}(\sigma) = -\tfrac{1}{2}\left\{J_-{}^\alpha, \left[J_{-\alpha}, \frac{\delta}{\delta B}\right]\right\}$$
$$= -i(\tfrac{1}{2}\widehat{P}^2 - 1) + C^{\alpha\prime}\frac{\delta}{\delta C^\alpha} + \left(C^\alpha\frac{\delta}{\delta C^\alpha}\right)' + B'\frac{\delta}{\delta B} + \left(B\frac{\delta}{\delta B}\right)' . \quad (8.3.5)$$

The analog in the usual OSp(1,1|2) formalism is

$$\tfrac{1}{2}\{J_-{}^\alpha, [J_{-\alpha}, p_+{}^2]\} = p^A p_A = 2p_+p_- + p^\alpha p_\alpha = \Box - M^2 \quad . \quad (8.3.6)$$

This differs from the usual light-cone gauge-fixed hamiltonian p_-, which is not OSp(1,1|2) invariant. Unlike the ordinary BRST case (but like the light-cone formalism), this operator is invertible, since it's of the standard form $K = \frac{1}{2}p^2 + \cdots$. This was made possible by the appearance of an even number (2) of anticommuting zero-modes. In ordinary BRST (sect. 4.4), the kinetic operator is fermionic: $c(\Box - M^2) - 2\frac{\partial}{\partial c}M^+$ is not invertible because M^+ is not invertible.

As usual, the propagator can be converted into a form amenable to first-quantized path-integral techniques by first introducing the Schwinger proper-time parameter:

$$\frac{1}{K} = \int_0^\infty d\tau \; e^{-\tau K} \quad , \quad (8.3.7)$$

where τ is identified with the (Wick-rotated) world-sheet time. At the free level, the analysis of this propagator corresponds to solving the Schrödinger equation or, in the Heisenberg picture (or classical mechanics of the string), to solving for the time dependence of the coordinates which follow from treating K as the hamiltonian:

$$[K, Z] = iZ' \quad , \quad \dot{Z} - [K, Z] = 0 \quad \to \quad Z = Z(\sigma + i\tau) \quad (8.3.8)$$

for $Z = P, C, B, \delta/\delta C, \delta/\delta B$. Thus in the mode expansion $Z = z_0 + \sum_1^\infty (z_n{}^\dagger e^{-in(\sigma+i\tau)} + z_n e^{in(\sigma+i\tau)})$ the positive-energy $z_n{}^\dagger$'s are creation operators while the negative-energy z_n's are annihilation operators. (Remember active vs. passive transformations: In the Schrödinger picture coordinates are constant while states have time dependence e^{-tH}; in the Heisenberg picture states are constant while coordinates have time dependence $e^{tH}(\;)e^{-tH}$.)

When doing string field theory, in order to define real string fields we identify complex conjugation of the fields as usual with proper-time reversal in the mechanics, which, in order to preserve handedness in the world sheet, means reversing σ as well as τ. As a result, all reparametrization-covariant variables with an even number of 2D vector indices are interpreted as string-field coordinates, while those with an odd number are momenta. (See sect. 8.1.) This means that X is a coordinate, while B and C are momenta. Therefore, we should define the string field as $\Phi[X(\sigma), G(\sigma), F_\alpha(\sigma)]$, where $B = i\delta/\delta G$ and $C^\alpha = i\delta/\delta F_\alpha$. This field is real under a combined complex conjugation and "twist" ($\sigma \to -\sigma$), and Q^α is odd in the number of functional plus σ derivatives. (Note that the corresponding replacement of B with G and C with F would not be required if the $\mathcal{G}_i$'s had been associated with a Yang-Mills symmetry rather than general coordinate transformations, since in that case B and C carry no vector indices.)

This OSp(1,1|2) algebra can also be derived from the classical mechanics action. The 2D general coordinate transformations (6.1.2) generated by $\delta = \int \frac{d\sigma}{2\pi}\, \epsilon^{\mathbf{m}}(\sigma)\mathcal{G}_{\mathbf{m}}(\sigma)$ determine the BRST transformations by (3.3.2):

$$\begin{aligned}
Q^\alpha X &= C^{\mathbf{m}\alpha}\partial_{\mathbf{m}}X \quad , \\
Q^\alpha g^{\mathbf{mn}} &= \partial_{\mathbf{p}}(C^{\mathbf{p}\alpha}g^{\mathbf{mn}}) - g^{\mathbf{p}(\mathbf{m}}\partial_{\mathbf{p}}C^{\mathbf{n})\alpha} \quad , \\
Q^\alpha C^{\mathbf{m}\beta} &= \tfrac{1}{2}C^{\mathbf{n}(\alpha}\partial_{\mathbf{n}}C^{\mathbf{m}\beta)} - C^{\alpha\beta}B^{\mathbf{m}} \quad , \\
Q^\alpha B^{\mathbf{m}} &= \tfrac{1}{2}(C^{\mathbf{n}\alpha}\partial_{\mathbf{n}}B^{\mathbf{m}} - B^{\mathbf{n}}\partial_{\mathbf{n}}C^{\mathbf{m}\alpha}) \\
&\quad - \tfrac{1}{8}\left[2C^{\mathbf{n}\beta}(\partial_{\mathbf{n}}C^{\mathbf{p}\alpha})\partial_{\mathbf{p}}C^{\mathbf{m}}{}_\beta + C^{\mathbf{n}\beta}C^{\mathbf{p}}{}_\beta\partial_{\mathbf{n}}\partial_{\mathbf{p}}C^{\mathbf{m}\alpha}\right] \ .
\end{aligned} \tag{8.3.9}$$

We then redefine

$$\widetilde{B}^{\mathbf{m}} = B^{\mathbf{m}} - \tfrac{1}{2}C^{\mathbf{n}\alpha}\partial_{\mathbf{n}}C^{\mathbf{m}}{}_\alpha \quad \to$$

$$\begin{aligned}
Q^\alpha C^{\mathbf{m}\beta} &= C^{\mathbf{n}\alpha}\partial_{\mathbf{n}}C^{\mathbf{m}\beta} - C^{\alpha\beta}\widetilde{B}^{\mathbf{m}} \quad , \\
Q^\alpha \widetilde{B}^{\mathbf{m}} &= C^{\mathbf{n}\alpha}\partial_{\mathbf{n}}\widetilde{B}^{\mathbf{m}} \quad .
\end{aligned} \tag{8.3.10}$$

The rest of the OSp(1,1|2) follows from (3.3.7):

$$J_{+\alpha}(X, g^{\mathbf{mn}}, C^{\mathbf{m}\beta}) = 0 \quad , \quad J_{+\alpha}\widetilde{B}^{\mathbf{m}} = 2C^{\mathbf{m}}{}_\alpha \quad ;$$

$$J_{\alpha\beta}(X, g^{\mathbf{mn}}, \widetilde{B}^{\mathbf{m}}) = 0 \quad , \quad J_{\alpha\beta}C^{\mathbf{m}\gamma} = \delta_{(\alpha}{}^{\gamma}C^{\mathbf{m}}{}_{\beta)} \quad ;$$

$$J_{-+}(X, g^{\mathbf{mn}}) = 0 \quad , \quad J_{-+}\widetilde{B}^{\mathbf{m}} = 2\widetilde{B}^{\mathbf{m}} \quad , \quad J_{-+}C^{\mathbf{m}\alpha} = C^{\mathbf{m}\alpha} \quad . \tag{8.3.11}$$

An ISp(2)-invariant gauge-fixing term is (dropping boundary terms)

$$\mathcal{L}_1 = Q_\alpha{}^2\tfrac{1}{2}\eta_{\mathbf{pq}}g^{\mathbf{pq}} = -\eta_{\mathbf{pq}}\left[\widetilde{B}^{\mathbf{p}}\partial_{\mathbf{m}}g^{\mathbf{qm}} + \tfrac{1}{2}g^{\mathbf{mn}}(\partial_{\mathbf{m}}C^{\mathbf{p}\alpha})(\partial_{\mathbf{n}}C^{\mathbf{q}}{}_\alpha)\right] \quad , \tag{8.3.12}$$

where η is the flat world-sheet metric. This expression is the analog of the gauge-fixing term $Q^2\frac{1}{2}A^2$ for Lorentz gauges in Yang-Mills [3.6,12]. Variation of $\widetilde{B}$ gives the condition for harmonic coordinates. The ghosts have the same form of lagrangian as X, but not the same boundary conditions: At the boundary, any variable with an even number of 2D vector indices with the value 1 has its σ-derivative vanish, while any variable with an odd number vanishes itself. These are the only boundary conditions consistent with Poincaré and BRST invariance. They are preserved by the redefinitions below.

Rather than this Landau-type harmonic gauge, we can also define more general Lorentz-type gauges, such as Fermi-Feynman, by adding to (8.3.12) a term proportional to $Q^2\frac{1}{2}\eta_{mn}C^{m\alpha}C^{n}{}_{\alpha} = -\frac{1}{2}\eta_{mn}\widetilde{B}^{m}\widetilde{B}^{n} + \cdots$. We will not consider such terms further here.

Although the hamiltonian quantum mechanical form of Q^α (8.3.3) also follows from (3.3.2) (with the functional derivatives now with respect to functions of just σ instead of both σ and τ), the relation to the lagrangian form follows only after some redefinitions, which we now derive. The hamiltonian form that follows directly from (8.3.9,10) can be obtained by applying the Noether procedure to $\mathcal{L} = \mathcal{L}_0 + \mathcal{L}_1$: The BRST current is

$$J^{m\alpha} = (\mathcal{g}^{np}C^{m\alpha} - \mathcal{g}^{m(n}C^{p)\alpha})\tfrac{1}{2}\left[(\partial_n X)\cdot(\partial_p X) + (\partial_n C^{q\beta})(\partial_p C_{q\beta})\right] - \widetilde{B}_n\partial_p(\mathcal{g}^{n[m}C^{p]\alpha} - \mathcal{g}^{mp}C^{n\alpha}) \quad , \qquad (8.3.13)$$

where (2D) vector indices have been raised and lowered with the flat metric. Canonically quantizing, with

$$\frac{1}{\alpha'}P^0 = i\frac{\delta}{\delta X} \quad , \quad \frac{1}{\alpha'}\widetilde{B}_m = -i\frac{\delta}{\delta \mathcal{g}^{0m}} \quad , \quad \frac{1}{\alpha'}\Pi_{m\alpha} = i\frac{\delta}{\delta C^{m\alpha}} \quad , \qquad (8.3.14a)$$

we apply

$$\dot{X} = -\frac{1}{\mathcal{g}^{00}}(P^0 + \mathcal{g}^{01}X') \quad , \quad \dot{C}^{m\alpha} = -\frac{1}{\mathcal{g}^{00}}(\Pi^{m\alpha} + \mathcal{g}^{01}C^{m\alpha\prime}) \quad , \qquad (8.3.14b)$$

to the first term in (8.3.13) and

$$\partial_m \mathcal{g}^{mn} = 0 \quad , \quad \mathcal{g}^{0m}\partial_m C^{n\alpha} = -\Pi^{n\alpha} \quad , \qquad (8.3.14c)$$

to the second to obtain

$$J^{0\alpha} = \left(-\frac{1}{\mathcal{g}^{00}}C^{0\alpha}\right)\tfrac{1}{2}\left[(P^{02} + X'^2) + (\Pi^{m\beta}\Pi_{m\beta} + C^{m\beta\prime}C_{m\beta}{}')\right]$$

$$+\left(C^{1\alpha}-\frac{\mathscr{g}^{01}}{\mathscr{g}^{00}}C^{0\alpha}\right)(X'\cdot P^{0}+C^{\mathbf{m}\beta'}\Pi_{\mathbf{m}\beta})-\widetilde{B}_{\mathbf{m}}\left[\Pi^{\mathbf{m}\alpha}+(\mathscr{g}^{\mathbf{m}[0}C^{1]\alpha})'\right], \tag{8.3.15}$$

where $Q^{\alpha}\sim\int d\sigma\ J^{0\alpha}$.

By comparison with (8.3.3), an obvious simplification is to absorb the $\mathscr{g}$ factors into the C's in the first two terms. This is equivalent to

$$C^{\mathbf{m}\alpha}\ \rightarrow\ \delta_{1}{}^{\mathbf{m}}C^{1\alpha}-\mathscr{g}^{0\mathbf{m}}C^{0\alpha}\quad , \tag{8.3.16}$$

and the corresponding redefinitions (unitary transformation) of Π and $\widetilde{B}$. This puts $\mathscr{g}$-dependence into the $\widetilde{B}\Pi$ terms,

$$\Pi_{\mathbf{m}}\widetilde{B}^{\mathbf{m}}\quad\rightarrow\quad\Pi_{0}\left(-\frac{1}{\mathscr{g}^{00}}\widetilde{B}^{0}\right)+\Pi_{1}\left(\widetilde{B}^{1}-\frac{\mathscr{g}^{01}}{\mathscr{g}^{00}}\widetilde{B}^{0}\right)+\cdots\quad , \tag{8.3.17}$$

unlike (8.3.3), so we remove it by the $\mathscr{g}$ redefinition

$$\mathscr{g}^{01}=-g^{1}\quad ,\quad \mathscr{g}^{00}=-\sqrt{1+2g^{0}+(g^{1})^{2}}\quad ;$$
$$\frac{1}{\alpha'}B_{\mathbf{m}}=i\frac{\delta}{\delta g^{\mathbf{m}}}\quad . \tag{8.3.18}$$

These redefinitions give

$$\begin{aligned}J^{0\alpha}=\ &\left[C^{0\alpha}\tfrac{1}{2}(P^{02}+X'^{2})+C^{1\alpha}X'\cdot P^{0}\right]\\&+\left\{C^{0\alpha}(C^{0\beta}\Pi_{1\beta})'+C^{1\alpha}\left[C^{\mathbf{m}\beta'}\Pi_{\mathbf{m}\beta}+(C^{0\beta}\Pi_{0\beta})'\right]\right\}\\&-\left\{C^{0\alpha}(B^{1}+g_{1}B^{0})'+C^{1\alpha}\left[-B^{0'}+g_{\mathbf{m}}B^{\mathbf{m}'}+(g_{0}B^{0})'\right]\right\}\\&+C^{0\alpha}\left[\tfrac{1}{2}C^{\mathbf{m}\beta'}C_{\mathbf{m}\beta}{}'+(g_{\mathbf{m}}C^{0\beta})'C^{\mathbf{m}}{}_{\beta}{}'\right]-\Pi_{\mathbf{m}}{}^{\alpha}B^{\mathbf{m}}\quad .\end{aligned} \tag{8.3.19}$$

The quadratic terms CB' don't appear in (8.3.3), and can be removed by the unitary transformation

$$Q^{\alpha\prime}=UQ^{\alpha}U^{-1}\quad ,\quad ln\ U=-i\frac{1}{\alpha'}\int C^{0\alpha}C^{1}{}_{\alpha}{}'\quad , \tag{8.3.20}$$

giving

$$\begin{aligned}J^{0\alpha}=\ &\left[C^{0\alpha}\tfrac{1}{2}(P^{02}+X'^{2})+C^{1\alpha}X'\cdot P^{0}\right]\\&+\left\{C^{0\alpha}(C^{0\beta}\Pi_{1\beta})'+C^{1\alpha}\left[C^{\mathbf{m}\beta'}\Pi_{\mathbf{m}\beta}+(C^{0\beta}\Pi_{0\beta})'\right]\right\}\\&-\left\{C^{0\alpha}(g_{1}B^{0})'+C^{1\alpha}\left[g_{\mathbf{m}}B^{\mathbf{m}'}+(g_{0}B^{0})'\right]\right\}\\&+C^{0\alpha}(g_{\mathbf{m}}C^{0\beta})'C^{\mathbf{m}}{}_{\beta}{}'-\Pi_{\mathbf{m}}{}^{\alpha}B^{\mathbf{m}}\quad .\end{aligned} \tag{8.3.21}$$

Finally, the remaining terms can be fixed by the transformation

$$ln\ U = i\frac{1}{\alpha'}\int C^{0\alpha}(g_1 C^0{}_\alpha{}' + g_0 C^1{}_\alpha{}') \tag{8.3.22}$$

to get (8.3.3), after extending σ to $[-\pi, \pi]$ by making the definitions

$$\hat{P} = \frac{1}{\sqrt{2\alpha'}}(P^0 + X') \quad ,$$

$$C^\alpha = \frac{1}{\sqrt{2\alpha'}}(C^{0\alpha} + C^{1\alpha}) \quad , \quad i\frac{\delta}{\delta C^\alpha} = \frac{1}{\sqrt{2\alpha'}}(\Pi_{0\alpha} + \Pi_{1\alpha}) \quad ,$$

$$G = \frac{1}{\sqrt{2\alpha'}}(g_0 + g_1) \quad , \quad i\frac{\delta}{\delta G} = \frac{1}{\sqrt{2\alpha'}}(B^0 + B^1) \quad , \tag{8.3.23}$$

where the previous coordinates, defined on $[0, \pi]$, have been extended to $[-\pi, \pi]$ as

$$Z(-\sigma) = \pm Z(\sigma) \quad , \tag{8.3.24}$$

with "+" if Z has an even number of vector indices with the value 1, and "−" for an odd number, in accordance with the boundary conditions, so that the new coordinates will be periodic in σ with period 2π.

To describe the closed string with the world-sheet metric, we again use 2 sets of open-string operators, as in (7.1.17), with each set of open-string operators given as in (8.3.3,4), and the translations are now, in terms of the zero-modes b and c of B and C,

$$\begin{aligned} p_+ &= \sqrt{-2i\frac{\partial}{\partial b}} \quad , \\ p_\alpha &= \frac{1}{p_+}i\frac{\partial}{\partial c^\alpha} \quad , \\ p_- &= -\frac{1}{p_+}\left(K + p_\alpha{}^2\right) \quad . \end{aligned} \tag{8.3.25}$$

The OSp(1,1|2) subgroup of the resulting IOSp(1,1|2), after the use of the constraints $\Delta p = 0$, reduces to what would be obtained from applying the general result (3.3.2) to closed string mechanics. ((3.2.6a), without the $B\partial/\partial\tilde{C}$ term, gives the usual BRST operator when applied to closed-string mechanics, which is the same as the sum of two open-string ones with the two sets of physical zero-modes identified.)

We now put the OSp(1,1|2) generators in a form analogous to those derived from the light cone [3.13]. Let's first consider the open string. Separating out the dependence on the zero-modes g and f^α in the OSp(1,1|2) operators,

$$
\begin{aligned}
J_{\alpha\beta} &= f_{(\alpha}\frac{\partial}{\partial f^{\beta)}} + \widetilde{M}_{\alpha\beta} \quad , \\
J_{-+} &= -2g\frac{\partial}{\partial g} - f^\alpha\frac{\partial}{\partial f^\alpha} + \widetilde{M}_{-+} \quad , \\
J_{+\alpha} &= -2g\frac{\partial}{\partial f^\alpha} + \widetilde{M}_{+\alpha} \quad , \\
J_{-\alpha} &= f_\alpha\frac{\partial}{\partial g} - ig\mathcal{B}_\alpha + K\frac{\partial}{\partial f^\alpha} + i\mathcal{C}_\alpha{}^\beta f_\beta + \widetilde{M}_{-\alpha} \quad ,
\end{aligned}
\tag{8.3.26}
$$

where the $\widetilde{M}$'s are the parts of the J's not containing these zero-modes, K is given in (8.3.4), and

$$
\begin{aligned}
\mathcal{B}_\alpha &= \int BC_\alpha{}' \quad , \\
\mathcal{C}_{\alpha\beta} &= \int C_\alpha C_\beta{}' = \mathcal{C}_{\beta\alpha} \quad .
\end{aligned}
\tag{8.3.27}
$$

From (3.3.8) and (8.3.26) we find the commutators of $\widetilde{M}_{AB}$, $\mathcal{B}_\alpha$, $\mathcal{C}_{\alpha\beta}$, and K; the nonvanishing ones are:

$$
\begin{aligned}
[\widetilde{M}_{\alpha\beta}, \widetilde{M}_{\gamma\delta}] &= -C_{(\gamma(\alpha}\widetilde{M}_{\beta)\delta)} \quad , \\
[\widetilde{M}_{\alpha\beta}, \widetilde{M}_{\pm\gamma}] &= -C_{\gamma(\alpha}\widetilde{M}_{\pm\beta)} \quad , \\
\{\widetilde{M}_{-\alpha}, \widetilde{M}_{+\beta}\} &= -C_{\alpha\beta}\widetilde{M}_{-+} - \widetilde{M}_{\alpha\beta} \quad , \\
[\widetilde{M}_{-+}, \widetilde{M}_{\pm\alpha}] &= \mp\widetilde{M}_{\pm\alpha} \quad , \\
\{\widetilde{M}_{-\alpha}, \widetilde{M}_{-\beta}\} &= 2iK\mathcal{C}_{\alpha\beta} \quad ,
\end{aligned}
$$

$$
\begin{gathered}
[\widetilde{M}_{\alpha\beta}, \mathcal{B}_\gamma] = -C_{\gamma(\alpha}\mathcal{B}_{\beta)} \quad , \quad [\widetilde{M}_{\alpha\beta}, \mathcal{C}_{\gamma\delta}] = -C_{(\gamma(\alpha}\mathcal{C}_{\beta)\delta)} \quad , \\
[\widetilde{M}_{-+}, \mathcal{B}_\alpha] = 3\mathcal{B}_\alpha \quad , \quad [\widetilde{M}_{-+}, \mathcal{C}_{\alpha\beta}] = 2\mathcal{C}_{\alpha\beta} \quad , \\
\{\widetilde{M}_{+\alpha}, \mathcal{B}_\beta\} = 2\mathcal{C}_{\alpha\beta} \quad , \quad [\widetilde{M}_{-\alpha}, \mathcal{C}_{\beta\gamma}] = -C_{\alpha(\beta}\mathcal{B}_{\gamma)} \quad .
\end{gathered}
\tag{8.3.28}
$$

(To show $\{\widetilde{M}_{-[\alpha}, \mathcal{B}_{\beta]}\} = 0$ requires explicit use of (8.3.3), but it won't be needed below.)

We then make the redefinition (see (3.6.8))

$$g = \tfrac{1}{2}h^2 \tag{8.3.29}$$

as for the particle. We next redefine the zero-modes as in (3.5.2) by first making the unitary transformation

$$ln\ U = (ln\ h)\left(\tfrac{1}{2}\left[\frac{\partial}{\partial f_\alpha}, f_\alpha\right] - \widetilde{M}_{-+}\right) \tag{8.3.30a}$$

to redefine h and then

$$\ln U = -f^\alpha \widetilde{M}_{+\alpha} \tag{8.3.30b}$$

to redefine c^α. The net result is that the transformed operators are

$$J_{-+} = -\frac{\partial}{\partial h}h \quad , \quad J_{+\alpha} = -h\frac{\partial}{\partial f^\alpha} \quad , \quad J_{\alpha\beta} = f_{(\alpha}\frac{\partial}{\partial f^{\beta)}} + \widetilde{M}_{\alpha\beta} \quad ,$$

$$J_{-\alpha} = f_\alpha \frac{\partial}{\partial h} + \frac{1}{h}\left[\frac{\partial}{\partial f^\alpha}(K + f^2) - \widetilde{M}_\alpha{}^\beta f_\beta + \widehat{Q}_\alpha\right] \quad , \tag{8.3.31a}$$

where

$$\widehat{Q}_\alpha = \widetilde{M}_{-\alpha} - i\tfrac{1}{2}B_\alpha + K\widetilde{M}_{+\alpha} \quad . \tag{8.3.31b}$$

We also have

$$p_+ = h \quad , \quad p_\alpha = -f_\alpha \quad , \quad p_- = -\frac{1}{h}(K + f^2) \quad . \tag{8.3.32}$$

These expressions have the canonical form (3.4.2a), with the identification

$$\widetilde{M}_{\alpha\beta} \leftrightarrow M_{\alpha\beta} \quad , \quad \widehat{Q}_\alpha \leftrightarrow Q_\alpha \quad . \tag{8.3.33}$$

From (8.3.28) we then find

$$\{\widehat{Q}_\alpha, \widehat{Q}_\beta\} = -2K\widetilde{M}_{\alpha\beta} \quad , \tag{8.3.34}$$

consistent with the identification (8.3.33). Thus the IOSp(1,1|2) algebra (8.3.31,32) takes the canonical form of chapt. 3. This also allows the closed string formalism to be constructed.

For expanding the fields, it's more convenient to expand the coordinate in *real* oscillators (preserving the reality condition of the string field) as

$$\hat{P} = p + \sum_1^\infty \sqrt{n}(-ia_n{}^\dagger e^{-in\sigma} + ia_n e^{in\sigma}) \quad ,$$

$$G = g + \sum_1^\infty \sqrt{n}(g_n{}^\dagger e^{-in\sigma} + g_n e^{in\sigma}) \quad ,$$

$$B = b + \sum_1^\infty \frac{1}{\sqrt{n}}(-ib_n{}^\dagger e^{-in\sigma} + ib_n e^{in\sigma}) \quad ,$$

$$F^\alpha = f^\alpha + \sum_1^\infty \sqrt{n}(f^\alpha{}_n{}^\dagger e^{-in\sigma} + f^\alpha{}_n e^{in\sigma}) \quad ,$$

$$C^\alpha = c^\alpha + \sum_1^\infty \frac{1}{\sqrt{n}}(-ic^\alpha{}_n{}^\dagger e^{-in\sigma} + ic^\alpha{}_n e^{in\sigma}) \quad . \tag{8.3.35}$$

(With our conventions, always $z_n{}^{\alpha\dagger} \equiv (z_n{}^\alpha)^\dagger$, and thus $z_{n\alpha}{}^\dagger = -(z_{n\alpha})^\dagger$.) The commutation relations are then

$$[a_m, a_n{}^\dagger] = [b_m, g_n{}^\dagger] = [g_m, b_n{}^\dagger] = \delta_{mn} \quad ,$$
$$\{c_{m\alpha}, f_{n\beta}{}^\dagger\} = \{f_{m\alpha}, c_{n\beta}{}^\dagger\} = \delta_{mn} C_{\alpha\beta} \quad , \tag{8.3.36}$$

We then have

$$\hat{Q}_\alpha = \sum_1^\infty [(\mathcal{O}_n{}^\dagger c_{n\alpha} - c_{n\alpha}{}^\dagger \mathcal{O}_n) - (b_n{}^\dagger f_{n\alpha} - f_{n\alpha}{}^\dagger b_n)] \quad ,$$

$$\mathcal{O}_n = \frac{1}{\sqrt{n}} \widetilde{L}_n + \tfrac{1}{2} b_n - 2g_n K \quad ,$$

$$\begin{aligned} K = \widetilde{L}_0 &= -\tfrac{1}{2}\Box - 1 + \sum_1^\infty n(a_n{}^\dagger \cdot a_n + b_n{}^\dagger g_n + g_n{}^\dagger b_n + c_n{}^{\alpha\dagger} f_{n\alpha} + f_n{}^{\alpha\dagger} c_{n\alpha}) \\ &= -\tfrac{1}{2}\Box - 1 + N \quad , \end{aligned}$$

$$\begin{aligned} \widetilde{L}_n = &- \sqrt{n}\partial \cdot a_n \\ &+ \sum_{m=1}^\infty \sqrt{m(n+m)}(a_m{}^\dagger \cdot a_{n+m} + b_m{}^\dagger g_{n+m} + c_m{}^{\alpha\dagger} f_{n+m,\alpha} \\ &\qquad + g_m{}^\dagger b_{n+m} + f_m{}^{\alpha\dagger} c_{n+m,\alpha}) \\ &+ \sum_{m=1}^{n-1} \sqrt{m(n-m)}(-\tfrac{1}{2} a_m \cdot a_{n-m} + b_m g_{n-m} + c_m{}^\alpha f_{n-m,\alpha}) \; . \end{aligned} \tag{8.3.37}$$

The Lorentz-gauge OSp(1,1|2) algebra can also be derived by the method of sect. 3.6. In the GL(1) case, we get the same algebra as above, while in the U(1) case we get a different result using the same coordinates, suggesting a similar first-quantization with the world-sheet metric. For the GL(1) case we define new coordinates at the GL(2|2) stage of reduction:

$$X^A = \tilde{x}^A + \widetilde{C}^A \quad , \quad P_A = \tilde{p}_A + \widetilde{C}'_A \quad , \tag{8.3.38}$$

where C is the generalization of C^α of (8.2.3) to arbitrary index. (Remember that in GL(2|2) a lower primed index can be converted into an upper unprimed index.) P_A is then canonically conjugate to X^A. We also have relations between the coordinates such as

$$\widetilde{P}_A = P_A \quad , \quad \widetilde{P}^A = \check{p}^A + X'^A \quad ,$$

$$\widetilde{X}^A = \breve{p}^A\sigma + X^A \quad , \tag{8.3.39}$$

where $\widehat{P}_A = (\widetilde{P}_A, \widetilde{P}_{A'})$, etc. However, $\widetilde{X}_A$ can be expressed in terms of P_A only with an inverse σ-derivative. As a result, when reexpressed in terms of these new coordinates, of all the IOSp(2,2|4) generators only the IGL(2|2) ones have useful expressions. Of the rest, $\breve{J}^{AB}$ is nonlocal in σ, while $\breve{J}_{AB}$ and $\breve{p}_A$ have explicit separate terms containing x^A and p_A. Explicitly, the local generators are p_A, $\breve{p}^A$, and

$$\breve{J}^A{}_B = i(-1)^{AB}\breve{x}_B\breve{p}^A - i\int X^A P_B \quad , \quad \breve{J}^{AB} = -ix^{[A}\breve{p}^{B)} - i\int X^A X'^B \quad . \tag{8.3.40}$$

In these expressions we also use the light-cone constraint and gauge condition as translated into the new coordinates:

$$P_- = -\frac{1}{2\breve{p}^-}\left[\widehat{P}_a{}^2 - 2 + 2(\breve{p}^+ + X'^+)P_+ + 2(\breve{p}^\alpha + X'^\alpha)P_\alpha\right] \quad , \quad X^- = 0 \quad . \tag{8.3.41}$$

After following the procedure of sect. 3.6 (or simply comparing expressions to determine the M_{ij}), the final $OSp(1,1|2)$ generators (3.6.7c) are (dropping the primes on the J's)

$$J_{+\alpha} = 2i\int X_\alpha P_+ \quad , \quad J_{-+} = -i\int (2X_-P_+ + X^\alpha P_\alpha) \quad , \quad J_{\alpha\beta} = -i\int X_{(\alpha}P_{\beta)} \quad ,$$

$$J_{-\alpha} = -i\int [X_-P_\alpha + X_\alpha(\tfrac{1}{2}\widehat{P}_a{}^2 - 1 + X'_-P_+ + X'^\beta P_\beta)] \quad . \tag{8.3.42}$$

This is just (8.3.3), with the identification $P_+ = G$ and $X^\alpha = C^\alpha$.

In the U(1) case there is no such redefinition possible (which gives expressions local in σ). The generators are

$$J_{+\alpha} = ix_\alpha p_+ + i\int \hat{X}_{\alpha'}\widehat{P}_{+'} \quad , \quad J_{-+} = -ix_-p_+ - i\int \hat{X}_{-'}\widehat{P}_{+'} \quad ,$$

$$J_{\alpha\beta} = -ix_{(\alpha}p_{\beta)} - i\int' \hat{X}_\alpha\widehat{P}_\beta - i\int \hat{X}_{\alpha'}\widehat{P}_{\beta'} \quad ,$$

$$\begin{aligned} J_{-\alpha} = & -ix_-p_\alpha + ix_\alpha\int \widehat{P}_- + i\int' \hat{X}_\alpha\widehat{P}_- \\ & -i\frac{1}{p_+}\left(p^a\int' \hat{X}_\alpha\widehat{P}_a - p^\beta\int' \hat{X}_\alpha\widehat{P}_\beta\right) - i\int \hat{X}_{-'}\widehat{P}_{\alpha'} \quad , \end{aligned}$$

$$-2p_+\widehat{P}_- = \widehat{P}_a{}^2 - 2 + \widehat{P}^\alpha\widehat{P}_\alpha + \widehat{P}^{\alpha'}\widehat{P}_{\alpha'} + 2\widehat{P}_{+'}\widehat{P}_{-'} \quad , \tag{8.3.43}$$

where $x_{A'} = p_{A'} = 0$. After performing the unitary transformations (3.6.13), they become

$$J_{+\alpha} = ix_\alpha p_+ \quad , \quad J_{-+} = -ix_- p_+ \quad , \quad J_{\alpha\beta} = -ix_{(\alpha}p_{\beta)} - i\int' \hat{X}_\alpha \widehat{P}_\beta - i\int \hat{X}_{\alpha'}\widehat{P}_{\beta'} \quad ,$$

$$\begin{aligned} J_{-\alpha} = & -ix_- p_\alpha + ix_\alpha \int \widehat{P}_- + i\int' \hat{X}_\alpha \widehat{P}_- \\ & - i\frac{1}{p_+}\Big[p^a \int' \hat{X}_\alpha \widehat{P}_a - p^\beta \left(\int' \hat{X}_\alpha \widehat{P}_\beta + \int \hat{X}_{\alpha'}\widehat{P}_{\beta'}\right) \\ & + \int \hat{X}_{-'}\widehat{P}_{\alpha'} + \tfrac{1}{2}\int \hat{X}_{+'}\widehat{P}_{\alpha'}\left(p_a{}^2 - 2p_+ \int' \widehat{P}_-\right)\Big] \quad . \end{aligned} \tag{8.3.44}$$

We can still interpret the new coordinates as the world-sheet metric, but different redefinitions would be necessary to obtain them from those in the mechanics action, and the gauge choice will probably differ with respect to the zero-modes.

Introducing extra coordinates has also been considered in [8.10], but in a way equivalent to using bosonized ghosts, and requiring imposing $D = 26$ by hand instead of deriving it from unitarity. Adding 4+4 extra dimensions to describe bosonic strings with enlarged BRST algebras $OSp(1,1|4)$ and $OSp(2,2|4)$ has been considered by Aoyama [8.11].

The use of such extra coordinates may also prove useful in the study of loop diagrams: In particular, harmonic-type gauges can be well-defined globally on the world sheet (unlike conformal gauges), and consequently certain parameters (the world-sheet generalization of proper-time parameters for the particle, (5.1.13)) appear automatically [8.12]. This suggests that such coordinates may be useful for closed string field theory (or superstring field theory, sect. 7.3).

The gauge-invariant action, its component analysis, and its comparison with that obtained from the other OSp(1,1|2) will be made in sect. 11.2.

Exercises

(1) Prove that the operators (8.1.9) satisfy commutation relations like (7.1.7a). Prove that they are conformally invariant.

(2) Derive (8.1.20). Verify that the usual fermionic anticommutation relations for (8.1.18) then follow from (8.1.19). Derive (8.1.22).

(3) Derive (8.1.15). Derive (8.1.23bc). Prove $Q^2 = 0$.

(4) Derive (8.1.23ac) from the energy-momentum tensors of (8.1.27,28ac).

(5) Find the commutation relations of $\frac{1}{2}\hat{D}d\hat{D}$ of (7.2.2), generalizing (8.1.2). Find the BRST operator. Derive the gauge-fixed Virasoro operators, and show the conformal weights of $\hat{\Psi}$ is $1/2$, and of its ghosts are $-1/2$ and $+3/2$. Use (8.1.23) to show the anomaly cancels for $D = 10$.

(6) Find an alternate first-order form of (8.1.28b) by rewriting the last term in terms of F and ω, and show how (8.1.28ac) follow.

(7) Show from the explicit expression for ω_a that (8.1.27) can be made vielbein-independent by field redefinition in the conformal gauge.

(8) Explicitly prove the equivalence of the IGL(1)'s derived in sect. 8.1 and from the light cone, using the analysis of sect. 8.2.

(9) Derive (8.3.1,3). Derive (8.3.9).

(10) Derive (8.3.28).

9. GRAPHS

9.1. External fields

One way to derive Feynman graphs is by considering a propagator in an external field:

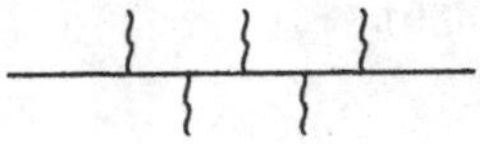

For example, for a scalar particle in an external scalar field,

$$L = \tfrac{1}{2}\dot{x}^2 - m^2 - \phi(x) \quad \rightarrow \tag{9.1.1a}$$

$$[p^2 + m^2 + \phi(x)]\psi(x) = 0 \quad \rightarrow \tag{9.1.1b}$$

$$\textit{propagator} \qquad \frac{1}{p^2 + m^2 + \phi(x)} = \frac{1}{p^2 + m^2} - \frac{1}{p^2 + m^2}\phi\frac{1}{p^2 + m^2} + \cdots \tag{9.1.1c}$$

For the (open, bosonic) string, it's useful to use the "one-handed" version of $X(\sigma)$ (as in (7.1.7b)) so that $\hat{X}$ and $\widehat{P}$ can be treated on an (almost) equal footing, so we will switch to that notation hand and foot. Then the generalization of (9.1.1b) (again jumping directly to the first-quantized form for convenience) is [7.5]

$$\left\langle \chi \middle| [\tfrac{1}{2}\widehat{P}^2(\sigma) - 1 + \mathcal{V}(\sigma)] \middle| \psi \right\rangle = 0 \quad . \tag{9.1.2}$$

(The OSp(1,1|2) algebra can be similarly generalized.) However, for consistency of these equations of motion, they must satisfy the same algebra as $\frac{1}{2}\widehat{P}^2 - 1$ ((8.1.2)). (In general, this is only true including ghost contributions, which we will ignore for the examples considered here.) Expanding this condition order-by-order in $\mathcal{V}$, we get the new relations

$$
\begin{aligned}
[\tfrac{1}{2}\hat{P}^2(\sigma_1) - 1, \mathcal{V}(\sigma_2)] &= 2\pi i \delta'(\sigma_2 - \sigma_1)\mathcal{V}(\sigma_1) \\
&= 2\pi i \left[\delta(\sigma_2 - \sigma_1)\mathcal{V}'(\sigma_2) + \delta'(\sigma_2 - \sigma_1)\mathcal{V}(\sigma_2)\right] , \quad (9.1.3a) \\
[\mathcal{V}(\sigma_1), \mathcal{V}(\sigma_2)] &= 0 \quad . \quad (9.1.3b)
\end{aligned}
$$

The first condition gives the conformal transformation properties of $\mathcal{V}$ (it transforms covariantly with conformal weight 1, like $\hat{P}$), and the second condition is one of "locality". A simple example of such a vertex is a photon field coupled to one end of the string:

$$
\mathcal{V}(\sigma) = 2\pi\delta(\sigma)A(X(0)) \cdot \hat{P} \quad , \quad (9.1.4)
$$

where $X(0) = \hat{X}(0)$. Graphs are now given by expanding the propagator as the inverse of the hamiltonian

$$
H = \int \frac{d\sigma}{2\pi}\left(\tfrac{1}{2}\hat{P}^2 - 1 + \mathcal{V}\right) = \mathbf{L}_0 + \int d\sigma\, \mathcal{V} \equiv \mathbf{L}_0 + V \quad . \quad (9.1.5)
$$

More general vertices can be found when normal ordering is carefully taken into account [1.3,4], and one finds that (9.1.3a) can be satisfied when the external field is on shell. For example, consider the scalar vertex

$$
\mathcal{V}(\sigma) = -2\pi\delta(\sigma)\phi(X(0)) \quad . \quad (9.1.6)
$$

Classically, scalar fields have the wrong conformal weight (zero):

$$
\frac{1}{2\pi}\left[\tfrac{1}{2}\hat{P}^2(\sigma_1) - 1, \phi\left(\hat{X}(\sigma_2)\right)\right] = i\delta(\sigma_2 - \sigma_1)\frac{\partial}{\partial\sigma_2}\phi\left(\hat{X}(\sigma_2)\right) \quad ; \quad (9.1.7a)
$$

but quantum mechanically they have weight "$-\frac{1}{2}\Box$":

$$
\begin{aligned}
\frac{1}{2\pi}\left[\tfrac{1}{2}\hat{P}^2(\sigma_1) - 1, \phi\left(\hat{X}(\sigma_2)\right)\right] &= i\delta(\sigma_2 - \sigma_1)\frac{\partial}{\partial\sigma_2}\phi\left(\hat{X}(\sigma_2)\right) \\
&+ \left[-\frac{1}{2}\frac{\partial^2}{\partial\hat{X}^2(\sigma_2)}\right] i\delta'(\sigma_2 - \sigma_1)\phi\left(\hat{X}(\sigma_2)\right) \quad .
\end{aligned}
\quad (9.1.7b)
$$

Therefore $\phi\left(\hat{X}(\sigma_2)\right)$, and thus $\mathcal{V}$ of (9.1.6), satisfies (9.1.3a) if ϕ is the on-shell ground state (tachyon): $-\frac{1}{2}\Box\phi = 1 \cdot \phi$. (Similar remarks apply quantum mechanically for the masslessness of the photon in the vertex (9.1.4).)

As an example of an S-matrix calculation, consider a string in an external plane-wave tachyon field, where the initial and final states of the string are also tachyons:

$$\phi(x) = e^{-ik\cdot x} \quad \to \quad : e^{-ik\cdot X(0)} : \tag{9.1.8}$$

We then find

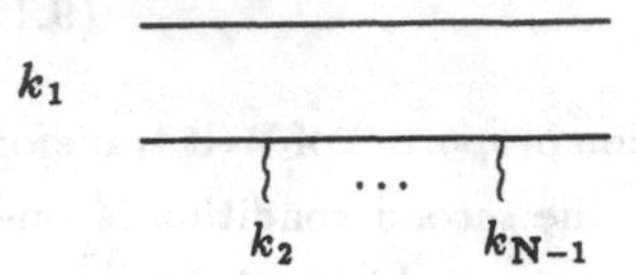

$$= g^{N-2} \langle k_N | V(k_{N-1}) \cdots \Delta(p) V(k_3) \Delta(p) V(k_2) | k_1 \rangle$$

$$= g^{N-2} \left\langle 0 \left| \widetilde{V}(k_{N-1}) \cdots \Delta(k_3 + k_4 + \cdots + k_N) \widetilde{V}(k_2) \right| 0 \right\rangle \quad ,$$

$$V(k) = : e^{-ik\cdot X(0)} : = \widetilde{V}(k) e^{-ik\cdot x} \quad , \quad X(0) = \sum_1^\infty \frac{1}{\sqrt{n}} (a_n{}^\dagger + a_n) \quad ,$$

$$\Delta(p) = \frac{1}{\frac{1}{2}p^2 + (N-1)} \quad , \quad N = \sum_1^\infty n a_n{}^\dagger \cdot a_n \quad , \tag{9.1.9}$$

where g is the coupling constant, and we have pulled the x pieces out of the X's and pushed them to the right, causing shifts in the arguments of the Δ's (which were originally p, the momentum operator conjugate to x, not to be confused with the constants k_i). We use Schwinger-like parametrizations (5.1.13) for the propagators:

$$\frac{1}{\frac{1}{2}p^2 + (N-1)} = \int_0^\infty dt\ e^{-t[\frac{1}{2}p^2 + (N-1)]} = \int_0^1 \frac{dx}{x}\ x^{\frac{1}{2}p^2 + (N-1)} \quad (x = e^{-t}) \quad , \tag{9.1.10}$$

where we use t_i for $\Delta(k_i + \cdots + k_N)$, as the difference in proper time between $\widetilde{V}(k_{i-1})$ and $\widetilde{V}(k_i)$. Plugging (9.1.10) into (9.1.9), the amplitude is

$$g^{N-2} \left(\prod_{i=3}^{N-1} \int_0^1 \frac{dx_i}{x_i}\ x_i{}^{\frac{1}{2}(k_i + \cdots + k_N)^2 - 1} \right) \prod_n \left\langle 0 \left| \cdots x_3{}^{na^\dagger \cdot a} e^{-ik_2 \cdot \frac{1}{\sqrt{n}} a^\dagger} \right| 0 \right\rangle \quad . \tag{9.1.11}$$

To evaluate matrix-elements of harmonic oscillators it's generally convenient to use coherent states:

$$|z\rangle \equiv e^{za^\dagger} |0\rangle \quad \to$$

$$a|z\rangle = z|z\rangle \quad , \quad a^\dagger \Big| z \Big\rangle = \frac{\partial}{\partial z} \Big| z \Big\rangle \quad , \quad e^{z'a^\dagger} |z\rangle = |z + z'\rangle \quad , \quad x^{a^\dagger a} |z\rangle = |xz\rangle \quad ,$$

$$\langle z | z' \rangle = e^{\bar{z}z'} \quad , \quad 1 = \int \frac{d^2 z}{\pi}\ e^{-|z|^2} |z\rangle \langle z| \quad ,$$

$$tr(\mathcal{O}) = \int \frac{d^2 z}{\pi}\, e^{-|z|^2} \langle z|\mathcal{O}|z\rangle \quad . \tag{9.1.12}$$

Using (9.1.12) and the identity $\prod_1^\infty e^{-cx^n/n} = (1-x)^c$, (9.1.11) becomes

$$g^{\mathrm{N}-2}\left(\prod \int \frac{dx_i}{x_i} x_i^{\frac{1}{2}(k_i+\cdots+k_{\mathrm{N}})^2-1}\right) \prod_{2\le i<j\le \mathrm{N}-1} \left(1 - \prod_{k=i+1}^{j} x_k\right)^{k_i\cdot k_j} \quad . \tag{9.1.13}$$

We next make the change of variables

$$\tau_i = \sum_{j=3}^{i} t_j \quad , \quad \tau_2 = 0 \quad , \quad \tau_{\mathrm{N}} = \infty \quad \to \quad 0 = \tau_2 \le \tau_3 \le \cdots \le \tau_{\mathrm{N}} = \infty \quad , \tag{9.1.14a}$$

or

$$z_i = \prod_{j=3}^{i} x_j \quad , \quad z_2 = 1 \quad , \quad z_{\mathrm{N}} = 0 \quad \to \quad 1 = z_2 \ge z_3 \ge \cdots \ge z_{\mathrm{N}} = 0 \quad , \tag{9.1.14b}$$

with

$$z_i = e^{-\tau_i} \quad , \tag{9.1.14c}$$

where τ_i is the absolute proper time of the corresponding vertex. Using the mass-shell condition $k_i{}^2 = 2$, the final result is then [9.1]

$$g^{\mathrm{N}-2}\left(\int \prod_{i=3}^{\mathrm{N}-1} dz_i\right) \prod_{2\le i<j\le \mathrm{N}} (z_i - z_j)^{k_i\cdot k_j} \quad . \tag{9.1.15}$$

The simplest case is the 4-point function (N= 4) [9.2]

$$g^2 \int_0^1 dz\, z^{-\frac{1}{2}s-2}(1-z)^{-\frac{1}{2}t-2} \qquad (s = -(k_1+k_2)^2 \quad , \quad t = -(k_2+k_3)^2)$$

$$= g^2 B(-\tfrac{1}{2}s - 1, -\tfrac{1}{2}t - 1) = g^2 \frac{\Gamma(-\frac{1}{2}s-1)\Gamma(-\frac{1}{2}t-1)}{\Gamma(-\frac{1}{2}s-\frac{1}{2}t-2)} \tag{9.1.16a}$$

$$= g^2 \sum_{j=0}^{\infty} \left[\frac{(\frac{1}{2}t+1+j)(\frac{1}{2}t+j)\cdots(\frac{1}{2}t+1)}{j!}\right] \frac{1}{j-(\frac{1}{2}s+1)} \tag{9.1.16b}$$

$$= \left(in \lim_{\substack{s\to-\infty \\ t\ fixed}}\right) \quad g^2\Gamma(-\tfrac{1}{2}s-1)(-\tfrac{1}{2}s-1)^{\frac{1}{2}t+1} \quad . \tag{9.1.16c}$$

(9.1.16b) shows that the amplitude can be expressed as a sum of poles in the s channel with squared masses $2(j-1)$, with maximum spin j (represented by the coefficient with leading term t^j). Since the amplitude is symmetric in s and t, it can also be expressed as a sum of poles in the t channel, and thus summing over poles in one channel generates poles in the other. (It's not necessary to sum over both.) This property is called "duality". (9.1.16c) shows that the high-energy behavior goes like $s^{\frac{1}{2}t+1}$ instead of the usual fixed-power behavior s^j due to the exchange of a spin j particle, which can be interpreted as the exchange of a particle with effective spin $j = \frac{1}{2}t + 1$. This property is known as "Regge behavior", and $j(t) = \frac{1}{2}t + 1$ is called the "leading Regge trajectory", which not only describes the high-energy behavior but also the (highest) spin at any given mass level (the mass levels being given by integral values of $j(t)$).

Instead of using operators to evaluate propagators in the presence of external fields, we can also use the other approach to quantum mechanics, the Feynman path integral formalism. In particular, for the calculation of purely tachyonic amplitudes considered above, we evaluate (9.1.9) directly in terms of V (rather than $\widetilde{V}$), after using (9.1.10):

$$g^{\mathbf{N}-2} \int_0^\infty dt_3 \cdots dt_{\mathbf{N}-1} \left\langle k_{\mathbf{N}} \left| V(k_{\mathbf{N}-1}) \cdots e^{-t_3(\frac{1}{2}p^2+N-1)} V(k_2) \right| k_1 \right\rangle \quad . \tag{9.1.17}$$

Using (9.1.14a), we can rewrite this as

$$g^{\mathbf{N}-2} \left(\prod_{i=3}^{\mathbf{N}-1} \int_{\tau_{i-1}}^{\tau_{i+1}} d\tau_i \right) \langle k_{\mathbf{N}} | \, V(k_{\mathbf{N}-1}, \tau_{\mathbf{N}-1}) \cdots V(k_3, \tau_3) V(k_2, \tau_2) \, | k_1 \rangle \quad , \tag{9.1.18a}$$

where

$$V(k,\tau) = \, : e^{-ik\cdot X(0,\tau)} : \quad , \quad X(0,\tau) = e^{\tau(\frac{1}{2}p^2+N-1)} X(0) e^{-\tau(\frac{1}{2}p^2+N-1)} \quad , \tag{9.1.18b}$$

is the vertex which has been (proper-)time-translated from 0 to τ. (Remember that in the Heisenberg picture operators have time dependence $\mathcal{O}(t) = e^{tH}\mathcal{O}(0)e^{-tH}$, whereas in the Schrödinger picture states have time dependence $|\psi(t)\rangle = e^{-tH} |\psi(0)\rangle$, so that time-dependent matrix elements are the same in either picture. This is equivalent to the relation between first- and second-quantized operators.) External states can also be represented in terms of vertices:

$$|k\rangle = \lim_{\tau \to -\infty} V(k,\tau) e^{\tau} |0\rangle \quad , \quad \langle k| = \lim_{\tau \to \infty} \langle 0| \, e^{-\tau} V(k,\tau) \quad . \tag{9.1.19}$$

The amplitude can then be represented as, using (9.1.14c),

$$g^{\mathbf{N}-2} \left(\prod_{i=3}^{\mathbf{N}-1} \int_{z_{i-1}}^{z_{i+1}} dz_i \right) \lim_{\substack{z_1 \to \infty \\ z_{\mathbf{N}} \to 0}} \frac{z_1}{z_{\mathbf{N}}} \langle 0| \, V'(k_{\mathbf{N}}, z_{\mathbf{N}}) \cdots V'(k_1, z_1) \, |0\rangle \quad , \tag{9.1.20a}$$

where

$$V'(k,z) = \left(-\frac{1}{z}\right) V(k,\tau(z)) \tag{9.1.20b}$$

according to (8.1.6), since vertices have weight $w = 1$. The amplitude with this form of the external lines can be evaluated by the same method as the previous calculation. (In fact, it directly corresponds to the calculation with 2 extra external lines and vanishing initial and final momenta.) However, being a vacuum matrix element, it is of the same form as those for which path integrals are commonly used in field theory. (Equivalently, it can also be evaluated by the operator methods commonly used in field theory before path integral methods became more popular there.) More details will be given in the following section, where such methods will be generalized to arbitrary external states.

Coupling the superstring to external super-Yang-Mills is analogous to the bosonic string and superparticle [2.6]: Covariantize $D_\alpha \to D_\alpha + \delta(\sigma)\Gamma_\alpha$, $P_a \to P_a + \delta(\sigma)\Gamma_a$, $\Omega^\alpha \to \Omega^\alpha + \delta(\sigma)W^\alpha$. Assuming $\int d\sigma\ \mathcal{A}$ as kinetic operator (again ignoring ghosts), the vertex becomes

$$V = W^\alpha D_\alpha + \Gamma^a P_a - \Gamma_\alpha \Omega^\alpha \tag{9.1.21}$$

evaluated at $\sigma = 0$. Solving the constraints (5.4.8) in a Wess-Zumino gauge, we find

$$\begin{aligned} W^\alpha &\approx \lambda^\alpha \quad , \\ \Gamma^a &\approx A^a + 2\gamma^a{}_{\alpha\beta}\Theta^\alpha\lambda^\beta \quad , \\ \Gamma_\alpha &\approx \gamma^a{}_{\alpha\beta}\Theta^\beta A_a + \tfrac{4}{3}\gamma^a{}_{\alpha\beta}\gamma_{a\gamma\delta}\Theta^\beta\Theta^\gamma\lambda^\delta \quad , \end{aligned} \tag{9.1.22}$$

evaluated at $\sigma = 0$, where "$\approx$" means dropping terms involving x-derivatives of the physical fields A_a and λ^α. Plugging (7.3.5) and (9.1.22) into (9.1.21) gives

$$V \approx A^a \hat{P}_a + \lambda^\alpha \left(\frac{\delta}{\delta\Theta^\alpha} - \gamma^a{}_{\alpha\beta}\hat{P}_a\Theta^\beta - \tfrac{1}{6}i\gamma^a{}_{\alpha\beta}\gamma_{a\gamma\delta}\Theta^\beta\Theta^\gamma\Theta'^\delta \right) \quad . \tag{9.1.23}$$

Comparing with (7.3.6), we see that the vertices, in this approximation, are the same as the integrands of the supersymmetry generators, evaluated at $\sigma = 0$. (In the case of ordinary field theory, the vertices *are* just the supersymmetry generators p_a and q_α, to this order in θ.) Exact expressions can be obtained by expansion of the superfields Γ_α, Γ_a, and W^α in (9.1.21) to all orders in Θ [7.6]. In practice, superfield techniques should be used even in the external field approach, so such explicit expansion (or even (9.1.22) and (9.1.23)) is unnecessary. It's interesting to note that, if we generalize D, P, and Ω to gauge-covariant derivatives $\nabla_\alpha = D_\alpha + \Gamma_\alpha$,

$\nabla_a = D_a + \Gamma_a$, $\nabla^\alpha = \Omega^\alpha + W^\alpha$, with Γ_α, Γ_a, and W^α now functions of σ, describing the vector multiplets of all masses, then the fact that the only mode of the ∇'s missing is the zero-mode of Ω^α ($\int d\sigma\ \Omega^\alpha = 0$) directly corresponds to the fact that the only gauge-invariant mode of the connections is the zero-mode of W^α (the massless spinor, the massive spinors being Stueckelberg fields).

The external field approach has also been used in the string mechanics lagrangian method to derive field theory lagrangians (rather than just S-matrices) for the lower mass levels (tachyons and massless particles) [9.3,1.16]. Since arbitrary external fields contain arbitrary functions of the coordinates, the string mechanics lagrangian is no longer free, and loop corrections give the field theory lagrangian including effective terms corresponding to eliminating the higher-mass fields by their classical field equations. Thus, calculating all mechanics-loop corrections gives an effective field theory lagrangian whose S-matrix elements are the tree graphs of the string field theory with external lines corresponding to the lower mass levels. Such effective lagrangians are useful for studying tree-level spontaneous breakdown due to these lower-mass fields (vacua where these fields are nontrivial). Field-theory-loop corrections can be calculated by considering more general topologies for the string (mechanics-loops are summed for one given topology).

9.2. Trees

The external field approach is limited by the fact that it treats ordinary fields individually instead of treating the string field as a whole. In order to treat general string fields, a string graph can be treated as just a propagator with funny topology: For example,

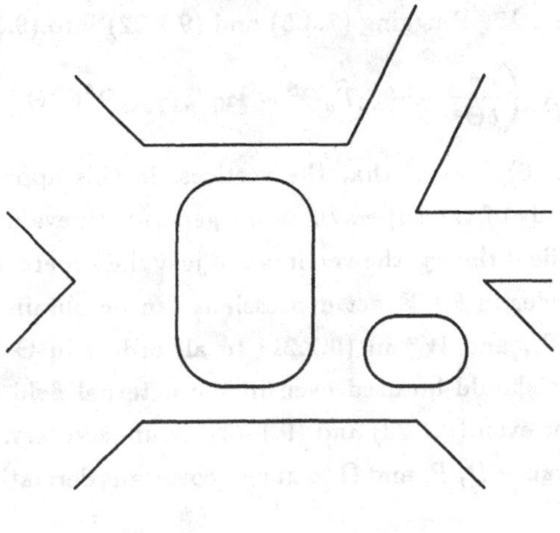

can be considered as a propagator where the initial and final "one-string" states just happen to be disconnected. The holes in the world sheet represent loops. When group theory indices are associated with the ends of the lines, the values of the indices are required to be the same along the entire line, which corresponds to tracing in the matrices associated with the string states. (The ends of the strings can be interpreted as "quarks" which carry the "flavor" quantum numbers, bound by a string of "gluons" which carry only "color" canceled by that of the quarks.) Such an approach is limited to perturbation theory, since the string is necessarily gauge-fixed, and any one graph has a fixed number of external lines and loops, i.e., a fixed topology. The advantage of this approach to graphs is that they can be evaluated by first-quantization, analogously to the free theory. (Even the second-quantized coupling constant can be included in the first-quantized formalism by noting that the power of the coupling constant which appears in a graph, up to wave function normalizations, is just the Euler number, and then adding the corresponding curvature integral to the mechanics action.)

We first consider the light-cone formalism. We Wick rotate the proper time $\tau \to i\tau$ (see sects. 2.5-6), so now conformal transformations are arbitrary reparametrizations of $\rho = -\tau + i\sigma$ (and the complex conjugate transformation on $\bar{\rho}$) instead of $\tau + \sigma$ (and of $\tau - \sigma$ independently: see (6.2.7)), since the metric is now $d\sigma^2 = d\rho d\bar{\rho}$. There are three parts to the graph calculation: (1) expressing the S-matrix in terms of the Green function for the 2D Laplace equation, (2) finding an explicit expression for the Green function for the 2D surface for that particular graph, by conformally transforming the ρ plane to the upper-half complex (z) plane where the Green function takes a simple form, and (3) finding the measure for the integration over the positions of the interaction points.

The first step is easy, and can be done using functional integration [9.4,1.4] or solving functional differential equations (the string analog of Feynman path integrals and the Schrödinger equation, respectively). Since all but the zero-mode (the usual spacetime coordinate) of the free string is described by an infinite set of harmonic oscillators, the most convenient basis is the "number" basis, where the nonzero-modes are represented in terms of creation and annihilation operators. The basic idea is then to represent S-matrix elements as

$$A = \langle ext|V\rangle = \langle ext|\, e^{\Delta}\, |0\rangle \quad , \tag{9.2.1}$$

where $|V\rangle$ represents the interaction and $\langle ext|$ represents all the states (initial and final) of the external strings, in the interaction picture. This is sort of a spacetime symmetric version of the usual picture, where an initial state propagates into a final state: Instead, the vacuum propagates into an "external" state. The exponential e^{Δ} is then the analog of the S-matrix $exp(-H_{INT}t)$, which propagates the vacuum at time 0 to external states at time t. It thus converts annihilation operators on its left (external, "out" states) into creation operators (for the "in" state, the vacuum, at "time" $x_+ = 0$). Δ itself is then the "connected" S-matrix: In this first-quantized picture, which looks like a free 2D theory in a space with funny geometry, it corresponds directly to the free propagator in this space. Since we work in the interaction picture, we subtract out terms corresponding to propagation in an "ordinary" geometry.

In the former approach, the amplitude can be evaluated as the Feynman path integral

$$A = \int \left(\prod_{i=3}^{N-1} d\tau_i \right) \int \mathcal{D}X^i(\sigma,\tau)$$
$$\cdot \left[\prod_r \int \mathcal{D}P_r(\sigma)\ \Psi[P_r] e^{-i\frac{1}{\alpha'}\int \frac{d\sigma}{2\pi} P_r(\sigma)\cdot X_r(\sigma,\tau_1)} \right]$$
$$\cdot\, e^{-\sum p_{-r}\tau_{1r} - \frac{1}{\alpha'}\int \frac{d^2\sigma}{2\pi}\left[\frac{1}{2}(\dot{X}^2 + X'^2) + constant\right]} \quad , \qquad (9.2.2)$$

corresponding to the picture (e.g., for N= 5)

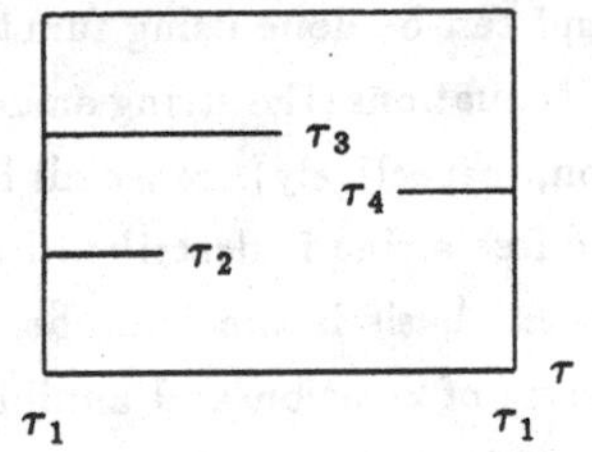

where the τ's have been Wick-rotated, τ_{1r} is the end of the rth string (to be taken to $\pm\infty$), the $p_-\tau_1$ factor amputates external lines (converts from the Schrödinger picture to the interaction picture), the factor in large brackets is the external-line wave function $\Psi[X]$ (or $\bar{\Psi}[X]$ for outgoing states) written as a Fourier transform, and the constant corresponds to the usual normal-ordering constant in the free hamiltonian. For explicitness, we have written the integration over interaction points ($\prod d\tau$) for the simple case of open-string tree graphs. Planar graphs always appear as rectangles due to the string lengths being $2\pi\alpha' p_+$, which is conserved. The functional integral (9.2.2) is gaussian, so, making the definition

$$J(\sigma,\tau) = i\delta(\tau-\tau_1)\frac{1}{\sqrt{\alpha'}}P(\sigma) \quad , \tag{9.2.3}$$

we find

$$-\int \frac{d^2\sigma}{2\pi}\left[\frac{1}{\alpha'}\tfrac{1}{2}(\partial X)^2 + \frac{1}{\sqrt{\alpha'}}JX\right] \quad \rightarrow \quad -\tfrac{1}{2}\int \frac{d^2\sigma}{2\pi}\,\frac{d^2\sigma'}{2\pi}\,J(\sigma)G(\sigma,\sigma')J(\sigma') \quad ,$$

$$\partial^2 G(\sigma,\sigma') = 2\pi\delta^2(\sigma'-\sigma) \quad , \quad \frac{\partial}{\partial n}G(\sigma,\sigma') = f(\sigma) \quad \left(\int d^2\sigma\, J \sim \sum p = 0\right) \quad ,$$

$$G = \sum(2-\delta_{m0})(2-\delta_{n0})G^{rs}{}_{mn}\, cos\left(m\frac{\sigma_r}{p_{+r}}\right) cos\left(n\frac{\sigma_s{}'}{p_{+s}}\right) e^{m\tau_r/p_{+r}+n\tau_s{}'/p_{+s}}$$
$$+\, G_{free} + (zero\text{-}mode)^2 \ terms$$

$$\rightarrow \quad A = \int (\prod d\tau_i)V(\tau)\left\langle \Psi \left| e^{\Delta} \right| 0\right\rangle \quad , \quad \Delta = \tfrac{1}{4}\sum G^{rs}{}_{mn}\alpha^r{}_m\cdot\alpha^s{}_n \quad , \tag{9.2.4}$$

where G is the 2D Green function for the kinetic operator (laplacian) $\partial^2/\partial\tau^2 + \partial^2/\partial\sigma^2$ of that particular surface, and $V(\tau)$ comes from $det(G)$. We have used Neumann boundary conditions (corresponding to (6.1.5)), where the ambiguity contained in the arbitrary function f (necessary in general to allow a solution) is harmless because of the conservation of the current J (i.e., the momentum p). The G_{free} term is dropped in converting to the interaction picture: In functional notation (see (9.2.2)), it produces the ground-state wave function. The (zero-mode)^2 terms are due to boundary conditions at ∞, and appear when the map to the upper-half plane is chosen so that the end of one string goes to ∞, giving a divergence. They correspond to the factor $1/z_N$ in the similar map used for (9.1.20a). The factors $(2-\delta_{m0})$, which don't appear in the naive Green function, are to correct for the fact that the figure above is not quite the correct one: The boundaries of the initial and final strings should not go to $\pm\infty$ before the source terms (9.2.3) (i.e., the wave functions) do, because of the boundary conditions. The net result is that nonzero-modes appear with an extra factor of 2 due to reflections from the boundary. However, these relative factors of 2 are canceled in the σ-integration, since $\int_0^\pi \frac{d\sigma}{2\pi} cos^2 m\sigma = \frac{1}{4}(1+\delta_{m0})$. Explicitly, after transforming the part of the ρ

plane corresponding to the string to the whole of the upper-half z plane, the Green function becomes

$$G(z,z') = ln|z-z'| + ln|z-\bar{z}'| \quad . \tag{9.2.5}$$

The first term is the usual Green function without boundaries, whose use in the z plane (and not just the ρ plane) follows from the fact that the Laplace equation is conformally invariant. The second term, which satisfies the homogeneous Laplace equation in the upper-half plane, has been added according to the method of images in order to satisfy the boundary conditions at the real axis, and gives the reflections which contribute the factors of 2.

In the latter (Schrödinger equation/operator) approach, it all boils down to using the general expression

$$\check{\Psi}(z) = \oint_z \frac{dz'}{2\pi i}\frac{1}{z'-z}\check{\Psi}(z') = -\sum_r \oint_{z_r} \frac{dz'}{2\pi i}\frac{1}{z'-z}\check{\Psi}(z') \quad , \tag{9.2.6}$$

where z_r are the points in the z-plane representing the ends of the strings (at $\rho = \pm\infty$). $\check{\Psi}(z)$ is an arbitrary operator which has been conformally transformed to the z plane:

$$\check{\Psi}_r(z) = \left(\frac{\partial\rho}{\partial z}\right)^w \tilde{\Psi}_r(\rho) \quad , \quad \tilde{\Psi}_r(\rho) = (p_{+r})^{-w}\Psi_r\left(\frac{\tilde{\rho}_r}{p_{+r}}\right) \quad , \quad \Psi_r(\zeta) = \sum_{-\infty}^{\infty} \psi_{rn}e^{-n\zeta} \quad , \tag{9.2.7a}$$

$$\tilde{\rho}_r = \rho - i\pi\sum_{s=1}^{r-1} p_{+s} \quad , \tag{9.2.7b}$$

with $\Psi(i\sigma) = \hat{\Psi}(\sigma)$ in terms of $\widehat{P}(\sigma)$ (so the ψ_n's are the α_n's of (7.1.7a)), and the conformal transformations (9.2.7a) (cf. (8.1.6)) are determined by the conformal weights w ($= 1$ for $\check{P}$). The $\rho \to z$ map is the map from the above figure to the upper-half plane. The $\zeta \to \rho$ map is the map from the free-string $\sigma \in [0,\pi]$ to the rth interacting-string $\sigma \in [\pi\sum_{s=1}^{r-1} p_{+s}, \pi\sum_{s=1}^{r} p_{+s}]$. All σ integrals from $-\pi$ to π become contour integrals in the z plane. (The upper-half z plane corresponds to $\sigma \in [0,\pi]$, while the lower half is $\sigma \in [-\pi, 0]$.) Since the string (including its extension to $[-\pi, 0]$) is mapped to the entire z plane without boundaries, (9.2.6) gets contributions from only the ends, represented by z_r. We work directly with $\widehat{P}(\sigma)$, rather than $X(\sigma)$, since $\widehat{P}$ depends only on ρ, while X depends on both ρ and $\bar{\rho}$. ($\hat{X}$ has a cut in the z plane, since $p\sigma$ isn't periodic in σ.) The open string results can also be applied directly to the closed string, which has separate operators which depend on $\bar{z}$ instead of z (i.e., $+$ and $-$ modes, in the notation of sect. 6.2). Actually, there is a bit of a cheat, since $\widehat{P}(\sigma)$ doesn't contain the zero-mode x. However, this zero-mode needs special care in any method: Extra factors of 2 appeared in the path-integral approach because of the boundary conditions along the real z-axis and at infinity. In fact, we'll see that the lost zero-mode terms can be found from the same calculation generally used in both operator and path integral approaches to find the integration measure of step 3 above, and thus requires no extra effort.

We therefore look directly for a propagator e^{Δ} that gives

$$\check{\Psi}(z) = e^{\Delta}\check{\Psi}_r(z)e^{-\Delta} = \check{\Psi}_r(z) + [\Delta, \check{\Psi}_r(z)] + \cdots \quad , \tag{9.2.8}$$

where $\check{\Psi}_r$ corresponds to a free in-field for the rth string in the interaction picture, and $\check{\Psi}$ to the interacting field. To do this, we first find a $\tilde{\Delta}$ for which

$$[\tilde{\Delta}, \check{\Psi}_r] = \check{\Psi} \quad . \tag{9.2.9}$$

We next subtract out the free part of $\tilde{\Delta}$ (external-line amputation):

$$\tilde{\Delta} = \Delta + \Delta_{free} \quad , \quad [\Delta_{free}, \check{\Psi}_r] = \check{\Psi}_r \quad . \tag{9.2.10}$$

This gives a Δ which is quadratic in operators, but contains no annihilation operators (which are irrelevant anyway, since the $|0\rangle$ will kill them). As a result, there are no terms with multiple commutators in the expansion of the exponential. We therefore obtain (9.2.8). When we subtract out free parts below, we will include the parts of the external-line amputation which compensate for the fact that z and z_r are not at the same time.

We first consider applying this method to operators of arbitrary conformal weight, as in (8.1.23). The desired form of Δ which gives (9.2.8) for both f and $\delta/\delta f$ is

$$\Delta_0 = -\sum_{r,s} \oint_{z_r} \frac{dz}{2\pi i} \oint_{z_s} \frac{dz'}{2\pi i} \frac{1}{z-z'} \check{f}_r(z) \frac{\delta}{\delta \check{f}_s(z')} - \mathit{free\text{-}string\ terms} \quad , \tag{9.2.11a}$$

where

$$\left[\frac{\delta}{\delta \check{f}_r(z_1)}, \check{f}_s(z_2)\right\} = 2\pi i \delta_{rs}\delta(z_2 - z_1) \quad \leftrightarrow \quad \left[\frac{\delta}{\delta \hat{f}_r(\sigma_1)}, \hat{f}_s(\sigma_2)\right\} = 2\pi\delta_{rs}\delta(\sigma_2 - \sigma_1) \quad , \tag{9.2.11b}$$

as follows from the fact that the conformal transformations preserve the commutation relations of f and $\delta/\delta f$. The integration contours are oriented so that

$$\oint_{\rho_r} \frac{d\rho}{2\pi i p_{+r}} = \int_{-\pi p_{+r}}^{\pi p_{+r}} \frac{d\sigma}{2\pi p_{+r}} = 1 \quad . \tag{9.2.12}$$

(9.2.11) can easily be shown to satisfy (9.2.8). The value of τ_r ($\to \pm\infty$) for the integration contour is fixed, so the δ in ζ in these commutation relations is really just a δ in σ of that contour. The free-string terms are subtracted as explained above. (In fact, they are poorly defined, since the integration contours for $r = s$ fall on top of each other.)

Unfortunately, Δ will prove difficult to evaluate in this form. We therefore rewrite it by expressing the $1/(z - z')$ as the derivative with respect to either z or z' of a ln and perform an integration by parts. The net result can be written as

$$\Delta(\check{\Psi}_1, \check{\Psi}_2) = \sum_{r,s} \left(\oint_{z_r} \frac{dz}{2\pi i} \oint_{z_s} \frac{dz'}{2\pi i} \right)' ln(z - z') \check{\Psi}_{1r}(z) \check{\Psi}_{2s}(z')$$
$$- \textit{free-string terms} \quad + (\textit{zero-mode})^2 \textit{ terms} \ , \qquad (9.2.13a)$$

$$[\check{\Psi}_{2r}(z_2) \ , \ \check{\Psi}_{1s}(z_1)\} = -2\pi i \delta_{rs} \delta'(z_2 - z_1) \quad , \qquad (9.2.13b)$$

The $'$ on the contour integration is because the integration is poorly defined due to the cut for the ln: We therefore define it by integration by parts with respect to either z or z', dropping surface terms. This also kills the constant part of the ln which contributes the (zero-mode)^2 terms, which we therefore add back in. Actually, (9.2.11) has no (zero-mode)^2 terms, but in the case $\Psi_1 = \Psi_2 = P$, these terms determine the evolution of the zero-mode x, which doesn't appear in $\widehat{P}$, and thus could not be determined by (9.2.8) anyway. (x does appear in X and $\hat{X}$, but they're less convenient to work with, as explained above.) These (quadratic-in-)zero-mode contributions are most easily calculated separately by considering the case when all external states are ground states of nonvanishing momentum (see below). In order for the commutation relations (9.2.13b) to be preserved by the conformal transformations, it's necessary that the conformal weights w_1 and w_2 of Ψ_1 and Ψ_2 both be 1. In that case, $\check{\Psi}$ can be replaced with $\tilde{\Psi}$ in (9.2.13a) while replacing dz with $d\rho$. However, one important use of this equation is for the evaluation of vertices (S-matrices with no internal propagators). Since these vertices are just δ functionals in the string field coordinates (see below), and δ functionals are independent of conformal weight except for the $\zeta \to \rho$ transformation (since that transformation appears explicitly in the argument of the δ functionals), we can write this result, for the cases of S-matrices with $w_1 = w_2 = 1$ or vertices (with $w_1 + w_2 = 2$) as

$$\Delta(\check{\Psi}_1, \check{\Psi}_2) = \sum_{r,s} \left(\oint_{\rho_r} \frac{d\rho}{2\pi i} \oint_{\rho_s} \frac{d\rho'}{2\pi i} \right)' ln(z - z') \check{\Psi}_{1r}(\rho) \check{\Psi}_{2s}(\rho')$$
$$- \textit{free-string terms} \quad + (\textit{zero-mode})^2 \textit{ terms} \ , \qquad (9.2.14a)$$

$$[\Psi_{2r}(\zeta_2)\ ,\ \Psi_{1s}(\zeta_1)\} = -2\pi i\delta_{rs}\delta'(\zeta_2-\zeta_1)\ , \tag{9.2.14b}$$

where the appropriate Δ for X $(w_1 = w_2 = 1)$ is

$$\Delta = \tfrac{1}{2}\Delta(\check{P},\check{P}) \quad . \tag{9.2.15}$$

These ′ and free-string corrections may seem awkward, but they will automatically be fixed by the same method which gives a simple evaluation of the contour integrals: i.e., the terms which are difficult to evaluate are exactly those which we don't want. (For non-vertex S-matrices with $w_1 \neq 1 \neq w_2$, (9.2.11) or (9.2.13) can be used, but their evaluated forms are much more complicated in the general case.) In general (for covariant quantization, supersymmetry, etc.) we also need extra factors which are evaluated at infinitesimal separation from the interaction (splitting) points, which follow from applying the conformal transformation (9.2.7), and (9.2.6) with $z = z_{INT}$. Only creation operators contribute.

The contour-integral form (9.2.14) can also be derived from the path-integral form (9.2.4). For the open string [9.5], these contour integrals can be obtained by either combining integrals over semicircles in the upper-half z plane (σ integrals from 0 to π) with their reflections [9.6], or more directly by reformulating the open string as a closed string with modes of one handedness only, and with interactions associated with just the points $\sigma = 0, \pi$ rather than all σ.

For the second step, $G^{rs}{}_{mn}$ unfortunately is hard to calculate in general. For open-string tree graphs, we perform the following conformal mapping to the upper-half complex plane [9.4], where Δ is easy to calculate:

$$\rho = \sum_{r=2}^{N} p_{+r}\, ln(z - z_r) \quad . \tag{9.2.16}$$

The boundary of the (interacting) string is now the real z axis, and the interior is the upper half of the complex z-plane. (The branches in the ln's in (9.2.16) are thus chosen to run down into the lower-half plane. When we use the whole plane for contour integrals below, we'll avoid integrals with contours with cuts inside them.) As a result, operators such as $\widehat{P}$, which were periodic in σ, are now meromorphic at z_r, so the contour integrals are easy to evaluate. Also, extending σ from $[0,\pi]$ to $[-\pi,\pi]$ extends the upper-half plane to the whole complex plane, so there are no boundary conditions to worry about. To evaluate (9.2.14), we note that, since we are neglecting (zero-modes)^2 and free string terms ($r = s$, $m = -n$), we can replace

$$ln(z-z')e^{-m\rho/p_{+r}-n\rho'/p_{+s}} \quad \to \quad \frac{1}{\frac{m}{p_{+r}}+\frac{n}{p_{+s}}}\left[\left(\frac{\partial}{\partial\rho}+\frac{\partial}{\partial\rho'}\right)ln(z-z')\right]e^{-m\rho/p_{+r}-n\rho'/p_{+s}} \tag{9.2.17}$$

by integration by parts. We then use the identity, for the case of (9.2.16),

$$\left(\frac{\partial}{\partial\rho}+\frac{\partial}{\partial\rho'}\right) ln(z-z') = \sum_{r=2}^{N} p_{+r}\left[\frac{\partial}{\partial\rho}ln(z-z_r)\right]\left[\frac{\partial}{\partial\rho'}ln(z'-z_r)\right] \quad . \qquad (9.2.18)$$

Then we can trivially change completely to z coordinates by using $d\rho\partial/\partial\rho = dz\partial/\partial z$, and converting the ρ exponentials into products of powers of z monomials. Differentiating the ln's gives (products of) single-variable contour integrals which can easily be evaluated as multiple derivatives:

$$\Delta = \sum_{rsmn} \psi_{1rm}\psi_{2sn}(p_{+r})^{1-w_1}(p_{+s})^{1-w_2}\frac{1}{np_{+r}+mp_{+s}}\sum_{t=2}^{N} p_{+t}A_{rtm}A_{stn}+(zero\text{-}mode)^2 \quad ,$$

$$A_{rtm} = \oint_{z_r}\frac{dz}{2\pi i}\frac{1}{z-z_t}\left[(z-z_r)\prod_{s=2}^{r-1}(z_s-z)^{p_{+s}/p_{+r}}\prod_{s=r+1}^{N}(z-z_s)^{p_{+s}/p_{+r}}\right]^{-m} \quad . \qquad (9.2.19)$$

For the third step, for open-string trees, we also need the Jacobian from $\prod d\tau_i \rightarrow (\prod dz_i)V(z)$, which for trees can easily be calculated by considering the graph where all external states are tachyons and all but 2 strings (one incoming and one outgoing) have infinitesimal length. We can also restrict all transverse momenta to vanish, and determine dependence on them at the end of the calculation by the requirement of Lorentz covariance. (Alternatively, we could complicate the calculation by including transverse momenta, and get a calculation more similar to that of (9.1.9).) We then have the amplitude (from nonrelativistic-style quantum mechanical arguments, or specializing (9.2.2))

$$A = g^{N-2}f(p_{+r})\int\left(\prod_{i=3}^{N-1} d\tau_i\right)e^{-\sum_{r=2}^{N-1} p_{-r}\tau_r} \quad , \qquad (9.2.20)$$

where f is a function to be determined by Lorentz covariance, the τ's are the interaction points, the strings 1 and N are those whose length is not infinitesimal, and we also choose $z_1=\infty$, $z_N=0$ in the transformation (9.2.16). We then solve for τ_r $(=-Re(\rho_r))$, in terms of z_r (in this approximation of infinitesimal lengths for all but 2 of the strings), as the finite values of ρ where the boundary "turns around":

$$\left.\frac{\partial\rho}{\partial z}\right|_{\rho_r} = 0 \quad \rightarrow$$

$$\rho_r = p_{+N}ln\, z_r + p_{+r}\left[ln\left(-\frac{p_{+r}}{p_{+N}}z_r\right)-1\right]+\sum_{s=2,s\neq r}^{N-1} p_{+s}ln(z_r-z_s)+\mathcal{O}\left[\left(\frac{p_{+r}}{p_{+N}}\right)^2\right] \quad . \qquad (9.2.21)$$

We then find, using the mass-shell condition $p_- = 1/p_+$ for the tachyon (p_- is $-H$ in nonrelativistic-style calculations)

$$A = g^{\mathrm{N}-2}\left[f p_{+\mathrm{N}}{}^{-1}\left(\prod_{r=2}^{\mathrm{N}-1}\frac{p_{+r}}{e}\right)\right]\int\left(\prod_{i=3}^{\mathrm{N}-1}dz_i\right) z_2 \prod_{s>r=2}^{\mathrm{N}}(z_r - z_s)^{p_{+r}/p_{+s}+p_{+s}/p_{+r}}$$
$$\cdot e^{\sum p_{ir} n_{rs}(p_+, z) p_{is}} \quad , \tag{9.2.22a}$$

where we have now included the transverse-momentum factor n_{rs}, whose exponential form follows from previous arguments. Its explicit value, as well as that of f, can now be determined by the manifest covariance of the tachyonic amplitude. However, (9.2.22a) is also the correct measure for the z-integration to be applied to (9.2.1) (or (9.2.4)), using Δ from (9.2.19) (which is expressed in terms of the same transformation (9.2.16)). At this point we can see that Lorentz covariance determines f to be such that the factor in brackets is a constant. We then note that, using $p_- = -H = -(\frac{1}{2}p_i{}^2 + N - 1)/p_+$, we have $p_r \cdot p_s = p_{ir}p_{is} - [(p_{+r}/p_{+s})(\frac{1}{2}p_{is}{}^2 + N - 1) + r \leftrightarrow s]$. This determines the choice of n_{rs} which makes the amplitude manifestly covariant for tachyons:

$$\int dz\, V(z) = g^{\mathrm{N}-2}\int\left(\prod_{i=3}^{\mathrm{N}-1}dz_i\right) z_2 \prod_{s>r=2}^{\mathrm{N}}(z_r - z_s)^{p_r\cdot p_s-(p_{+r}/p_{+s})(\frac{1}{2}p_s{}^2-1)-(p_{+s}/p_{+r})(\frac{1}{2}p_r{}^2-1)} . \tag{9.2.22b}$$

Note that p_- dependence cancels, so $p_r \cdot p_s$ and $p_r{}^2$ can be chosen to be the covariant ones. Taking N=1 to compare with the tachyonic particle theory, we see this agrees with the result (9.1.15) (after choosing also $z_2 = 1$). It also gives the (zero-mode)^2 terms which were omitted in our evaluation of Δ. (That is, we have determined both of these factors by considering this special case.) For the case of the tachyon, we could have obtained the covariant result (9.2.22b) more directly by using covariant amputation factors, i.e., by using p_- as an independent momentum instead of as the hamiltonian (see sect. 2.5). However, the result loses its manifest covariance, even on shell, for excited states because of the usual $1/p_+$ interactions which result in the light-cone formalism after eliminating auxiliary fields.

As mentioned in sect. 8.1, there is an Sp(2) invariance of free string theory. In terms of the tree graphs, which were calculated by performing a conformal map to the upper-half complex plane, it corresponds to the fact that this is the subgroup of the conformal group which takes the upper-half complex plane to itself. Explicitly, the transformation is $z \to (m_{11}z + m_{12})/(m_{21}z + m_{22})$, where the matrix m_{ij} is real, and without loss of generality can be chosen to have determinant 1. This transformation also takes the real line to itself, and when combined with (9.2.16) modifies it only by adding a constant and changing the values of the z_r (but not their order). In particular, the 3 arbitrary parameters allow arbitrary values (subject to ordering) for z_1, z_2, and z_N, which were ∞, 1, and 0. (This adds a term for z_1 to (9.2.16) which was previously dropped as an infinite constant. $\rho \to \infty$ as $z \to \infty$ in (9.2.16) corresponds to the end of the first string.) Because of the Sp(2) invariance, (9.2.22) can be rewritten in a form with all z's treated symmetrically: The tachyonic amplitude is then

$$A = g^{N-2} \int \frac{\prod_{r=1}^{N} dz_r}{dz_i dz_j dz_k} (z_i - z_j)(z_i - z_k)(z_j - z_k) \prod_{1 \le r < s \le N} (z_r - z_s)^{p_r \cdot p_s} \quad , \quad (9.2.22c)$$

where z_i, z_j, z_k are any 3 z's, which are not integrated over, and with all z's cyclically ordered as in (9.1.14).

Closed-string trees are similar, but whereas open-string interaction points occur anywhere on the *boundary*, closed-string interaction points occur anywhere on the *surface*. (Light-cone coordinates are chosen so that these points always occur for those values of τ where the strings split or join.) Thus, for closed strings there are also integrations over the σ's of the interaction points. The amplitude corresponding to (9.1.15) or (9.2.22) for the closed string, since it has both clockwise and counterclockwise modes, has the product of the integrand for the open string (for one set of modes) with its complex conjugate (for the modes propagating in the opposite direction), and the integral is over both z's and $\bar{z}$'s (i.e., both τ's and σ's). (There are also additional factors of 1/4 in the exponents due to the different normalization of the zero modes.) However, whereas the integral in (9.1.15) for the open string is restricted by (9.1.14) so that the z's (interaction points) lie on the boundary (the real axis), and are ordered, in the closed string case the z's are anyplace on the surface (arbitrary complex).

We next consider the evaluation of the open-string 3-point function, which will be needed below as the vertex in the field theory action. The 3-string vertex for the open string can be written in functional form as a δ-functional equating the coordinates of a string to those of the strings into which it splits. In the case of general string coordinates $\mathbf{Z}$:

$$S_{INT} = g \int d\tau\, \mathcal{D}^3\mathbf{Z}\, d^3p_+\, \delta\left(\sum p_+\right) \delta[\tilde{\mathbf{Z}}_1(\sigma) - \tilde{\mathbf{Z}}_3(\sigma)]\delta[\tilde{\mathbf{Z}}_2(\sigma) - \tilde{\mathbf{Z}}_3(\sigma)]\Phi[1]\Phi[2]\Phi[3] \quad , \quad (9.2.23)$$

with $\tilde{\mathbf{Z}}$ as in (9.2.7). We now use (9.2.6-16) for $\mathbf{N}= 3$, with $z_r = \infty, 1, 0$ in (9.2.16), and p_{+1} with the opposite sign to p_{+2} and p_{+3}. The splitting point is

$$\frac{\partial\rho}{\partial z}=0 \quad\rightarrow\quad z=z_0=-\frac{p_{+3}}{p_{+1}} \quad , \quad \rho=\tau_0+i\pi p_{+2} \quad , \tag{9.2.24a}$$

$$\tau_0=\tfrac{1}{2}\sum p_+\, ln\,({p_+}^2) \quad . \tag{9.2.24b}$$

For (9.2.19), we use the integral

$$\oint_0 \frac{dz}{2\pi i}\, z^{-n-1}(z+1)^u=\frac{1}{n!}\frac{d^n}{dz^n}(z+1)^u|_{z=0}=\frac{u(u-1)\cdots(u-n+1)}{n!}=\binom{u}{n} \quad ,$$

$$\binom{-u+n-1}{n}=(-1)^n\binom{u}{n} \tag{9.2.25}$$

to evaluate

$$m>0: \quad A_{r2m}=p_{+3}\mathcal{N}_{rm} \quad , \quad A_{r3m}=-p_{+2}\mathcal{N}_{rm} \quad ; \quad \mathcal{N}_{rm}=\frac{1}{p_{+,r+1}}\binom{-m\frac{p_{+,r+1}}{p_{+r}}}{m} \quad ;$$
$$m=0: \quad A_{rt0}=\delta_{rt}-\delta_{r1} \quad . \tag{9.2.26}$$

The result is then (see, e.g., [9.4]):

$$\Delta(\Psi_1,\Psi_2)=-\psi_1\mathfrak{N}\psi_2-\widetilde{\psi}_1\mathcal{N}\frac{1}{n}\psi_2-\psi_1\frac{1}{n}\mathcal{N}\widetilde{\psi}_2-\tau_0\sum\frac{p^2+M^2}{2p_+} \quad ,$$

$$\mathfrak{N}_{rsmn}=\frac{p_{+1}p_{+2}p_{+3}}{np_{+r}+mp_{+s}}\mathcal{N}_{rm}\mathcal{N}_{sn} \quad , \quad \widetilde{\psi}=p_{+[r}(\psi_0)_{r+1]} \quad ,$$

$$S=\int d^3p_+\, d^3\psi\, \delta\left(\sum p_+\right)\delta\left(\sum \psi_0\right)\langle\Phi_1\Phi_2\Phi_3\left|e^{\Delta}\right|0\rangle \quad . \tag{9.2.27}$$

(In some places we have used matrix notation with indices $r,s=1,2,3$ and $m,n=1,2,...,\infty$ implicit.) The ψ's include the p's. For simplicity, we have assumed the ψ's have $w=1$; otherwise, each ψ should be replaced with ${p_+}^{1-w}\psi$. The τ_0 term comes from shifting the value of τ at which the vertex is evaluated from $\tau=0$ to the interaction time $\tau=\tau_0$ (it gives just the propagator factor $e^{-\tau_0\sum H_r}$, where H_r is the free hamiltonian on each string). In more general cases we'll also need to evaluate a regularized $\widehat{\Psi}$ at the splitting point, which is also expressed in terms of the mode expansion of $ln(z-z_r)$ (actually its derivative $1/(z-z_r)$) which was used in (9.2.14) to obtain (9.2.27):

$$\widehat{\Psi}(z_0)\rightarrow\frac{1}{\sqrt{p_{+1}p_{+2}p_{+3}}}\widetilde{\psi}+\sqrt{p_{+1}p_{+2}p_{+3}}{p_+}^{-w}\mathcal{N}\psi \quad . \tag{9.2.28}$$

(Again, matrix notation is used in the second term.) We have arbitrarily chosen a convenient normalization factor in the regularization. (A factor which diverges as $z\rightarrow z_0$ must be divided out anyway.) The vertex is cyclically symmetric in the 3 strings (even though some strings have $p_+ < 0$). Besides the conservation law $\sum p = 0$, we also have $\sum p_+ x = 0$, which is actually the conservation law for angular momentum J_{+i}. These are special cases of the ψ_0 conservation law indicated above by the δ function, after including the ${p_+}^{1-w}$. (Remember that a coordinate of weight w is conjugate to one with weight $1-w$.) This conservation law makes the definition of $\tilde{\psi}$ above τ-independent. However, such conservation laws may be violated by additional vertex factors (9.2.28).

The 3-string vertex for the closed string in operator form is essentially just the product of open-string vertices for the clockwise and counterclockwise modes, since the δ functionals can be written as such a product, except for the zero modes. However, whereas open strings must join at their ends, closed strings may join anywhere, and the σ parametrizing this joining is then integrated over. Equivalently, the vertex may include projection operators $\delta_{\Delta N,0} = \int \frac{d\sigma}{2\pi}\, e^{i\sigma\Delta N}$ which perform a σ translation equivalent to the integration. (The former interpretation is more convenient for a first-quantized approach, whereas the latter is more convenient in the operator formalism.) These projection operators are redundant in a "Landau gauge," where the residual $\sigma \to \sigma + constant$ gauge invariance is fixed by introducing such projectors into the propagator.

In the covariant first-quantized formalism one can consider more general gauges for the σ-τ reparametrization invariance and local scale invariance than $g_{mn} = \eta_{mn}$. Changing the gauge has the effect of "stretching" the surface in σ-τ space. Since the 2D metric can always be chosen to be flat in any small region of the surface, it's clear that the only invariant quantities are global. These are topological quantities (some integers describing the type of surface) and certain proper-length parameters (such as the proper-length of the propagator in the case of the particle, as in (5.1.13)). In particular, this applies to the light-cone formalism, which is just a covariant gauge with stronger gauge conditions (and some variables removed by their equations of motion). Thus, the planar light-cone tree graph above is essentially a flat disc; and the proper-length parameters are the τ_i, $i = 3, \ldots, N-1$. However, there are more general covariant gauges even for such surfaces with just straight-line boundaries: For example, we can identify

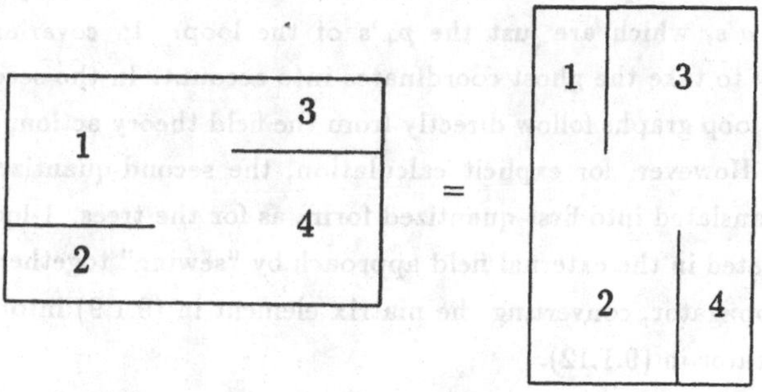

with the proper-length parameter being the relative position of the 2 splitting points (horizontal or vertical displacement, respectively for the 2 graphs, with the value of the parameter being positive or negative). More generally, the only invariants in this graph are the proper length distances measured along the boundary between the *end*points (the points associated with the external particles), less 3 which can be eliminated by remaining projective invariance (consider, e.g., the surface as a disc, with the endpoints on the circular boundary). Thus, we can keep the splitting points in the positions in the figures and vary the proper-length parameters by moving the ends instead. If this is interpreted in terms of ordinary Feynman graphs, the first graph seems to have intermediate states formed by the collision of particles 1 and 2, while the second one is from 1 and 3. The identity of these 2 graphs means that the same result can be obtained by summing over intermediate states in only 1 of these 2 channels as in the other, as we saw for the case of external tachyons in (9.1.16). Thus, duality is just a manifestation of σ-τ reparametrization invariance and local scale invariance.

9.3. Loops

Here we will only outline the procedure and results of loop calculations (for details see [0.1,1.3-5,9.7-10,5.4] and the shelf of this week's preprints in your library). In the first-quantized approach to loops the only essential difference from trees is that the topology is different. This means that: (1) It's no longer possible to conformally map to the upper-half plane, although one can map to the upper-half plane with certain lines identified (e.g., for the planar 1-loop graph, which is topologically a cylinder, we can choose the region between 2 concentric semicircles, with the semicircles identified). (2) The integration variables include not only the

τ's of the interaction points which define the position of the loop in the string, but also the σ's, which are just the p_+'s of the loop. In covariant gauges it's also necessary to take the ghost coordinates into account. In the second-quantized approach the loop graphs follow directly from the field theory action, as in ordinary field theory. However, for explicit calculation, the second-quantized expressions need to be translated into first-quantized form, as for the trees. 1-loop graphs can also be calculated in the external field approach by "sewing" together the 2 ends of the string propagator, converting the matrix element in (9.1.9) into a trace, using the trace operator in (9.1.12).

An interesting feature of open string theories is that closed strings are generated as bound states. This comes from stretching the one-loop graph with intermediate states of 2 180°-twisted open strings:

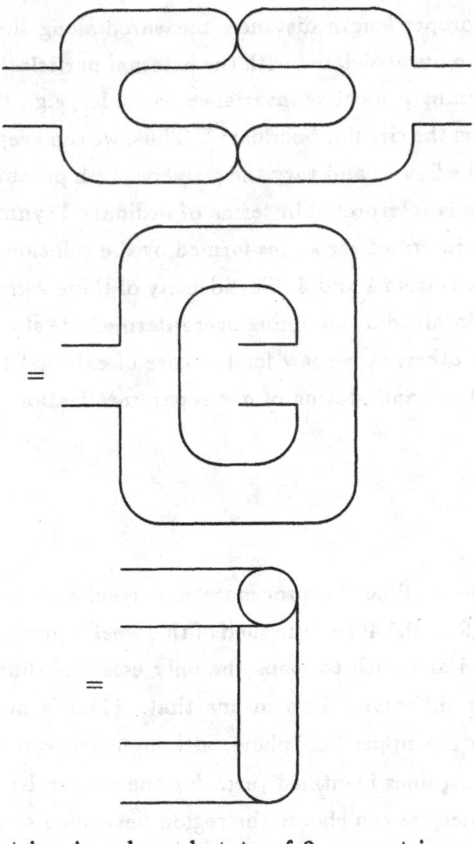

Thus, a closed string is a bound state of 2 open strings. The closed-string coupling can then be related to the open-string coupling, either by examining more general graphs, or by noticing that the Gauss-Bonnet theorem says that twice the number of "handles" (closed-string loops) plus the number of "windows" (open string loops) is a topological invariant (the Euler number, up to a constant), and thus 1 closed-string loop can be converted into 2 open-string loops. Specifically,

$$\hbar g_{closed} = (\hbar g_{open})^2 \quad \rightarrow \quad g_{closed} = \hbar g_{open}{}^2 \quad . \tag{9.3.1}$$

Thus, for consistent $\hbar$ counting the open strings must be thought of as fundamental is such a theory (which so far means just the SO(32) superstring), and the closed strings as bound states. Since (known) closed strings always contain gravitons, this makes the SO(32) superstring the only known example of a theory where the graviton appears as a bound state. The graviton propagator is the result of ultraviolet divergences due to particles of arbitrarily high spin which sum to diverge only at the pole:

$$\int_0^\infty dk^2\, (1 - \alpha'^2 p^2 k^2 + \tfrac{1}{2}\alpha'^4 p^4 k^4 - \cdots) = \int dk^2\, e^{-\alpha'^2 p^2 k^2} = \frac{1}{\alpha'^2 p^2} \quad . \tag{9.3.2}$$

(In general, $p^2 + M^2$ appears instead of p^2, so the entire closed-string spectrum is generated.)

As mentioned in the introduction, the topology of a 2D surface is defined by a few integers, corresponding to, e.g., the number of holes. By choosing the coordinates of the surface appropriately ("stretching" it in various ways), the surface takes the form of a string tree graph with one-loop insertions, each one loop insertion having the value 1 of one of the topological invariants (e.g., 1 hole). (Actually, some of these insertions are 1-loop closed-string insertions, and therefore are counted as 2-loop in an open-string theory due to (9.3.1).) For example, a hole in an open-string sheet may be pushed around so that it represents a loop as in a box graph, a propagator correction, a tadpole, or an external line correction. Such duality transformations can also be represented in Feynman graph notation as a consequence of the duality properties of simpler graphs such as the 4-point tree graph (9.1.16):

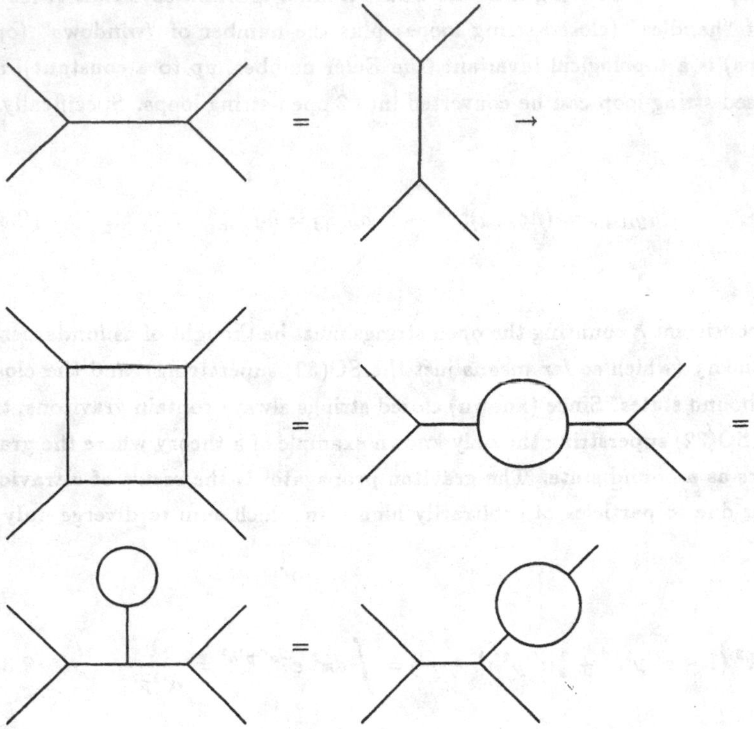

By doing stretching of such planar graphs out of the plane, these loops can even be turned into closed-string tadpoles:

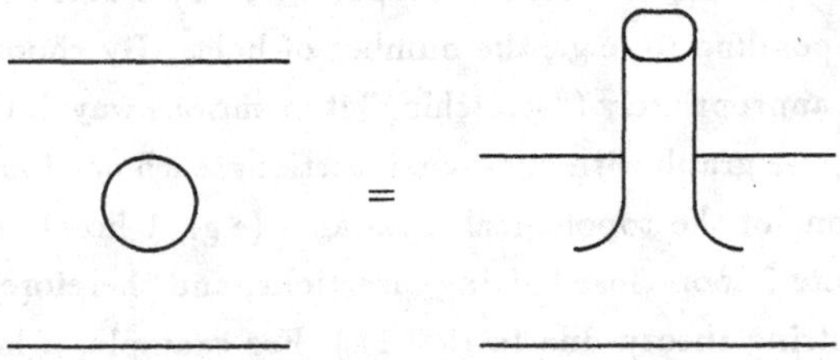

Stretching represents continuous world-sheet coordinate transformations. However, there are some coordinate transformations which can't be obtained by combining infinitesimal transformations, and thus must be considered separately in

analyzing gauge fixing and anomalies [9.11]. The simplest example is for a closed-string loop (vacuum bubble), which is a torus topologically. The group of general coordinate transformations has as a subgroup conformal transformations (which can be obtained as a residual gauge invariance upon covariant gauge fixing, sect. 6.2). Conformal transformations, in turn, have as a subgroup the (complex) projective group Sp(2,C): The defining representation of this group is given by 2×2 complex matrices with determinant 1, so the corresponding representation space consists of pairs of complex numbers. If we consider the transformation property of a complex variable which is the ratio of the 2 numbers of the pair, we then find:

$$\begin{pmatrix} z_1 \\ z_2 \end{pmatrix}' = \begin{pmatrix} a & b \\ c & d \end{pmatrix} \begin{pmatrix} z_1 \\ z_2 \end{pmatrix} \quad , \quad z_0 = \frac{z_1}{z_2} \tag{9.3.3a}$$

$$\to \quad z_0{}' = \frac{az_0 + b}{cz_0 + d} \quad , \tag{9.3.3b}$$

where $ad - bc = 1$. Finally, the projective group has as a discrete subgroup the "modular" group Sp(2,Z), where a, b, c, d are (real) integers (still satisfying $ad - bc = 1$). To see how this relates to the torus, define the torus as the complex plane with the identification of points

$$z \to z + n^1 z_1 + n^2 z_2 \tag{9.3.4}$$

for any integers n^1, n^2, for 2 particular complex numbers z_1, z_2 which point in different directions in the complex plane. We can then think of the torus as the parallelogram with corners $0, z_1, z_2, z_1 + z_2$, with opposite sides identified, and the complex plane can be divided up into an infinite number of equivalent copies, as implied by (9.3.4). The conformal structure of the torus can be completely described by specifying the value of $z_0 = z_1/z_2$. (E.g., z_1 and z_2 both change under a complex scale transformation, but not their ratio. Without loss of generality, we can choose the imaginary part of z_0 to be positive by ordering z_1 and z_2 appropriately.) However, if we transform (z_1, z_2) under the modular group as in (9.3.3a), then (9.3.4) becomes

$$z \to z + n^i z'_i \quad , \quad z'_i = g_i{}^j z_j \quad , \tag{9.3.5a}$$

where $g_i{}^j$ is the Sp(2,Z) matrix, or equivalently

$$z \to z + n'^i z_i \quad , \quad n'^i = n^j g_j{}^i \quad . \tag{9.3.5b}$$

In other words, an Sp(2,Z) transformation gives back the same torus, since the identification of points in the complex plane (9.3.4) and (9.3.5b) is the same (since it holds are all pairs of integers n^i). We therefore define the torus by the complex parameter z_0, modulo equivalence under the Sp(2,Z) transformation (9.3.3b). It turns out that the modular group can be generated by just the 2 transformations

$$z_0 \to -\frac{1}{z_0} \quad and \quad z_0 \to z_0 + 1 \ . \tag{9.3.6}$$

The relevance of the modular group to the 1-loop closed-string diagram is that the functional integral over all surfaces reduces (for tori) to an integral over z_0. Gauge fixing for Sp(2,Z) then means picking just one of the infinite number of equivalent regions of the complex plane (under (9.3.3b)). However, for a closed string in less than its critical dimension, there is an anomaly in the modular invariance, and the theory is inconsistent. Modular invariance also restricts what types of compactification are allowed.

If the 2D general coordinate invariance is not violated by anomalies, it's then sufficient to consider these 1-loop objects to understand the divergence structure of the quantum string theory. However, while the string can be stretched to separate any two 1-loop divergences, we know from field theory that overlapping divergences can't be factored into 1-loop divergences. This suggests that any 1-loop divergences, since they would lead to overlapping divergences, would violate the 2D reparametrization invariance which would allow the 1-loop divergences to be disentangled. Hence, it seems that a string theory must be finite in order to avoid such anomalies. Conversely, we expect that finiteness at 1 loop implies finiteness at all loops. Some direct evidence of this is given by the fact that all known string theories with fermions have 1-loop anomalies in the usual gauge invariances of the massless particles if and only if they also have 1-loop divergences. After the restrictions placed by tree-level duality (which determines the ground-state mass and restricts the open-string gauge groups to U(N), USp(N), and SO(N)), supersymmetry in the presence of massless spin-3/2 particles, and 1-loop modular invariance, this last anomaly restriction allows only SO(32) as an open-string gauge group (although it doesn't restrict the closed-string theories) [1.11].

Of the finite theories, the closed-string theories are finite graph-by-graph, whereas the open-string theory requires cancellation between pairs of 1-loop graphs, with the exception of the nonplanar loop discussed above. The 1-loop closed-string graphs (corresponding to 2-loop graphs in the open-string theory) are (1) the torus ("handle") and (2) the Klein bottle, with external lines attached. The pairs of 1-loop open-string graphs are (1) the annulus (planar loop, or "window") + Möbius strip (nonorientable loop) with external open and/or closed strings, and (2) the disk + a graph with the topology of RP_2 (a disk with opposite points identified) with external closed strings. The Klein bottle is allowed only for nonoriented closed strings, and the Möbius strip and RP_2 are allowed only for nonoriented open strings.

It should be possible to simplify calculations and give simple proofs of finiteness by the use of background field methods similar to those which in gravity and supersymmetry made higher-loop calculations tractable and allowed simple derivations of no-renormalization theorems [1.2]. However, the use of arbitrary background string fields will require the development of gauge-covariant string field theory, the present status of which is discussed in the following chapters.

Exercises

(1) Generalize (9.1.3) to the spinning string, using (7.2.2) instead of $\frac{1}{2}\hat{P}^2$. Writing $\frac{1}{2}\hat{D}d\hat{D}(\sigma) \rightarrow \frac{1}{2}\hat{D}d\hat{D}(\sigma) + \delta(\sigma)\mathcal{V}$, $\mathcal{V} = W + \theta V$, show that V is determined explicitly by W.

(2) Derive (9.1.7b).

(3) Fill in all the steps needed to obtain (9.1.15) from (9.1.6,8). Derive all parts of (9.1.16). Derive (9.1.19).

(4) Evaluate (9.1.20) by using the result (9.1.15) with $\mathrm{N} \rightarrow \mathrm{N}+2$ (but dropping 2 $d\tau$ integrations) and letting $k_0 = k_{\mathrm{N}+1} = 0$.

(5) Derive (9.2.18).

(6) Generalize (9.2.16) to arbitrary z_1, z_2, z_{N} and derive (9.2.22c) by the method of (9.2.20,21). Take the infinitesimal form of the Sp(2) transformation and show that it's generated by $\mathbf{L}_0$, $\mathbf{L}_{\pm 1}$ with the correspondence (8.1.3), where $z = e^{i\sigma}$.

(7) Derive (9.2.27). Evaluate G_{rsmn} of (9.2.4) using (9.2.19,26,27).

10. LIGHT-CONE FIELD THEORY

In this chapter we extend the discussion of sect. 2.1 to the string and consider interacting contributions to the Poincaré algebra of sect. 7.1 along the lines of the Yang-Mills case treated in sect. 2.3.

For the string [10.1,9.5,10.2], it's convenient to use a field $\Phi[X_i(\sigma), p_+, \tau]$, since p_+ is the length of the string. This X for $\sigma \in [0, \pi p_+]$ is related to that in (7.1.7) for $p_+ = 1$ by $X(\sigma, p_+) = X(\sigma/p_+, 1)$. The *hermiticity* condition on the (open-string) field is

$$\Phi[X_i(\sigma), p_+, \tau] = \Phi^\dagger[X_i(\pi p_+ - \sigma), -p_+, \tau] \quad . \tag{10.1a}$$

The same relation holds for the closed string, but we may replace $\pi p_+ - \sigma$ with just $-\sigma$, since the closed string has the residual gauge invariance $\sigma \to \sigma + constant$, which is fixed by the constraint (or gauge choice) $\Delta N \Phi = 0$. (See (7.1.12). In loops, this gauge choice can be implemented either by projection operators or by Faddeev-Popov ghosts.) As described in sect. 5.1, this charge-conjugation condition corresponds to a combination of ordinary complex conjugation (τ reversal) with a twist (matrix transposition combined with σ reversal). The twist effectively acts as a charge-conjugation matrix in σ space, in the sense that expressions involving $tr\ \Phi^\dagger \Phi$ acquire such a factor if reexpressed in terms of just Φ and not $\Phi^\dagger$ (and (10.1a) looks like a reality condition for a group for which the twist is the group metric). Here Φ is an N×N matrix, and the odd mass levels of the string (including the massless Yang-Mills sector) are in the adjoint representation of U(N), SO(N), or USp(N) (for even N) [10.3], where in the latter 2 cases the field also satisfies the *reality* condition

$$\eta \Phi = (\eta \Phi)^* \quad , \tag{10.1b}$$

where η is the group metric (symmetric for SO(N), antisymmetric for USp(N)). The fact that the latter cases use the operation of τ reversal separately, or, by combining with (10.1a), the twist separately, means that they describe nonoriented strings: The string field is constrained to be invariant under a twist. The same is true for closed strings (although closed strings have no group theory, so the choice

of oriented vs. nonoriented is arbitrary, and $\eta = 1$ in (10.1b)). The twist operator can be defined similarly for superstrings, including heterotic strings. For general open strings the twist is most simply written in terms of the hatted operators, on which it acts as $\hat{\mathcal{O}}(\sigma) \to \hat{\mathcal{O}}(\sigma - \pi)$ (i.e., as $e^{i\pi N}$). For general closed strings, it takes $\hat{\mathcal{O}}^{(\pm)}(\sigma) \to \hat{\mathcal{O}}^{(\pm)}(-\sigma)$. (For closed strings, $\sigma \to \sigma - \pi$ is irrelevant, since $\Delta N = 0$.)

Light-cone superstring fields [10.2] also satisfy the reality condition (in place of (10.1b), generalizing (5.4.32)) that the Fourier transform with respect to Θ^a is equal to the complex conjugate (which is the analog of a certain condition on covariant superfields)

$$\int \mathcal{D}\Theta \; e^{\int \Theta^a \Pi_a p_+/2} \eta \Phi[\Theta^a] = (\eta \Phi[\Pi^a])^* \quad . \tag{10.1c}$$

Θ^a has a mode expansion like that of the ghost $\hat{C}$ or the spinning string's $\hat{\Psi}$ (of the fermionic sector). The ground-state of the open superstring is described by the light-cone superfield of (5.4.35), which is a function of the zero-modes of all the above coordinates. Thus, the lowest-mass (massless) sector of the open superstring is supersymmetric Yang-Mills.

The free action of the bosonic open string is [10.1,9.5]

$$S_0 = -\int \mathcal{D}X_i \int_{-\infty}^{\infty} dp_+ \int_{-\infty}^{\infty} d\tau \; tr \; \Phi^\dagger p_+ \left(i\frac{\partial}{\partial\tau} + H \right) \Phi \quad ,$$

$$H = \int_0^{2\pi\alpha' p_+} \frac{d\sigma}{2\pi} \left[\tfrac{1}{2} \left(-\alpha' \frac{\delta^2}{\delta X_i{}^2} + \frac{1}{\alpha'} X_i'^2 \right) - 1 \right] = \int_{-\pi p_+}^{\pi p_+} \frac{d\sigma}{2\pi} (\tfrac{1}{2} \hat{P}_i{}^2 - 1) = \frac{p_i{}^2 + M^2}{2p_+} \quad . \tag{10.2}$$

The free field equation is therefore just the quantum mechanical Schrödinger equation. (The p_+ integral can also be written as $2\int_0^\infty$, due to the hermiticity condition. This form also holds for closed strings, with H the sum of 2 open-string ones, as described in sec. 7.1.) Similar remarks apply to superstrings (using (7.3.13)).

As in the first-quantized approach, interactions are described by splitting and joining of strings, but now the graph gets chopped up into propagators and vertices:

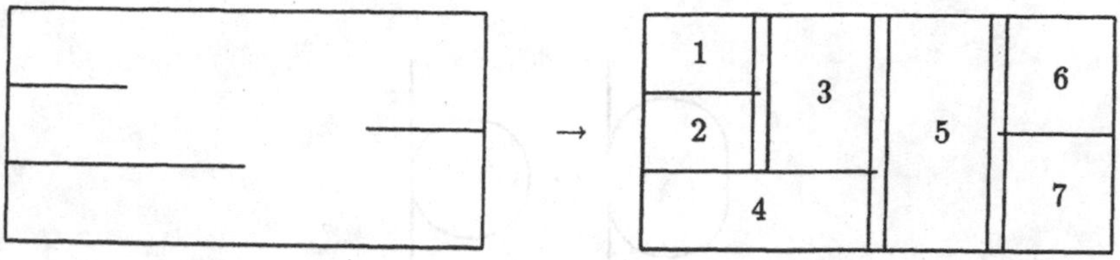

The 3-open-string vertex is then just a δ functional setting 1 string equal to 2 others, represented by an infinitesimal strip in the world sheet. The interaction term in the action is given by (9.2.23) for $Z = X$. The 3-closed-string vertex is a similar δ functional for 3 closed strings, which can be represented as the product of 2 open-string δ functionals, since the closed-string coordinates can be represented as the sum of 2 open-string coordinates (one clockwise and one counterclockwise, but with the same zero-modes). This vertex generally requires an integration over the σ of the integration point (since closed strings can join anywhere, not having any ends, corresponding to the gauge invariance $\sigma \to \sigma + constant$), but the equivalent operation of projection onto $\Delta N = 0$ can be absorbed into the propagators.

General vertices can be obtained by considering similar slicings of surfaces with general global topologies [10.2,9.7]. There are 2 of order g, corresponding ***locally*** to a splitting or joining:

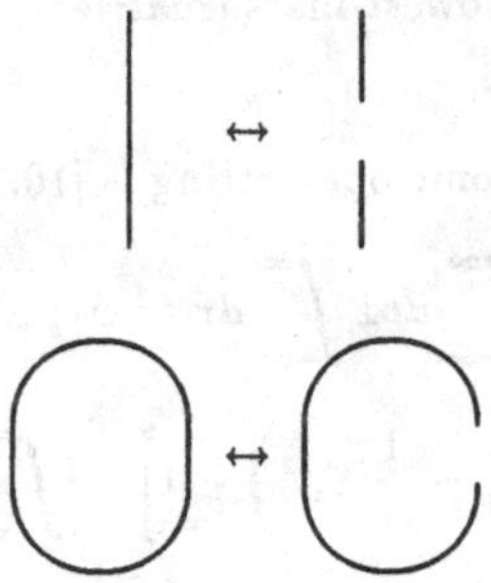

The former is the 3-open-string vertex. The existence of the latter is implied by the former via the nonplanar loop graph (see sect. 9.3). The rest are order g^2, and correspond locally to 2 strings touching their middles and switching halves:

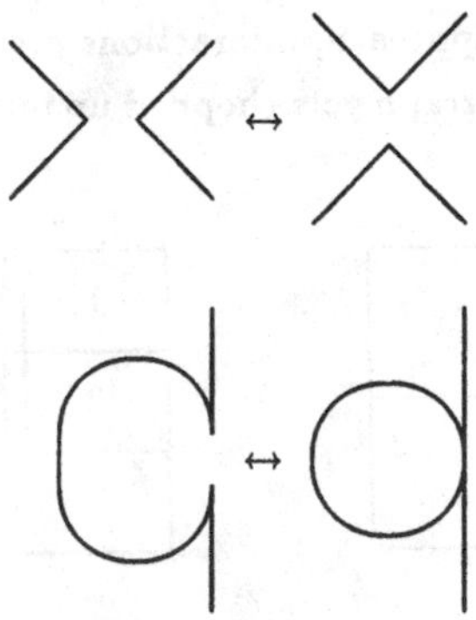

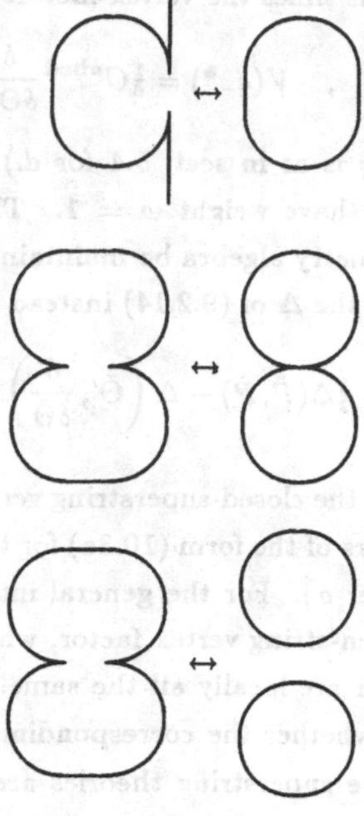

The type-I (SO(32)) open-closed) theory has all these vertices, but the type-IIAB and heterotic theories have only the last one, since they have only closed strings, and they are oriented (clockwise modes are distinguishable from counterclockwise). If the type-I theory is treated as a theory of fundamental open strings (with closed strings as bound states, so $\hbar$ can be defined), then we have only the first of the order-g vertices and the first of the order-g^2.

The light-cone quantization of the spinning string follows directly from the corresponding bosonic formalism by the 1D supersymmetrization described in sect. 7.2. In particular, in such a formalism the vertices require no factors besides the δ functionals [10.4]. (In converting to a non-superfield formalism, integration of the vertex over θ produces a vertex factor.) However, the projection (7.2.5) must be put in by hand. Also, the fact that boundary conditions can be either of 2 types (for bosons vs. fermions) must be kept in mind.

The interactions of the light-cone superstring [10.2] are done as for the light-cone bosonic string, but there are extra factors. For example, for the 3-open-string vertex (the interacting contribution to p_-), we have the usual δ-functionals times

$$V(p_-) = P_L + \tfrac{1}{2}\bar{P}^{ab}\frac{\delta}{\delta\Theta^a}\frac{\delta}{\delta\Theta^b} + \tfrac{1}{24}\bar{P}_L C^{abcd}\frac{\delta}{\delta\Theta^a}\frac{\delta}{\delta\Theta^b}\frac{\delta}{\delta\Theta^c}\frac{\delta}{\delta\Theta^d}, \qquad (10.3a)$$

evaluated at the splitting point. The interacting contributions to q_- are given by the same overlap δ-functionals times the vertex factors

$$V(q_{-\mathrm{a}}) = \frac{\delta}{\delta\Theta^{\mathrm{a}}} \quad , \quad V(\bar{q}_-{}^{\mathrm{a}}) = \tfrac{1}{6}C^{\mathrm{abcd}}\frac{\delta}{\delta\Theta^{\mathrm{b}}}\frac{\delta}{\delta\Theta^{\mathrm{c}}}\frac{\delta}{\delta\Theta^{\mathrm{d}}} \quad . \tag{10.3b}$$

(The euphoric notation for q is as in sect. 5.4 for d.) These are evaluated as in (9.2.28), where P and $\delta/\delta\Theta$ have weight $w = 1$. Their form is determined by requiring that the supersymmetry algebra be maintained. The δ functional part is given as in (9.2.27), but now the Δ of (9.2.14) instead of just (9.2.15) is

$$\Delta = \tfrac{1}{2}\Delta(\check{P},\check{P}) - \Delta\left(\check{\Theta}', \frac{\delta}{\delta\check{\Theta}}\right) \quad . \tag{10.4}$$

As for the bosonic string, the closed-superstring vertex is the product of 2 open-string ones (including 2 factors of the form (10.3a) for type I or II but just 1 for the heterotic, and integrated over σ). For the general interactions above, all order-g interactions have a single open-string vertex factor, while all order-g^2 have 2, since the interactions of each order are locally all the same. The vertex factor is either 1 or (10.3a), depending on whether the corresponding set of modes is bosonic or supersymmetric. When these superstring theories are truncated to their ground states, the factor (10.3a) keeps only the zero-mode contributions, which is the usual light-cone, 3-point vertex for supersymmetric Yang-Mills, and the product of 2 such factors (for closed strings) is the usual vertex for supergravity.

The second-quantized interacting Poincaré algebra for the light-cone string can be obtained perturbatively. For example,

$$[p_-, J_{-i}] = 0 \quad \rightarrow$$

$$[p^{(2)}{}_-, J^{(2)}{}_{-i}] = 0 \quad , \tag{10.5a}$$

$$[p^{(3)}{}_-, J^{(2)}{}_{-i}] + [p^{(2)}{}_-, J^{(3)}{}_{-i}] = 0 \quad , \tag{10.5b}$$

etc., where (n) indicates the order in fields (see sect. 2.4). The solution to (10.5a) is known from the free theory. The solution to (10.5b) can be obtained from known results for the first-quantized theory [1.4,10.5]: The first term represents the invariance of the 3-point interaction of the hamiltonian under free Lorentz transformations. The fact that this invariance holds only on-shell is an expected consequence of the fact that the second term in (10.5b) is simply the commutator of the free hamiltonian with the interaction correction to the Lorentz generator. Thus, the algebra of the complete interacting generators closes off shell as well as on, and the explicit form of $J^{(3)}{}_{-i}$ follows from the expression for nonclosure given in [1.4,10.5]:

$$\mathcal{J}^{(3)}{}_{-}{}^{i}(1,2,3) = -2igX_r{}^{i}(\sigma_{int})\delta\left(\sum p_+\right)\Delta[X^i] \quad , \tag{10.6}$$

where "r" denotes any of the three strings and Δ represents the usual overlap-integral δ-functionals with splitting point σ_{INT}. This is the analog of the generalization of (2.3.5) to the interacting scalar particle, where $p_-\phi \to -(1/2p_+)(p_i{}^2\phi + \phi^2)$. Since p_- also contains a 4-point interaction, there is a similar contribution to $J^{(2)}{}_{-i}$ (i.e., (10.6) with Δ replaced by the corresponding 3-string product derived from the 4-string light-cone vertex in p_-, and g replaced with g^2, but otherwise the same normalization). Explicit second-quantized operator calculations show that this closes the algebra [10.6]. Similar constructions apply to superstrings [10.7].

Covariant string rules can be obtained from the light-cone formalism in the same way as in sect. 2.6, and p_+ now also represents the string length [2.7]. Thus, from (10.2) we get the free action, in terms of a field $\Phi[X^a(\sigma), X^\alpha(\sigma), p_+, \tau]$,

$$S_0 = -\int \mathcal{D}X^a\,\mathcal{D}X^\alpha \int_{-\infty}^{\infty} dp_+ \int_{-\infty}^{\infty} d\tau\; tr\; \Phi^\dagger p_+\left(i\frac{\partial}{\partial\tau} + H\right)\Phi \quad ,$$

$$H = \int_0^{\pi p_+} \frac{d\sigma}{2\pi}\left\{\tfrac{1}{2}\left[-\alpha'\left(\frac{\delta^2}{\delta X^{a2}} + \frac{\delta}{\delta X_\alpha}\frac{\delta}{\delta X^\alpha}\right) + \frac{1}{\alpha'}\left(X^{a\prime 2} + X^{\alpha\prime}X_\alpha{}'\right)\right] - 1\right\} \quad . \tag{10.7}$$

The vertex is again a δ functional, in all variables.

The light-cone formalism for heterotic string field theory has also been developed, and can be extended to further types of compactifications [10.8].

Unfortunately, the interacting light-cone formalism is not completely understood, even for the bosonic string. There are certain kinds of contact terms which must be added to the superstring action and supersymmetry generators to insure lower-boundedness of the energy (supersymmetry implies positivity of the energy) and cancel divergences in scattering amplitudes due to coincidence of vertex operator factors [10.9], and some of these terms have been found. (Similar problems have appeared in the covariant spinning string formulation of the superstring: see sect. 12.2.) This problem is particularly evident for closed strings, which were thought to have only cubic interaction terms, which are insufficient to bound the potential in a formalism with only physical polarizations. Furthermore, the closed-string bound states which have been found to follow at one loop from open-string theories by explicitly applying unitarity to tree graphs do not seem to follow from the light-cone field theory rules [10.1]. Since unitarity requires that 1-loop corrections are uniquely determined by tree graphs, the implication is that the present light-cone field theory action is incomplete, or that the rules following from it have not been correctly applied. It is interesting to note that the type of graph needed to give the correct closed-string poles resembles the so-called Z-graph of ordinary light-cone field theory [10.10], which contains a line backward-moving in x_+ when the light-cone formalism is obtained as an ultrarelativistic limit, becoming an instantaneous line when the limit is reached.

Exercises

I can't think of any.

11. BRST FIELD THEORY

11.1. Closed strings

Since the gauge-invariant actions for free open strings follow directly from the methods of sects. 3.4-5 using the algebras of chapt. 8 (for the bosonic and fermionic cases, using either OSp(1,1|2) or IGL(1) algebras), we will consider here just closed strings, after a few general remarks.

Other string actions have been proposed which lack the complete set of Stueckelberg fields [11.1], and as a result they are expected to suffer from problems similar to those of covariant "unitary" gauges in spontaneously broken gauge theories: no simple Klein-Gordon-type propagator, nonmanifest renormalizability, and singularity of semiclassical solutions, including those representing spontaneous breakdown. Further attempts with nonlocal, higher-derivative, or incomplete actions appeared in [11.2]. Equivalent gauge-invariant actions for the free Ramond string have been obtained by several groups [4.13-15,11.3]. The action of [11.3] is related to the rest by a unitary transformation: It has factors involving coordinates which are evaluated at the midpoint of the string, whereas the others involve corresponding zero-modes.

For open or closed strings, the hermiticity condition (10.1a) now requires that the ghost coordinates also be twisted: In the IGL(1) formalism (where the ghost coordinates are momenta)

$$\Phi[X(\sigma), C(\sigma), \widetilde{C}(\sigma)] = \Phi^\dagger[X(\pi-\sigma), -C(\pi-\sigma), \widetilde{C}(\pi-\sigma)] \qquad (11.1.1a)$$

($\widetilde{C}$ gets an extra "−" because the twist is σ reversal, and $\widetilde{C}$ carries a σ index in the mechanics action), and in the OSp(1,1|2) formalism we just extend (10.1a):

$$\Phi[X^a(\sigma), X^\alpha(\sigma), p_+] = \Phi^\dagger[X^a(\pi p_+ - \sigma), X^\alpha(\pi p_+ - \sigma), -p_+] \quad . \qquad (11.1.1b)$$

As in the light-cone formalism (see sect. 10.1), for discussing free theories, we scale σ by p_+ in (11.1.1b), so the twist then takes $\sigma \to \pi - \sigma$. (For the closed string the twist is $\sigma \to -\sigma$, so in both (11.1.1a) and (11.1.1b) the arguments of the coordinates are just $-\sigma$ on the right-hand side.)

In the rest of this section we will consider closed strings only. First we show how to extend the OSp(1,1|2) formalism to closed strings [4.10]. By analogy to (4.1.1), the kinetic operator for the closed string is a δ function in IOSp(1,1|2):

$$S = \int d^D x d^2 x_\alpha dx_- d^2\Delta p_\alpha d\Delta p_+ \;\; \Phi^\dagger \; {p_+}^2 \delta(J_{AB})\delta(\Delta p_A)\,\Phi \quad ,$$

$$\delta\Phi = \tfrac{1}{2} J^{AB}\Lambda_{BA} + \Delta p^A \Lambda_A \quad , \qquad (11.1.2)$$

where, as in (7.1.17), the Poincaré generators J_{AB} and p_A are given as sums, and Δp_A as differences, of the left-handed and right-handed versions of the open-string generators of (7.1.14). The Hilbert-space metric necessary for hermiticity is now ${p_+}^2$ (or equivalently ${p_+}^{(+)}{p_+}^{(-)}$), since a factor of p_+ is needed for the open-string modes of each handedness. For simplicity, we do not take the physical momenta p_a to be doubled here, since the IOSp(1,1|2) algebra closes regardless, but they can be doubled if the corresponding δ functions and integrations are included in (11.1.2). More explicitly, the δ function in the Poincaré group is given by

$$\delta(J_{AB})\delta(\Delta p_A) = \delta({J_{\alpha\beta}}^2) i\delta(J_{-+})\delta^2(J_{+\alpha})\delta^2(J_{-\alpha})\delta(\Delta p_-)\delta^2(\Delta p_\alpha)\delta(\Delta p_+) \quad . \qquad (11.1.3)$$

To establish the invariance of (11.1.2), the fact that (4.1.1) is invariant indicates that it's sufficient to show that $\delta(J_{AB})$ commutes with $\delta(\Delta p_A)$. This follows from the fact that each of the J_{AB}'s commutes with $\delta(\Delta p_A)$. We interpret $\delta(\Delta p_-) = p_+\delta(\Delta N)$ in the presence of the other $\delta(\Delta p)$'s, where $\delta(\Delta N)$ is a Kronecker δ, and the other $\delta(\Delta p)$'s are Dirac δ's.

All the nontrivial terms are contained in the $\delta^2(J_{-\alpha})$. As in the open-string case, dependence on the gauge coordinates x_α and x_- is eliminated by the $\delta^2(J_{+\alpha})$ and $\delta(J_{-+})$ on the left, and further terms are killed by $\delta({J_{\alpha\beta}}^2)$. Similarly, dependence on Δp_α and Δp_+ is eliminated by the corresponding δ functions on the right, and further terms are killed by $\delta(\Delta p_-)$. For convenience, the latter elimination should be done before the former. After making the redefinition $\Phi \to \frac{1}{p_+}\Phi$ and integrating out the unphysical zero-modes, the action is similar to the OSp(1,1|2) case:

$$S = \int d^D x \; \tfrac{1}{2}\phi^\dagger \; \delta(\Delta N)\delta({M_{\alpha\beta}}^2)\left[\Box - M^2 + ({M_\alpha}^a p_a + M_{\alpha m}M)^2\right] \phi \quad ,$$

$$\delta\phi = (M^{\alpha a}p_a + {M^\alpha}_m M)\Lambda_\alpha + \tfrac{1}{2}M^{\alpha\beta}\Lambda_{\alpha\beta} + \Delta N\Lambda \quad . \qquad (11.1.4)$$

This is the minimal form of the closed-string action.

The nonminimal form is obtained by analogy to the IGL(1) formalism, in the same way OSp(1,1|2) was extended to IOSp(1,1|2): Using a δ function in the closed-string group GL(1|1) of sect. 8.2 (found from sums and differences of the expressions in (8.2.6), as in (7.1.17)), we obtain the action (with $\Delta p_- \to \Delta N$)

$$S = \int d^D x dc d\Delta p^c \, \Phi^\dagger \, iQ\delta(J^3)\delta(\Delta p^c)\delta(\Delta N) \, \Phi \quad ,$$

$$\delta\Phi = Q\Lambda + J^3\tilde{\Lambda} + \Delta N\tilde{\Lambda} + \Delta p^c\tilde{\Lambda} \quad , \qquad (11.1.5)$$

or, after integrating out Δp^c, with $\Phi = \phi + \Delta p^c \psi$,

$$S = \int d^D x dc \; \phi^\dagger \, i\widehat{Q}\delta(\widehat{J}^3)\delta(\Delta N) \, \phi \quad ,$$

$$\delta\phi = \widehat{Q}\Lambda + \widehat{J}^3\tilde{\Lambda} + \Delta N\tilde{\Lambda} \quad , \qquad (11.1.6)$$

where the $\widehat{\ }$'s indicate that all terms involving Δp^c and its canonical conjugate have been dropped. The field $\phi = \varphi + c\chi$ is commuting.

For the gauge-fixing in the GL(1|1) formalism above (or the equivalent one from first-quantization), we now choose [4.5]

$$\mathcal{O} = -2\Delta p^c \left[c, \frac{\partial}{\partial c}\right] \quad \to \quad K = \Delta p^c \left[c(\Box - M^2) - 4M^+ \frac{\partial}{\partial c}\right] - 2\Delta N \left[c, \frac{\partial}{\partial c}\right] \quad , \qquad (11.1.7)$$

where we have used

$$Q = -i\tfrac{1}{2}c(p^2 + M^2) + i\tfrac{1}{2}M^+ \frac{\partial}{\partial c} - i\Delta N \frac{\partial}{\partial \Delta p^c} + i\tfrac{1}{2}\Delta M^+ \Delta p^c + \mathcal{Q}^+ \quad . \qquad (11.1.8)$$

Expanding the string field over the ghost zero-modes,

$$\Phi = (\phi + ic\hat{\phi}) + i\Delta p^c(\hat{\psi} + c\psi) \quad , \qquad (11.1.9)$$

we substitute into the lagrangian $L = \frac{1}{2}\Phi^\dagger K \Phi$ and integrate over the ghost zero modes:

$$\frac{\partial}{\partial c}\frac{\partial}{\partial \Delta p^c} L = \tfrac{1}{2}\phi^\dagger(\Box - M^2)\phi + 2\psi^\dagger M^+ \psi + 4i(\hat{\phi}^\dagger \Delta N \phi + i\hat{\psi}^\dagger \Delta N \psi) \quad . \qquad (11.1.10)$$

ϕ contains propagating fields, ψ contains BRST auxiliary fields, and $\hat{\phi}$ and $\hat{\psi}$ contain lagrange multipliers which constrain $\Delta N = 0$ for the other fields. Although the propagating fields are completely gauge-fixed, the BRST auxiliary fields again have the gauge transformations

$$\delta\psi = \lambda \quad , \quad M^+\lambda = 0 \quad , \qquad (11.1.11a)$$

and the lagrange multipliers have the gauge transformations

$$\delta\hat{\phi} = \hat{\lambda}_\phi \quad , \quad \delta\hat{\psi} = \hat{\lambda}_\psi \quad ; \quad \Delta N\hat{\lambda}_\phi = \Delta N\hat{\lambda}_\psi = 0 \quad . \qquad (11.1.11b)$$

11.2. Components

To get a better understanding of the gauge-invariant string action in terms of more familiar particle actions, we now expand the string action over some of the lower-mass component fields, using the algebras of chapt. 8 in the formalism of sect. 4. All of these results can also be derived by simply identifying the reducible representations which appear in the light-cone, and then using the component methods of sect. 4.1. However, here we'll work directly with string oscillators, and not decompose the reducible representations, for purposes of comparison.

As an example of how components appear in the IGL(1) quantization, the massless level of the open string is given by (cf. (4.4.6))

$$\Phi = \left[(A^a a_a{}^\dagger + C^\alpha a_\alpha{}^\dagger) + icBa^{\tilde{c}\dagger} \right] |0\rangle \quad , \tag{11.2.1}$$

where $a^\alpha = (a^c, a^{\tilde{c}})$, all oscillators are for the first mode, and we have used (4.4.5). The lagrangian and BRST transformations then agree with (3.2.8,11) for $\zeta = 1$. In order to obtain particle actions directly without having to eliminate BRST auxiliary fields, from now on we work with only the OSp(1,1|2) formalism. (By the arguments of sect. 4.2, the IGL(1) formalism gives the same actions after elimination of BRST auxiliary fields.)

As described in sect. 4.1, auxiliary fields which come from the ghost sector are crucial for writing local gauge-invariant actions. (These auxiliary fields have the same dimension as the physical fields, unlike the BRST auxiliary fields, which are 1 unit higher in dimension and have algebraic field equations.) Let's consider the counting of these auxiliary fields. This requires finding the number of Sp(2) singlets that can be constructed out of the ghost oscillators at each mass level. The Sp(2) singlet constructed from two isospinor creation operators is $a_m{}^{\alpha\dagger} a_{n\alpha}{}^\dagger$, which we denote as (mn). A general auxiliary field is obtained by applying to the vacuum some nonzero number of these pairs together with an arbitrary number of bosonic creation operators. The first few independent products of pairs of fermionic operators, listed by eigenvalue of the number operator N, are:

$$
\begin{aligned}
0 &: \quad I \\
1 &: \quad - \\
2 &: \quad (11) \\
3 &: \quad (12)
\end{aligned}
$$

$$4 : \quad (13)\,,\,(22)$$
$$5 : \quad (14)\,,\,(23)$$
$$6 : \quad (15)\,,\,(24)\,,\,(33),\,(11)(22) \qquad (11.2.2)$$

where I is the identity and no operator exists at level 1. The number of independent products of singlets at each level is given by the partition function

$$\frac{1}{\prod_{n=2}^{\infty}(1-x^n)}$$
$$= 1+x^2+x^3+2x^4+2x^5+4x^6+4x^7+7x^8+8x^9+12x^{10}+14x^{11}+21x^{12}+\cdots \quad , (11.2.3)$$

corresponding to the states generated by a single bosonic coordinate missing its zeroth and first modes, as described in sect. 8.1.

We now expand the open string up to the third mass level (containing a massive, symmetric, rank-2 tensor) and the closed string up to the second mass level (containing the graviton) [4.1]. The mode expansions of the relevant operators are given by (8.2.1). Since the $\delta({M_{\alpha\beta}}^2)$ projector keeps only the Sp(2)-singlet terms, we find that up to the third mass level the expansion of ϕ is

$$\phi = [\phi_0 + A^a {a^\dagger}_{1a} + \tfrac{1}{2}h^{ab}{a^\dagger}_{1a}{a^\dagger}_{1b} + B^a {a^\dagger}_{2a} + \eta({a_{1\alpha}}^\dagger)^2]\,|0\rangle \quad . \qquad (11.2.4)$$

Here ϕ_0 is the tachyon, A^a is the massless vector, and (h^{ab}, B^a, η) describe the massive, symmetric, rank-two tensor. It's now straightforward to use (8.2.1) to evaluate the action (4.1.6):

$$\mathcal{L} = \mathcal{L}_{-1} + \mathcal{L}_0 + \mathcal{L}_1 \quad ; \qquad (11.2.5)$$

$$\mathcal{L}_{-1} = \tfrac{1}{2}\phi_0(\Box + 2)\phi_0 \quad , \qquad (11.2.6a)$$
$$\mathcal{L}_0 = \tfrac{1}{2}A\cdot\Box A + \tfrac{1}{2}(\partial\cdot A)^2 = -\tfrac{1}{4}F^2 \quad , \qquad (11.2.6b)$$
$$\mathcal{L}_1 = \tfrac{1}{4}h^{ab}(\Box - 2)h_{ab} + \tfrac{1}{2}B\cdot(\Box - 2)B - \tfrac{1}{2}\eta(\Box - 2)\eta + \tfrac{1}{2}(\partial^b h_{ab} + \partial_a\eta - B_a)^2 + \tfrac{1}{2}(\tfrac{1}{4}{h^a}_a + \tfrac{3}{2}\eta + \partial\cdot B)^2 \quad . \qquad (11.2.6c)$$

The gauge transformations are obtained by expanding (4.1.6). The pieces involving $\Lambda_{\alpha\beta}$ are trivial in the component viewpoint, since they are the ones that reduce the components of ϕ to the Sp(2) singlets given in (11.2.4). Since then only the Sp(2) singlet part of $Q_\alpha \Lambda^\alpha$ can contribute, only the $(M_{\alpha\beta})$ isospinor sector of Λ^α is relevant. We therefore take

$$\Lambda_\alpha = (\xi a_{1\alpha}{}^\dagger + \epsilon^a a_{1a}{}^\dagger a_{1\alpha}{}^\dagger + \epsilon a_{2\alpha}{}^\dagger)\,|0\rangle \quad . \tag{11.2.7}$$

Then the invariances are found to be

$$\delta A_a = \partial_a \xi \quad ; \tag{11.2.8a}$$

$$\begin{aligned} \delta h_{ab} &= \partial_{(a}\epsilon_{b)} - \frac{1}{\sqrt{2}}\eta_{ab}\epsilon \quad , \\ \delta B_a &= \partial_a \epsilon + \sqrt{2}\epsilon_a \quad , \\ \delta \eta &= -\partial \cdot \epsilon + \frac{3}{\sqrt{2}}\epsilon \quad . \end{aligned} \tag{11.2.8b}$$

(11.2.6b) and (11.2.8a) are the usual action and gauge invariance for a free photon; however, (11.2.6c) and (11.2.8b) are not in the standard form for massive, symmetric rank two. Letting

$$h_{ab} = \widehat{h}_{ab} + \tfrac{1}{10}\eta_{ab}\widehat{\eta} \quad , \quad \eta = -\tfrac{1}{2}\widehat{h}^a{}_a - \tfrac{3}{10}\widehat{\eta} \quad ; \tag{11.2.9}$$

one finds

$$\delta \widehat{h}_{ab} = \partial_{(a}\epsilon_{b)} \quad , \quad \delta B_a = \partial_a \epsilon + \sqrt{2}\epsilon_a \quad , \quad \delta\widehat{\eta} = -5\sqrt{2}\epsilon \quad . \tag{11.2.10}$$

In this form it's clear that $\widehat{\eta}$ and B_a are Stueckelberg fields that can be gauged away by ϵ and ϵ_a. (This was not possible for η, since the presence of the $\partial \cdot \epsilon$ term in its transformation law (11.2.8b) would require propagating Faddeev-Popov ghosts.) In this gauge $\mathcal{L}_1$ reduces to the Fierz-Pauli lagrangian for a massive, symmetric, rank-two tensor:

$$\mathcal{L} = \tfrac{1}{4}\widehat{h}^{ab}\Box\widehat{h}_{ab} + \tfrac{1}{2}(\partial_b \widehat{h}_{ab})^2 - \tfrac{1}{2}(\partial_b \widehat{h}^{ab})\partial_a \widehat{h}^c{}_c - \tfrac{1}{4}\widehat{h}^a{}_a \Box \widehat{h}^b{}_b - \tfrac{1}{4}(\widehat{h}^{ab}\widehat{h}_{ab} - \widehat{h}^a{}_a{}^2) \quad . \tag{11.2.11}$$

The closed string is treated similarly, so we'll consider only the tachyon and massless levels. The expansion of the physical closed-string field to the second mass level is

$$\phi = (\phi_0 + h^{ab} a^+{}_{1a}{}^\dagger a^-{}_{1b}{}^\dagger + A^{ab} a^+{}_{1a}{}^\dagger a^-{}_{1b}{}^\dagger + \eta a^+{}_1{}^{\alpha\dagger} a^-{}_{1\alpha}{}^\dagger)\,|0\rangle \quad , \tag{11.2.12}$$

where ϕ_0 is the tachyon and h_{ab}, A_{ab}, and η describe the massless sector, consisting of the graviton, an antisymmetric tensor, and the dilaton. We have also dropped fields which are killed by the projection operator for ΔN. We then find for the action (11.1.4):

$$\begin{aligned}\mathcal{L} &= \mathcal{L}_{-2} + \mathcal{L}_0 \quad ; \\ \mathcal{L}_{-2} &= \tfrac{1}{2}\phi_0(\Box + 4)\phi_0 \quad , \\ \mathcal{L}_0 &= \tfrac{1}{4}h^{ab}\Box h_{ab} + \tfrac{1}{4}A^{ab}\Box A_{ab} - \tfrac{1}{2}\eta\Box\eta \\ &\quad + \tfrac{1}{2}(\partial^b h_{ab} + \partial_a\eta)^2 + \tfrac{1}{2}(\partial^b A_{ab})^2 \quad .\end{aligned} \tag{11.2.13}$$

The nontrivial gauge transformations are found from (11.1.4):

$$\delta h_{ab} = \partial_{(a}\epsilon_{b)} \quad , \quad \delta A_{ab} = \partial_{[a}\zeta_{b]} \quad , \quad \delta\eta = \partial\cdot\epsilon \quad . \tag{11.2.14}$$

These lead to the field redefinitions

$$h_{ab} = \widehat{h}_{ab} + \eta_{ab}\widehat{\eta} \quad , \quad \eta = \widehat{\eta} + \tfrac{1}{2}\widehat{h}^a{}_a \quad ; \tag{11.2.15}$$

which result in the improved gauge transformations

$$\delta\widehat{h}_{ab} = \partial_{(a}\epsilon_{b)} \quad , \quad \delta A_{ab} = \partial_{[a}\zeta_{b]} \quad , \quad \delta\widehat{\eta} = 0 \quad . \tag{11.2.16}$$

Substituting back into (11.2.13), we find the covariant action for a tachyon, linearized Einstein gravity, an antisymmetric tensor, and a dilaton [4.10].

The formulation with the world-sheet metric (sect. 8.3) uses more gauge and auxiliary degrees of freedom than even the IGL(1) formulation. We begin with the open string [3.13]. If we evaluate the kinetic operator for the gauge-invariant action (4.1.6) between states without fermionic oscillators, we find $\widehat{Q}^2 \to \sum_1^\infty (b_n{}^\dagger \mathcal{O}_n + \mathcal{O}_n{}^\dagger b_n)$, so the kinetic operator reduces to

$$\delta(M_{\alpha\beta}{}^2)(-2K+\widehat{Q}^2) \quad \to \quad -2K(1-N_{GB})+\sum_1^\infty\left[-b_n{}^\dagger b_n + \frac{1}{\sqrt{n}}(b_n{}^\dagger\widetilde{L}_n + \widetilde{L}_n{}^\dagger b_n)\right] \quad ,$$

$$N_{GB} = \sum_1^\infty (b_n{}^\dagger g_n + g_n{}^\dagger b_n) \quad , \tag{11.2.17}$$

dropping the f and c terms in K and $\widetilde{L}_n$. The operator N_{GB} counts the number of $g^\dagger$'s plus $b^\dagger$'s in a state (without factors of n). We now evaluate the first few component levels. The evaluation of the tachyon action is trivial:

$$\phi = \varphi(x)\,|0\rangle \quad \to \quad \mathcal{L} = \tfrac{1}{2}\varphi(\Box + 2)\varphi \quad , \tag{11.2.18}$$

where $S=\int d^{26}x\ \mathcal{L}$. For the photon, expanding in only Sp(2) singlets,

$$\phi=(A\cdot a_1{}^\dagger+\mathsf{B}g_1{}^\dagger+\mathsf{G}b_1{}^\dagger)|0\rangle \quad\rightarrow$$

$$\mathcal{L}=\tfrac{1}{2}A\cdot\Box A-\tfrac{1}{2}\mathsf{B}^2-\mathsf{B}\partial\cdot A=-\tfrac{1}{4}F_{ab}{}^2-\tfrac{1}{2}(\mathsf{B}+\partial\cdot A)^2 \quad . \tag{11.2.19}$$

The disappearance of G follows from the gauge transformations

$$\Lambda_\alpha=(f_{1\alpha}{}^\dagger\lambda_c+c_{1\alpha}{}^\dagger\lambda_f)|0\rangle \quad\rightarrow$$

$$\delta(A,\mathsf{B},\mathsf{G})=(\partial\lambda_c,-\Box\lambda_c,\lambda_f-\tfrac{1}{2}\lambda_c) \quad . \tag{11.2.20}$$

Since G is the only field gauged by λ_f, no gauge-invariant can be constructed from it, so it must drop out of the action.

For the next level, we consider the gauge transformations first in order to determine which fields will drop out of the action so that its calculation will be simplified. The Sp(2) singlet part of the field is

$$\begin{aligned}\phi=(&\tfrac{1}{2}h^{ab}a_{1a}{}^\dagger a_{1b}{}^\dagger+h^a a_{2a}{}^\dagger+\mathsf{B}^a g_1{}^\dagger a_{1a}{}^\dagger+\mathsf{G}^a b_1{}^\dagger a_{1a}{}^\dagger\\&+hg_1{}^\dagger b_1{}^\dagger+\widehat{\mathsf{B}}g_1{}^{\dagger 2}+\mathsf{G}b_1{}^{\dagger 2}+\mathsf{B}g_2{}^\dagger+\widehat{\mathsf{G}}b_2{}^\dagger+\eta_+f_1{}^{\dagger 2}+\eta_-c_1{}^{\dagger 2}+\eta_0f_1{}^{\alpha\dagger}c_{1\alpha}{}^\dagger)|0\rangle \quad .\end{aligned} \tag{11.2.21}$$

The only terms in Λ_α which contribute to the transformation of the Sp(2) singlets are

$$\begin{aligned}\Lambda_\alpha=\Big[&(\lambda_{cp}\cdot a^\dagger{}_1+\lambda_{cb}g^\dagger{}_1+\lambda_{cg}b^\dagger{}_1)f^\dagger{}_{1\alpha}+(\lambda_{fp}\cdot a^\dagger{}_1+\lambda_{fb}g^\dagger{}_1+\lambda_{fg}b^\dagger{}_1)c^\dagger{}_{1\alpha}\\&+\lambda_cf^\dagger{}_{2\alpha}+\lambda_fc^\dagger{}_{2\alpha}\Big]\Big|0\Big\rangle \quad .\end{aligned} \tag{11.2.22}$$

The gauge transformations of the components are then

$$\begin{aligned}\delta h_{ab}&=\partial_{(a}\lambda^{cp}{}_{b)}-\eta_{ab}\frac{1}{\sqrt{2}}\lambda_c \quad ,\\ \delta h_a&=\sqrt{2}\lambda^{cp}{}_a+\partial_a\lambda_c \quad ,\\ \delta\mathsf{B}_a&=\partial_a\lambda_{cb}+2K\lambda^{cp}{}_a \quad ,\\ \delta\mathsf{G}_a&=-\tfrac{1}{2}\lambda^{cp}{}_a+\lambda^{fp}{}_a+\partial_a\lambda_{cg} \quad ,\\ \delta h&=\frac{1}{\sqrt{2}}\lambda_c-\tfrac{1}{2}\lambda_{cb}+\lambda_{fb}+2K\lambda_{cg} \quad ,\\ \delta\widehat{\mathsf{B}}&=2K\lambda_{cb} \quad ,\\ \delta\mathsf{G}&=-\tfrac{1}{2}\lambda_{cg}+\lambda_{fg} \quad ,\end{aligned}$$

$$\begin{aligned}
\delta\mathsf{B} &= \sqrt{2}\lambda_{cb} + 2K\lambda_c \quad , \\
\delta\widehat{\mathsf{G}} &= -\tfrac{1}{2}\lambda_c + \lambda_f + \sqrt{2}\lambda_{cg} \quad , \\
\delta\eta_+ &= \lambda_{cb} \quad , \\
\delta\eta_- &= \sqrt{2}\lambda_f - \tfrac{1}{2}\lambda_{fb} - \partial\cdot\lambda_{fp} + 2K\lambda_{fg} \quad , \\
\delta\eta_0 &= \sqrt{2}\lambda_c - \tfrac{1}{4}\lambda_{cb} + \tfrac{1}{2}\lambda_{fb} - \tfrac{1}{2}\partial\cdot\lambda_{cp} + K\lambda_{cg} \quad .
\end{aligned} \tag{11.2.23}$$

We then gauge away

$$\begin{aligned}
\mathsf{G}_a &= 0 \quad \rightarrow \quad \lambda^{fp}{}_a = \tfrac{1}{2}\lambda^{cp}{}_a - \partial_a\lambda_{cg} \quad , \\
\mathsf{G} &= 0 \quad \rightarrow \quad \lambda_{fg} = \tfrac{1}{2}\lambda_{cg} \quad , \\
\eta_+ &= 0 \quad \rightarrow \quad \lambda_{cb} = 0 \quad , \\
\eta_- &= 0 \quad \rightarrow \quad \lambda_f = \frac{1}{2\sqrt{2}}\lambda_{fb} + \frac{1}{2\sqrt{2}}\partial\cdot\lambda_{cp} - \frac{1}{\sqrt{2}}(\Box + K)\lambda_{cg} \quad , \\
\eta_0 &= 0 \quad \rightarrow \quad \lambda_{fb} = -2\sqrt{2}\lambda_c + \partial\cdot\lambda_{cp} - 2K\lambda_{cg} \quad .
\end{aligned} \tag{11.2.24}$$

The transformation laws of the remaining fields are

$$\begin{aligned}
\delta h_{ab} &= \partial_{(a}\lambda_{b)} - \eta_{ab}\lambda \quad , \\
\delta h_a &= \sqrt{2}(\lambda_a + \partial_a\lambda) \quad , \\
\delta h &= -3\lambda + \partial\cdot\lambda \quad , \\
\delta\mathsf{B}_a &= -(\Box - 2)\lambda_a \quad , \\
\delta\mathsf{B} &= -\sqrt{2}(\Box - 2)\lambda \quad , \\
\delta\widehat{\mathsf{B}} &= 0 \quad , \\
\delta\widehat{\mathsf{G}} &= \frac{1}{\sqrt{2}}(-3\lambda + \partial\cdot\lambda) \quad ,
\end{aligned} \tag{11.2.25}$$

where $\lambda_a = \lambda^{cp}{}_a$ and $\lambda = \lambda_c/\sqrt{2}$. λ_{cg} drops out of the transformation law as a result of the gauge invariance for gauge invariance

$$\delta\phi = \widehat{\mathcal{Q}}^\alpha\Lambda_\alpha \quad \rightarrow \quad \delta\Lambda_\alpha = \widehat{\mathcal{Q}}^\beta\Lambda_{(\alpha\beta)} \quad . \tag{11.2.26}$$

The lagrangian can then be computed from (4.1.6) and (11.2.17) to be

$$\begin{aligned}
\mathcal{L} = {} & \tfrac{1}{4}h^{ab}(\Box - 2)h_{ab} + \tfrac{1}{2}h^a(\Box - 2)h_a - \tfrac{1}{2}h(\Box - 2)h \\
& - \tfrac{1}{2}\mathsf{B}_a{}^2 - \tfrac{1}{2}\mathsf{B}^2 + 2\widehat{\mathsf{B}}(\sqrt{2}\widehat{\mathsf{G}} - h) \\
& + \mathsf{B}^a(-\partial^b h_{ab} + \partial_a h + \sqrt{2}h_a) + \mathsf{B}(-\partial\cdot h + \frac{3}{\sqrt{2}}h - \frac{1}{2\sqrt{2}}h^a{}_a) \, .
\end{aligned} \tag{11.2.27}$$

The lagrangian of (11.2.19) is equivalent to that obtained from the IGL(1) action, whereas the lagrangian of (11.2.27) is like the IGL(1) one but contains in addition the 2 gauge-invariant auxiliary fields $\widehat{\mathsf{B}}$ and $\sqrt{2}\widehat{\mathsf{G}} - h$. Both reduce to the OSp(1,1|2) lagrangians after elimination of auxiliary fields.

The results for the closed string are similar. Here we consider just the massless level of the nonoriented closed string (which is symmetric under interchange of + and − modes). The Sp(2) and ΔN invariant components are

$$\phi = (h^{ab}a^\dagger{}_{+a}a^\dagger{}_{-a} + B^a g^\dagger{}_{(+}a^\dagger{}_{-)a} + G^a b^\dagger{}_{(+}a^\dagger{}_{-)a} + hg^\dagger{}_{(+}b^\dagger{}_{-)} + Bg^\dagger{}_+ g^\dagger{}_- + Gb^\dagger{}_+ b^\dagger{}_- \\ + \eta_+ f^\dagger{}_+{}^\alpha f^\dagger{}_{-\alpha} + \eta_- c^\dagger{}_+{}^\alpha c^\dagger{}_{-\alpha} + \eta_0 f^\dagger{}_{(+}{}^\alpha c^\dagger{}_{-)\alpha})\,|0\rangle \quad , \qquad (11.2.28)$$

where all oscillators are from the first mode ($n = 1$), and h^{ab} is symmetric. The gauge parameters are

$$\Lambda_\alpha = \left[(\lambda_{cp}\cdot a^\dagger + \lambda_{cb}g^\dagger + \lambda_{cg}b^\dagger)_{(+}f^\dagger{}_{-)\alpha} + (\lambda_{fp}\cdot a^\dagger + \lambda_{fb}g^\dagger + \lambda_{fg}b^\dagger)_{(+}c^\dagger{}_{-)\alpha}\right]|0\rangle \quad . \qquad (11.2.29)$$

The component transformations are then

$$\begin{aligned}
\delta h_{ab} &= \partial_{(a}\lambda^{cp}{}_{b)} \quad , \\
\delta B_a &= \partial_a\lambda_{cb} - \tfrac{1}{2}\Box\lambda^{cp}{}_a \quad , \\
\delta G_a &= 2\lambda^{fp}{}_a - \lambda^{cp}{}_a + \partial_a\lambda_{cg} \quad , \\
\delta h &= 2\lambda_{fb} - \lambda_{cb} - \tfrac{1}{2}\Box\lambda_{cg} \quad , \\
\delta B &= -\Box\lambda_{cb} \quad , \\
\delta G &= 4\lambda_{fg} - 2\lambda_{cg} \quad , \\
\delta\eta_+ &= 2\lambda_{cb} \quad , \\
\delta\eta_- &= -\lambda_{fb} - \partial\cdot\lambda_{fp} - \tfrac{1}{2}\Box\lambda_{fg} \quad , \\
\delta\eta_0 &= \lambda_{fb} - \tfrac{1}{2}\lambda_{cb} - \tfrac{1}{2}\partial\cdot\lambda_{cp} - \tfrac{1}{4}\Box\lambda_{cg} \quad .
\end{aligned} \qquad (11.2.30)$$

We choose the gauges

$$\begin{aligned}
G_a &= 0 \quad \rightarrow \quad \lambda^{fp}{}_a = \tfrac{1}{2}\lambda^{cp}{}_a - \tfrac{1}{2}\partial_a\lambda_{cg} \quad , \\
G &= 0 \quad \rightarrow \quad \lambda_{fg} = \tfrac{1}{2}\lambda_{cg} \quad , \\
\eta_+ &= 0 \quad \rightarrow \quad \lambda_{cb} = 0 \quad , \\
\eta_0 &= 0 \quad \rightarrow \quad \lambda_{fb} = \tfrac{1}{2}\partial\cdot\lambda_{cp} + \tfrac{1}{4}\Box\lambda_{cg}
\end{aligned} \qquad (11.2.31)$$

The remaining fields transform as

$$\begin{aligned}
\delta h_{ab} &= \partial_{(a}\lambda_{b)} \quad , \\
\delta B_a &= -\tfrac{1}{2}\Box\lambda_a \quad , \\
\delta h &= \partial\cdot\lambda \quad , \\
\delta B &= 0 \quad , \\
\delta\eta_- &= -\partial\cdot\lambda \quad ,
\end{aligned} \qquad (11.2.32)$$

where $\lambda_a = \lambda^{cp}{}_a$, and λ_{cg} drops out. Finally, the lagrangian is

$$\mathcal{L} = \tfrac{1}{2}h^{ab}\Box h_{ab} - h\Box h - 4B_a{}^2 - 8B(\eta_- + h) - 4B^a(\partial^b h_{ab} - \partial_a h) \quad . \qquad (11.2.33)$$

Again we find the auxiliary fields B and $\eta_- + h$ in addition to the usual nonminimal ones. After elimination of auxiliary fields, this lagrangian reduces to that of (11.2.13) (for the nonoriented sector, up to normalization of the fields).

Exercises

(1) Derive the gauge-invariant actions (IGL(1) and OSp(1,1|2)) for free open strings. Do the same for the Neveu-Schwarz string. Derive the OSp(1,1|2) action for the Ramond string.

(2) Find all the Sp(2)-singlet fields at the levels indicated in (11.2.2). Separate them into sets corresponding to irreducible representations of the Poincaré group (including their Stueckelberg and auxiliary fields).

(3) Derive the action for the massless level of the open string using the bosonized ghosts of sect. 8.1.

(4) Derive the action for the next mass level of the closed string after those in (11.2.13). Do the same for (11.2.33).

(5) Rederive the actions of sect. 11.2 for levels which include spin 2 by first decomposing the corresponding light-cone representations into irreducible representations, and then using the Hilbert-space constructions of sect. 4.1 for each irreducible representation.

12. GAUGE-INVARIANT INTERACTIONS

12.1. Introduction

The gauge-invariant forms of the interacting actions for string field theories are far from understood. Interacting closed string field theory does not yet exist. (Although a proposal has been made [12.1], it is not even at the point where component actions can be examined.) An open-string formulation exists [4.8] (see the following section), but it does not seem to relate to the light-cone formulation (chapt. 10), and has more complicated vertices. Furthermore, all these formulations are in the IGL(1) formalism, so relation to particles is less direct because of the need to eliminate BRST auxiliary fields. Most importantly, the concept of conformal invariance is not clear in these formulations. If a formulation could be found which incorporated the world-sheet metric as coordinates, as the free theory of sect. 8.3, it might be possible to restore conformal transformations as a larger gauge invariance which allowed the derivation of the other formulations as (partial) gauge choices.

In this section we will mostly discuss the status of the derivation of an interacting gauge-covariant string theory from the light cone, with interactions similar to those of the light-cone string field theory. The derivation follows the corresponding derivation for Yang-Mills described in sects. 3.4 and 4.2 [3.14], but the important step (3.4.17) of eliminating p_+ dependence has not yet been performed.

As performed for Yang-Mills in sect. 3.4, the transformation (3.4.3a) with $\Phi \to U^{-1}\Phi$ is the first step in deriving an IGL(1) formalism for the interacting string from the light cone. Since the transformation is nonunitary, the factor of p_+ in (2.4.9) is canceled. In (2.4.7), using integration by parts, (3.4.3a) induces the transformation of the vertex function

$$\mathcal{V}^{(n)} \quad \to \quad \frac{1}{p_{+1}\cdots p_{+n}} U(1)\cdots U(n)\mathcal{V}^{(n)} \quad , \tag{12.1.1}$$

where we work in momentum space with respect to p_+. The lowest-order interacting contribution to Q then follows from applying this transformation to the

OSp-extended form of the light-cone vertex (10.6):

$$\mathcal{J}^{(3)}{}_{-}{}^{c} = -2ig\frac{p_{+r}}{p_{+1}p_{+2}p_{+3}}X_r{}^c(\sigma_{int})\,\delta\left(\sum p_+\right)\Delta[X^a]\,\Delta[p_+X^c]\,\Delta[p_+{}^{-1}X^{\tilde{c}}] \tag{12.1.2}$$

effectively giving conformal weight -1 to X^c and conformal weight 1 to $X^{\tilde{c}}$. A δ-functional that matches a coordinate must also match the σ-derivative of the coordinate, and with no zero-modes one must have

$$\Delta\left[p_+{}^{-1}X^{\tilde{c}}\left(\frac{\sigma}{p_+}\right)\right]\bigg|_{\tilde{c}=0} = \Delta\left[\partial_\sigma p_+{}^{-1}X^{\tilde{c}}\left(\frac{\sigma}{p_+}\right)\right] = \Delta\left[p_+{}^{-2}X^{\tilde{c}\prime}\left(\frac{\sigma}{p_+}\right)\right] \quad . \tag{12.1.3}$$

(Even the normalization is unambiguous, since without zero-modes Δ can be normalized to 1 between ground states.) We now recognize X^c and $X^{\tilde{c}\prime}$ to be just the usual Faddeev-Popov ghost $C(\sigma)$ of τ-reparametrizations and Faddeev-Popov *anti*ghost $\delta/\delta\widetilde{C}(\sigma)$ of σ-reparametrizations (as in (8.1.13)), of conformal weights -1 and 2, respectively, which is equivalent to the relation (8.2.2,3) (as seen by using (7.1.7b)). Finally, we can (functionally) Fourier transform the antighost so that it is replaced with the canonically conjugate ghost. Our final vertex function is therefore

$$\mathcal{J}^{(3)}{}_{-}{}^{c} = -2ig\frac{p_{+r}}{p_{+1}p_{+2}p_{+3}}C_r(\sigma_{int})\,\delta\left(\sum p_+\right)\Delta[X^a]\,\Delta[p_+C]\,\Delta[p_+\widetilde{C}] \quad , \tag{12.1.4a}$$

or in terms of momenta

$$\widetilde{\mathcal{J}}^{(3)}{}_{-}{}^{c} = -2igp_{+r}C_r(\sigma_{int})\,\delta\left(\sum p_+\right)\Delta[p_+{}^{-1}P^a]\,\Delta\left[p_+{}^{-2}\frac{\delta}{\delta C}\right]\Delta\left[p_+{}^{-2}\frac{\delta}{\delta\widetilde{C}}\right] \quad . \tag{12.1.4b}$$

The extra p_+'s disappear due to Fourier transformation of the zero-modes c:

$$\frac{1}{p_+}\int dc\, e^{-cp^{\tilde{c}}}f(p_+c) = \int dc\, e^{-cp^{\tilde{c}}/p_+}f(c) = \tilde{f}\left(\frac{p^{\tilde{c}}}{p_+}\right) \quad . \tag{12.1.5}$$

Equivalently, the exponent of U by (3.4.3a) is $c\partial/\partial c + M^3$, so the zero-mode part just scales c, but $c\partial/\partial c = 1 - (\partial/\partial c)c$, so besides scaling $\partial/\partial c$ there is an extra factor of p_+ for each zero-mode, canceling those in (12.1.1,2). There is no effect on the normalization with respect to nonzero-modes because of the above-mentioned normalization in the definition of Δ with respect to the creation and annihilation operators. A similar analysis applies to closed strings [12.2].

Hata, Itoh, Kugo, Kunitomo, and Ogawa [12.3] proposed an interacting BRST operator equivalent to this one, and corresponding gauge-fixed and gauge-invariant actions, but with p_+ treated as an extra coordinate as in [2.7]. (A similar earlier attempt appeared in [4.9,7.5], with p_+ fixed, as a consequence of which the BRST algebra didn't close to all orders. Similar attempts appeared in [12.4].) However, as explained in [2.7], such a formalism requires also an additional anticommuting coordinate in order for the loops to work (as easily checked for the planar 1-loop graph with external tachyons [2.7]), and can lead to problems with infrared behavior [4.4].

The usual four-point vertex of Yang-Mills (and even-higher-point vertices of gravity for the closed string) will be obtained only after field redefinitions of the massive fields. This corresponds to the fact that it shows up in the zero-slope limit of the S-matrix only after massive propagators have been included and reduced to points. In terms of the Lagrangian, for arbitrary massive fields μ and massless fields ν, the terms, for example,

$$L = \tfrac{1}{2}\mu[\Box + M^2 + M^2 U(\nu)]\mu + M^2 \mu V(\nu) \qquad (12.1.6a)$$

(where $U(\nu)$ and $V(\nu)$ represent some interaction terms) become, in the limit $M^2 \to \infty$,

$$L = M^2[\tfrac{1}{2}(1+U)\mu^2 + V\mu] \quad . \qquad (12.1.6b)$$

The corresponding field redefinition is

$$\mu = \tilde{\mu} - \frac{V}{1+U} \quad , \qquad (12.1.6c)$$

which modifies the Lagrangian to

$$L = \tfrac{1}{2}\tilde{\mu}(\Box + M^2 + M^2 U)\tilde{\mu} + \tfrac{1}{2}M^2 \frac{V^2}{1+U} + \mathcal{O}(M^0) \quad . \qquad (12.1.6d)$$

The redefined massive fields $\tilde{\mu}$ no longer contribute in the zero-slope limit, and can be dropped from the Lagrangian before taking the limit. However, the redefinition has introduced the new interaction term $\frac{1}{2}M^2 \frac{V^2}{1+U}$ into the ν-part of the Lagrangian.

12.2. Midpoint interaction

Witten has proposed an extension of the IGL(1) gauge-invariant open bosonic string action to the interacting case [4.8]. Although there may be certain limitations with his construction, it shares certain general properties with the light-cone (and covariantized light-cone) formalism, and thus we expect these properties will be common to any future approaches. The construction is based on the use of a vertex which consists mainly of δ-functionals, as in the light-cone formalism. Although the geometry of the infinitesimal surface corresponding to these δ-functionals differs from that of the light-cone case, they have certain algebraic features in common. In particular, by considering a structure for which the δ-functional (times certain vertex factors) is identified with the product operation of a certain algebra, the associativity of the product follows from the usual properties of δ-functionals. This is sufficient to define an interacting, nilpotent BRST operator (or Lorentz generators with $[J_{-i}, J_{-j}] = 0$), which in turn gives an interacting gauge-invariant (or Lorentz-invariant) action.

The string fields are elements of an algebra: a vector space with an outer product $*$. We write an explicit vector index on the string field Φ_i, where $i = \mathbf{Z}(\sigma)$ is the coordinates ($X, C, \widetilde{C}$ for the covariant formalism and $X_T, x_\pm$ for the light cone), and excludes group-theory indices. Then the product can be written in terms of a rank-3 matrix

$$(\Phi * \Psi)_i = f_i{}^{jk}\Phi_k\Psi_j \quad . \tag{12.2.1}$$

In order to construct actions, and because of the relation of a field to a first-quantized wave function, we require, in addition to the operations necessary to define an algebra, a Hilbert-space inner product

$$\langle\Phi|\Psi\rangle = \int \mathcal{D}Z \; tr\; \Phi^\dagger \Psi = tr\; \Phi^{i\dagger}\Psi_i \quad , \tag{12.2.2}$$

where tr is the group-theory trace. Furthermore, in order to give the hermiticity condition on the field we require an indefinite, symmetric charge-conjugation matrix Ω on this space:

$$\Phi_i = \Omega_{ij}\Phi^{j\dagger} \quad , \quad (\Omega\Phi)[X(\sigma), C(\sigma), \widetilde{C}(\sigma)] \equiv \Phi[X(\pi-\sigma), C(\pi-\sigma), -\widetilde{C}(\pi-\sigma)] \quad . \tag{12.2.3a}$$

Ω is the "twist" of (11.1.1). To allow a reality condition or, combining with (12.2.3a), a symmetry condition for real group representations (for SO(N) or USp(N)), we also require that indices can be freely raised and lowered:

$$(\eta\Phi)_i = \Omega_{ij}\delta^{jk}(\eta\Phi)^t{}_k \quad , \tag{12.2.3b}$$

where η is the group metric and t is the group-index transpose. In order to perform the usual graphical manipulations implied by duality, the twist must have the usual effect on vertices, and thus on the inner product:

$$\Omega(\Phi * \Psi) = (\Omega\Psi) * (\Omega\Phi) \quad . \tag{12.2.4}$$

Further properties satisfied by the product follow from the nilpotence of the BRST operator and integrability of the field equations $Q\Phi = 0$: Defining

$$Q\Phi = Q_0\Phi + \Phi * \Phi \quad \to \quad (Q\Phi)_i = Q_{0i}{}^j\Phi_j + f_i{}^{jk}\Phi_k\Phi_j \quad , \tag{12.2.5}$$

$$S = \int \mathcal{D}\mathbf{Z}\; tr\; \Phi^\dagger(\tfrac{1}{2}Q_0\Phi + \tfrac{1}{3}\Phi * \Phi) = tr\; \left[\tfrac{1}{2}(\Omega Q_0)^{ij}\Phi_j\Phi_i + \tfrac{1}{3}(\Omega f)^{ijk}\Phi_k\Phi_j\Phi_i\right] \quad , \tag{12.2.6}$$

we find that hermiticity requires

$$(\Omega Q_0)^{ij} = (\bar{Q}_0\Omega)^{ji} \quad , \quad (\Omega f)^{ijk} = (\bar{f}\Omega\Omega)^{kji} \quad , \tag{12.2.7}$$

integrability requires antisymmetry of Q_0 and cyclicity of $*$:

$$(\Omega Q_0)^{ij} = -(\Omega Q_0)^{ji} \quad , \quad (\Omega f)^{ijk} = (\Omega f)^{jki} \tag{12.2.8}$$

(where permutation of indices is in the "graded" sense, but we have omitted some signs: e.g., $(\Omega Q_0)^{ij}\Phi_j\Psi_i = +(\Omega Q_0)^{ij}\Psi_j\Phi_i$ when $\Phi[\mathbf{Z}]$ and $\Psi[\mathbf{Z}]$ are *anticommuting*, but Φ_i and Ψ_i include components of *either* statistics), and nilpotence requires, besides ${Q_0}^2 = 0$, that $*$ is BRST invariant (i.e., Q_0 is distributive over $*$) and associative:

$$Q_0(\Phi * \Psi) = (Q_0\Phi) * \Psi + \Phi * (Q_0\Psi)$$

$$\leftrightarrow \quad {Q_{0i}}^l(f\Omega\Omega)_{ljk} + {Q_{0j}}^l(f\Omega\Omega)_{ilk} + {Q_{0k}}^l(f\Omega\Omega)_{ijl} = 0 \quad , \tag{12.2.9}$$

$$\Phi * (\Psi * \Upsilon) = (\Phi * \Psi) * \Upsilon \quad \leftrightarrow \quad (\Omega f)^{ijm}{f_m}^{kl} = (\Omega f)^{jkm}{f_m}^{li} \tag{12.2.10}$$

(where we have again ignored some signs due to grading). $*$ should also be invariant under all transformations under conserved quantities, and thus the operators $\partial/\partial\mathbf{z} \sim \int d\sigma\; \partial/\partial\mathbf{Z}$ must also be distributive over $*$.

At this point we have much more structure than in an ordinary algebra, and only one more thing needs to be introduced in order to obtain a matrix algebra: an identity element for the outer product

$$\Phi * I = I * \Phi = \Phi \quad \leftrightarrow \quad {f_i}^{jk} I_k = {\delta_i}^j \quad . \tag{12.2.11}$$

It's not clear why string field theory must have such an object, but both the light-cone approach and Witten's approach have one. In the light-cone approach the identity element is the ground state (tachyon) at vanishing momentum (including the string length, $2\pi\alpha' p_+$), which is related to the fact that the vertex for an external tachyon takes the simple form $: e^{-ik\cdot X(0)} :$. We now consider the fields as being matrices in $\mathbf{Z}(\sigma)$-space as well as in group space, although the vector space on which such matrices act might not be (explicitly) defined. (Such a formalism might be a consequence of the same duality properties that require general matrix structure, as opposed to just adjoint representation, in the group space.) $*$ is now the matrix product. (12.2.7,10) then express just the usual hermiticity and associativity properties of the matrix product. The trace operation Tr of these matrices is implied by the Hilbert-space inner product (12.2.2):

$$\langle\Phi|\Psi\rangle = Tr\ \Phi^\dagger\Psi \quad \leftrightarrow \quad Tr\ \Phi = \langle I|\Phi\rangle \quad . \tag{12.2.12}$$

(12.2.8) states the usual cyclicity of the trace. Finally, the twist metric (12.2.3) is identified with the matrix transpose, in addition to transposition in the group space, as implied by (12.2.4). Using this transposition in combination with the usual hermitian conjugation to define the matrix complex conjugate, the hermiticity condition (12.2.3a) becomes just hermiticity in the group space: Denoting the group-space matrix indices as $\Phi_{\mathbf{a}}{}^{\mathbf{b}}$,

$$\Phi_{\mathbf{a}}{}^{\mathbf{b}} = \Phi_{\mathbf{b}}{}^{\mathbf{a}*} \quad . \tag{12.2.13}$$

As a result of

$$I^t = I \quad \leftrightarrow \quad \Omega_{ij} I_j = I_i \tag{12.2.14}$$

and the fact that Q_0 and $\partial/\partial\mathbf{z}$ are distributive as well as being "antisymmetric" (odd under simultaneous twisting and integration by parts), we find

$$Q_0 I = \frac{\partial}{\partial\mathbf{z}} I = 0 \quad . \tag{12.2.15}$$

Given one $*$ product, it's possible to define other associative products by combining it with some operators d which are distributive over it. Thus, the condition of associativity of $\star$ and $*$ implies

$$A \star B = A * dB \quad \rightarrow \quad d^2 = 0$$

$$A \star B = (dA) * (dB) \quad \rightarrow \quad d^2 = d \ or\ d^2 = 0 \quad . \tag{12.2.16}$$

The former allows the introduction of conserved anticommuting factors (as for the BRST open-string vertex), while the latter allows the introduction of projection operators (as expected for closed strings with respect to ΔN).

The gauge transformations and action come directly from the interacting BRST operator: Using the analysis of (4.2.17-21),

$$\delta\Phi = \left[\left[\int \Lambda\Phi, Q\right]_c, \Phi\right]_c = Q_0\Lambda + [\Phi, *\Lambda] \quad , \tag{12.2.17a}$$

where the last bracket is the commutator with respect to the $*$ product, and

$$S = -iQ = \int \tfrac{1}{2}\Phi^\dagger Q_0 \Phi + \tfrac{1}{3}\Phi^\dagger(\Phi * \Phi) \quad , \tag{12.2.17b}$$

where for physical fields we restrict to

$$J^3\Phi = 0 \quad , \quad J^3\Lambda = -\Lambda \quad . \tag{12.2.17c}$$

Gauge invariance follows from $Q^2 = \frac{1}{2}[Q,Q]_c = 0$. Actually, the projection onto $J^3 = 0$ is somewhat redundant, since the other fields can be removed by gauge transformations or nondynamical field equations, at least at the classical level. (See (3.4.18) for Yang-Mills.)

A possible candidate for a gauge-fixed action can be written in terms of Q as

$$S = \left[Q, \tfrac{1}{2}\int \Phi^\dagger \mathcal{O}\Phi\right]_c + \tfrac{1}{6}\int \Phi^\dagger (Q_{INT}\Phi) \tag{12.2.18a}$$

$$= -\tfrac{1}{2}\int (Q_0\Phi)^\dagger \left[c, \frac{\partial}{\partial c}\right]\Phi - \tfrac{1}{2}\int (Q_{INT}\Phi)^\dagger \left(\left[c, \frac{\partial}{\partial c}\right] - \tfrac{1}{3}\right)\Phi \ , \tag{12.2.18b}$$

where Q_0 and Q_{INT} are the free and interaction terms of Q, and $\mathcal{O} = \frac{1}{2}[c, \partial/\partial c]$. Each term in (12.2.18a) is separately BRST invariant. The BRST invariance of the second term follows from the associativity property of the $*$ product. Due to the $-1/3$ in (12.2.18b) one can easily show that all $\phi^2\psi$ terms drop out, due just to the c dependence of Φ and the cyclicity of Q_{INT}. Such terms would contain auxiliary fields which drop out of the free action. We would like these fields to occur only in a way which could be eliminated by field redefinition, corresponding to maintaining a gauge invariance of the free action at the interacting level, so we could choose the gauge where these fields vanish. Unfortunately, this is not the case in (12.2.18), so allowing this abelian gauge invariance would require adding some additional cubic-interaction gauge-fixing term, each term of which contains auxiliary-field factors, such that the undesired auxiliary fields are eliminated from the action.

Whereas the δ-functionals used in the light-cone formalism correspond to a "flat" geometry (see chapt. 10), with all curvature in the boundary (specifically, the splitting point) rather than the surface itself, those used in Witten's covariant formalism correspond to the geometry (with the 3 external legs amputated)

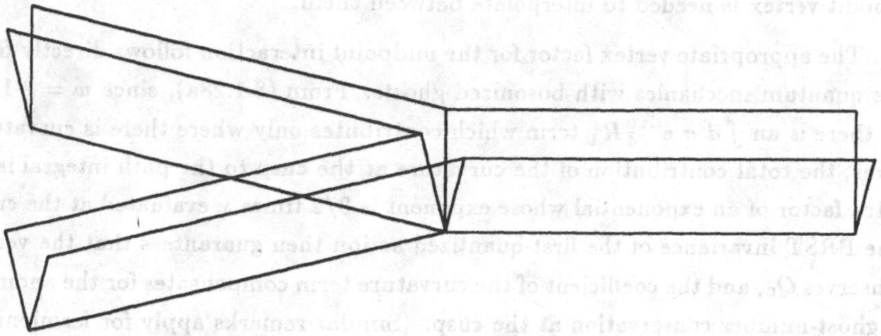

(The boundaries of the strings are on top; the folds along the bottom don't affect the intrinsic geometry.) All the curvature is concentrated in the cusp at the bottom (a small circle around it subtends an angle of 3π), with no intrinsic curvature in the boundary (all parts of the boundary form angles of π with respect to the surface). Each string is "folded" in the middle ($\pi/2$), and thus the vertex δ-functionals equate the coordinates $\mathbf{Z}(\sigma)$ of one string with $\mathbf{Z}(\pi-\sigma)$ for the next string for $\sigma \in [0,\pi/2]$ (and therefore with $\mathbf{Z}(\pi-\sigma)$ for the previous string for $\sigma \in [\pi/2,\pi]$). These δ functions are easily seen to define an associative product: Two successive $*$ products produce a configuration like the one above, but with 4 strings, and associativity is just the cyclicity of this 4-string object (see (12.2.10)). (However, vertex factors can ruin this associativity because of divergences of the coincidence of two such factors in the product of two vertices, as in the superstring: see below.)

Similar remarks apply to the corresponding product implied by the δ functionals of (9.2.23) of the light-cone formalism, but there associativity is violated by an amount which is fixed by the light-cone 4-point interaction vertex. This is due to the fact that in the light-cone formalism there is a 4-point graph where a string has a string split off from one side, followed by an incoming string joining onto the same side. If no conformal transformation is made, this graph is nonplanar, unlike the graph where this splitting and joining occur on opposite sides. There is a similar graph where the splitting and joining occur on the opposite side from the first graph, and these 2 graphs are continuously related by a graph with a 4-point vertex as described in chapt. 10. The σ-position of the interaction point in the surface of the string varies from one end of the string to the other, with the vertex having this point at an end being the same as the limit of the 1 of the other 2 graphs where the propagator has vanishing length. On the other hand, in the covariant formalism the limits of the 2 corresponding graphs are the same, so no 4-point vertex is needed to interpolate between them.

The appropriate vertex factor for the midpoint interaction follows directly from the quantum mechanics with bosonized ghosts. From (8.1.28a), since $w = -1$ for $\widehat{C}$, there is an $\int d^2\sigma\, e^{-1}\frac{3}{2}R\chi$ term which contributes only where there is curvature. Thus, the total contribution of the curvature at the cusp to the path integral is an extra factor of an exponential whose exponent is 3/2 times χ evaluated at the cusp. The BRST invariance of the first-quantized action then guarantees that the vertex conserves Q_0, and the coefficient of the curvature term compensates for the anomaly in ghost-number conservation at the cusp. (Similar remarks apply for fermionized ghosts using the Lorentz connection term.) Alternatively, the coefficient follows from considering the ghost number of the fields and what ghost number is required for the vertex to give the same matrix elements for physical polarizations as in the light-cone formalism: In terms of bosonized-ghost coordinates, any physical state must be $\sim e^{-\hat{q}/2}$ by (8.1.19,21a). Since the δ functional and functional integration in χ have no such factors, and the vertex factor can be only at the cusp (otherwise it destroys the above properties of the δ functionals), it must be $e^{3\chi/2}$ evaluated at the cusp to cancel the $\hat{q}$-dependence of the 3 fields. In terms of the original fermionic ghosts, we use the latter argument, since the anomalous curvature term doesn't show up in the classical mechanics lagrangian (although a similar argument could be made by considering quantum mechanical corrections). Then the physical states have no dependence on c, while the vertex has a dc integration for each of the 3 coordinates, and a single δ function for overall conservation of the "momentum" c. (A similar argument follows from working in terms of Fourier transformed fields which depend instead on the "coordinate" $\partial/\partial c$.) The appropriate vertex factor is thus

$$C(\tfrac{\pi}{2})\tilde{C}(\tfrac{\pi}{2}) \sim \hat{C}(\tfrac{\pi}{2})\hat{C}(-\tfrac{\pi}{2}) \quad , \qquad (12.2.19a)$$

or, in terms of bosonized ghosts (but still for fields with fermionic ghost coordinates)

$$e^{2\chi(\pi/2)} \sim e^{\hat{\chi}(\pi/2)}e^{\hat{\chi}(-\pi/2)} \quad , \qquad (12.2.19b)$$

where $\pi/2$, the midpoint of each string, is the position of the cusp. (The difference in the vertex factor for different coordinates is analogous to the fact that the "scalar" $(-g)^{-1/2}\delta^D(x-x')$ in general relativity has different expressions for g in different coordinate systems.) The vertex, including the factor (12.2.19), can be considered a Heisenberg-picture vacuum in the same way as in the light-cone formalism, where the vertex $|V\rangle = e^{\Delta}\,|0\rangle$ in (9.2.1) was the effect of acting on the interaction-picture vacuum with the S-matrix (of the first-quantized theory). However, in this case the vacuum includes vertex factors because the appropriate vacuum is not the tachyon one but rather the one left invariant by the Sp(2) subalgebra of the Virasoro algebra [12.5] (see sect. 8.1). This is a consequence of the Sp(2) symmetry of the tree graphs.

Because of the midpoint form of the interaction there is a global symmetry corresponding to conformal transformations which leave the midpoint fixed. In second-quantized notation, any operator which is the bracket of (the interacting) Q with something itself has a vanishing bracket with Q, and is therefore simultaneously BRST invariant and generates a global symmetry of the action (because Q is the action). In particular, we can consider

$$\left[Q, \int \Phi^\dagger \left(\frac{\delta}{\delta\widehat{C}(\sigma)} - \frac{\delta}{\delta\widehat{C}(\pi-\sigma)}\right)\Phi\right]_c \sim \int \Phi^\dagger \left(\widehat{\mathcal{G}}(\sigma) - \widehat{\mathcal{G}}(\pi-\sigma)\right)\Phi \quad , \qquad (12.2.20)$$

where the 2 $\delta/\delta\widehat{C}$ terms cancel in the interaction term because of the form of the overlap integral for the vertex (and the location of the vertex factor (12.2.19) at the midpoint), and the surviving free term comes from the first-quantized expression for $\widehat{\mathcal{G}} = \{Q_0, \delta/\delta\widehat{C}\}$. Thus, this subalgebra of the Virasoro algebra remains a global invariance at the interacting level (without becoming inhomogeneous in the fields).

The mode expansion of Witten's vertex can be evaluated [12.6,7] as in the light-cone case (sect. 9.2). (Partial evaluations were given and BRST invariance was also studied in [12.8].) Now

$$\Delta = \tfrac{1}{2}\Delta(\check{P},\check{P}) - \Delta\left(\check{C}', \frac{\delta}{\delta\check{C}}\right) \quad . \qquad (12.2.21)$$

($\widehat{C}$ has weight $w = 2$.) The map from the ρ plane to the z plane can be found from the following sequence of conformal transformations:

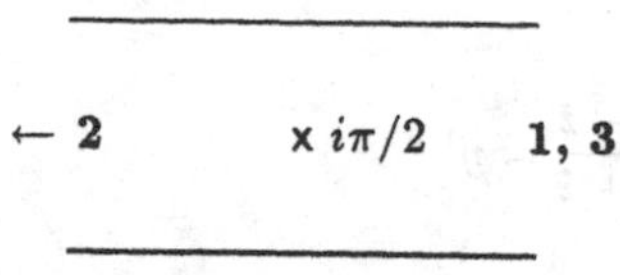

$\rho = ln\,\zeta$

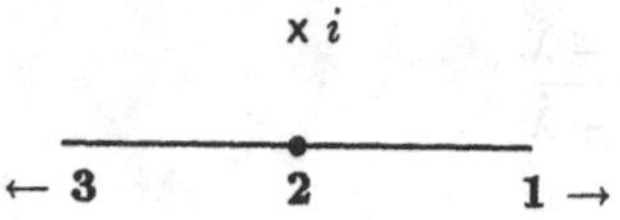

$$\zeta = i\frac{1-\eta}{1+\eta} = e^\rho$$

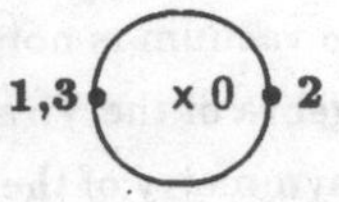

$$\eta = \lambda^{3/2} = \frac{i-\zeta}{i+\zeta}$$

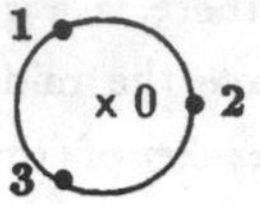

$$\lambda = \frac{i-z}{i+z} = \eta^{2/3}$$

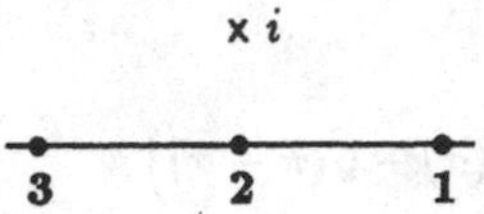

$$z = i\frac{1-\lambda}{1+\lambda}$$

(The bold-face numbers label the ends of the strings.) This maps the string from an infinite rectangle (ρ) to the upper-half plane (ζ) to the interior of the unit circle (η) to a different circle with all three strings appearing on the same sheet (λ) to the upper-half plane with all strings on one sheet (z). If the cut for $\lambda(\eta)$ is chosen appropriately (the positive imaginary axis of the η plane), the cut under which the third string is hidden is along the part of the imaginary ρ axis below $i\pi/2$. (More conveniently, if the cut is taken in the negative real direction in the η plane, then it's in the positive real direction in the ρ plane, with halves of 2 strings hidden under the cut.) Since the last transformation is projective, we can drop it. (Projective transformations don't affect equations like (9.2.14).)

Unfortunately, although the calculation can still be performed [12.7], there is now no simple analog to (9.2.18). It's easier to use instead a map similar to (9.2.16) by considering a 6-string δ functional with pairs of strings identified [12.6]: Specifically, we replace the last 2 maps above with

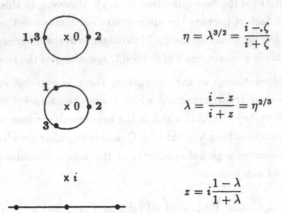

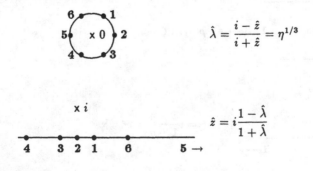

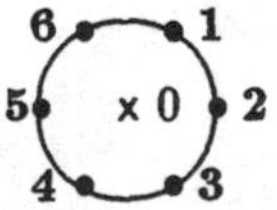

$$\hat{\lambda} = \frac{i-\hat{z}}{i+\hat{z}} = \eta^{1/3}$$

$$\hat{z} = i\frac{1-\hat{\lambda}}{1+\hat{\lambda}}$$

and thus, relabeling $r \to r+1$ (or performing an equivalent rotation of the $\hat{\lambda}$ circle)

$$\rho = \sum_{r=1}^{5} p_{+r} ln(\hat{z} - z_r) - ln\ 3 \quad ,$$

$$p_{+r} = (-1)^{r+1} \quad , \quad z_r = (\sqrt{3}, \frac{1}{\sqrt{3}}, 0, -\frac{1}{\sqrt{3}}, -\sqrt{3}, \infty) \quad . \tag{12.2.22}$$

We can then use the same procedure as the light cone. However, it turns out to be more convenient to evaluate the contour integrals in terms of ζ rather than $\hat{z}$. Also, instead of applying (9.2.18) to (12.2.22), we apply it to the corresponding expression for $\hat{\lambda}$:

$$\rho = \sum_{r=1}^{5} \alpha_r ln(\hat{\lambda} - \lambda_r) - \tfrac{1}{4} i\pi \quad ,$$

$$p_{+r} = (-1)^r \quad ; \quad \lambda_r = e^{-i\pi(r-2)/3} \quad . \tag{12.2.23}$$

Reexpressing (9.2.18) in terms of $\hat{\lambda}$, we find

$$\left(\frac{\partial}{\partial \rho} + \frac{\partial}{\partial \rho'}\right) ln(\hat{\lambda} - \hat{\lambda}') = \tfrac{1}{6}\left[(\hat{\lambda}^3 + \hat{\lambda}'^3) + (\hat{\lambda}\hat{\lambda}'^2 + \hat{\lambda}^2\hat{\lambda}') + \left(\frac{1}{\hat{\lambda}\hat{\lambda}'^2} + \frac{1}{\hat{\lambda}^2\hat{\lambda}'}\right)\right] \quad . \tag{12.2.24}$$

Using the conservation laws, the first set of terms can be dropped. Since it's actually $\lambda = \hat{\lambda}^2$ (or z), and not $\hat{\lambda}$, for which the string is mapped to the complex plane ($\hat{\lambda}$ describes a 6-string vertex, and thus double counts), the ln we actually want to evaluate is

$$ln(\lambda - \lambda') = ln(\hat{\lambda} - \hat{\lambda}') + ln(\hat{\lambda} + \hat{\lambda}') \quad . \tag{12.2.25}$$

This just says that the general coefficients N_{rs} in Δ multiplying oscillators from string r times those from string s is related to the corresponding fictitious 6-string coefficients $\tilde{N}_{rs}$ by

$$N_{rs} = \tilde{N}_{rs} + \tilde{N}_{r,s+3} \quad . \tag{12.2.26}$$

The contour integrals can now be evaluated over ζ in terms of

$$\left(\frac{1+x}{1-x}\right)^{1/3} = \sum_0^\infty a_n x^n \quad , \quad \left(\frac{1+x}{1-x}\right)^{2/3} = \sum_0^\infty b_n x^n \quad , \tag{12.2.27}$$

These coefficients satisfy the recursion relations

$$(n+1)a_{n+1} = \tfrac{2}{3}a_n + (n-1)a_{n-1} \quad , \quad (n+1)b_{n+1} = \tfrac{4}{3}b_n + (n-1)b_{n-1} \quad , \tag{12.2.28}$$

which can be derived by appropriate manipulations of the corresponding contour integrals: e.g.,

$$a_n = \oint_0 \frac{dx}{2\pi i x} \frac{1}{x^n} \left(\frac{1+x}{1-x}\right)^{1/3}$$

$$= \oint_0 \frac{dx}{2\pi i x} \frac{1}{x^n} \tfrac{3}{2}(1-x^2) \left[\left(\frac{1+x}{1-x} \right)^{1/3} \right]' \quad . \tag{12.2.29}$$

Because of i's relative to (12.2.27) appearing in the actual contour integrals, we use instead the coefficients

$$A_n = a_n \cdot \begin{cases} (-1)^{n/2} & (n\ even) \\ (-1)^{(n-1)/2} & (n\ odd) \end{cases} \quad , \tag{12.2.30}$$

and similarly for B_n. We finally obtain an expression similar to (9.2.27), except that we must use (12.2.26), and

$$\widetilde{N}_{rsmn} = \frac{1}{\frac{m}{p_{+r}} + \frac{n}{p_{+s}}} M_{rsmn} \quad ,$$

$$M_{r,r+t,mn} = \tfrac{1}{3} c_{mnt} \left[A_m B_n + (-1)^{m+n+t} B_m A_n \right] \quad ,$$

$$c_{mnt} = \begin{cases} (-1)^m Re(e^{it2\pi/3}) & (m+n\ even) \\ Im(e^{it2\pi/3}) & (m+n\ odd) \end{cases} \quad . \tag{12.2.31}$$

The terms for $n = m \neq 0$ or $n = 0 \neq m$ can be evaluated by taking the appropriate limit ($n \to m$ or $n \to 0$). $m = n = 0$ can then be evaluated separately, using (9.2.22b), (12.2.22), and (12.2.26). The final result is

$$\Delta(\check{\Psi}_1, \check{\Psi}_2) = -\sum{}' \psi_1 N \psi_2 - \tfrac{1}{4} ln \left(\frac{3^3}{2^4} \right) \sum p^2 \quad , \tag{12.2.32}$$

where $\sum'$ is over $r, s = 1, 2, 3$ and $m, n = 0, 1, \ldots, \infty$ except for the term $m = n = 0$. As for (9.2.27), ψ refers to all sets of oscillators, with ψ replaced with ${p_+}^{1-w}\psi$ for oscillators of weight w. In this case we use (12.2.21), and the p_+'s are all ± 1, so for the ghosts there is an extra sign factor $p_{+r} p_{+s}$ for $\widetilde{N}_{rsmn}$.

There are a number of problems to resolve for this formalism: (1) In calculating S-matrix elements, the 4-point function is considerably more difficult to calculate than in the light-cone formalism [12.9], and the conformal maps are so complicated that it's not yet known how to derive even the 5-point function for tachyons, although arguments have been given for equivalence to the light-cone/external-field result [12.10]. (2) It doesn't seem possible to derive an external-field approach to interactions, since the string lengths are all fixed to be π. In the light-cone formalism the external-field approach follows from choosing the Lorentz frame where all but 2 of the string lengths (i.e., p_+'s) vanish. (Thus, e.g., in the 3-string vertex 1 string reduces to a point on the boundary, reducing to a vertex as in sect. 9.1.) This is related to the fact that I of (12.2.11) is just the harmonic oscillator ground state at vanishing momentum (and length) for the light-cone formalism, but for this formalism it's $\sim \delta[X(\sigma) - X(\pi - \sigma)]$. (3) The fact that the gauge-invariant vertex is so different from the light-cone vertex indicates that gauge-fixing to the light-cone gauge should be difficult. Furthermore, the light-cone formalism requires a 4-point interaction in the action, whereas this covariant formalism doesn't. Perhaps a formalism with a larger gauge invariance exists such that these 2 formalisms are found by 2 different types of gauge choices. (4) There is some difficulty in extending the discussion of sect. 11.1 for the closed string to the interacting case, since the usual form of the physical-state vertex requires that the vertex be related to the product of open-string vertices for the clockwise and counterclockwise states, multiplied by certain vertex factors which don't exist in this formalism (although they would in a formalism more similar to the light-cone one, since the light-cone formalism has more zero-mode conservation laws). This is particularly confusing since open strings generate closed ones at the 1-loop level. However, some progress in understanding these closed strings has been made [12.11]. Also, a general analysis has been made of some properties of the 3-point closed-string vertex required by consistency of the 1-loop tadpole and 4-string tree graphs [12.12], using techniques which are applicable to vertices more general than δ-functionals [12.13].

The gauge-fixing of this formalism with a BRST algebra that closes on shell has been studied [12.14]. It has been shown both in the formalism of light-cone-like closed string theory [12.15] and for the midpoint-interaction open string theory [12.16] that the kinetic term can be obtained from an action with just the cubic term by giving an appropriate vacuum value to the string field. However, whereas in the former case (barring difficulties in loops mentioned above) this vacuum value is natural because of the vacuum value of the covariant metric field for the graviton, in the latter case there is no classical graviton in the open string theory, so the existence (or usefulness) of such a mechanism is somewhat confusing.

The midpoint-interaction formulation of the open superstring (as a truncated spinning string) has also been developed [11.3,12.17]. The supersymmetry algebra closes only on shell, and the action apparently also needs (at least) 4-point interactions to cancel divergences in 4-point amplitudes due to coincidence of vertex operator factors (both of which occur at the midpoint) [12.18]. Such interactions might be of the same type needed in the light-cone formulation (chapt. 10).

Exercises

(1) Check the BRST invariance of (12.2.18).

(2) Find the transformation of $ln(z - z')$ under the projective transformation $z \rightarrow \frac{az+b}{cz+d}$ (and similarly for z'). Use the conservation law $\sum_r \psi_{0r} = 0$ to show that (9.2.14) is unaffected.

(3) Derive the last term of (12.2.32).

REFERENCES

Preface

[0.1] M.B. Green, J.H. Schwarz, and E. Witten, Superstring theory, v. 1-2 (Cambridge University, Cambridge, 1987).

[0.2] V. Gates, E. Kangaroo, M. Roachcock, and W.C. Gall, The super G-string, *in* Unified string theories, eds. M. Green and D. Gross, Proc. of Santa Barbara Workshop, Jul. 29 - Aug. 16, 1985 (World Scientific, Singapore, 1986) p. 729;
Super G-string field theory, *in* Superstrings, Cosmology and Composite Structures, eds. S.J. Gates, Jr. and R.N. Mohapatra, Proc. of International Workshop on Superstrings, Composite Structures, and Cosmology, College Park, MD, March 11-18, 1987 (World Scientific, Singapore, 1987) p. 585.

Chapter 1

[1.1] P. van Nieuwenhuizen, Phys. Rep. 68 (1981) 189.

[1.2] M.T. Grisaru and W. Siegel, Nucl. Phys. B201 (1982) 292, B206 (1982) 496;
N. Marcus and A. Sagnotti, Phys. Lett. 135B (1984) 85;
M.H. Goroff and A. Sagnotti, Phys. Lett. 160B (1985) 81.

[1.3] Dual theory, ed. M. Jacob (North-Holland, Amsterdam, 1974);
J. Scherk, Rev. Mod. Phys. 47 (1975) 123;
P.H. Frampton, Dual resonance models and string theories (World Scientific, Singapore, 1986).

[1.4] S. Mandelstam, Phys. Rep. 13 (1974) 259.

[1.5] J.H. Schwarz, Phys. Rep. 89 (1982) 223.

[1.6] G. 't Hooft, Nucl. Phys. B72 (1974) 461.

[1.7] S. Weinberg, Critical phenomena for field theorists, *in* Understanding the fundamental constituents of matter, proc. of the International School of Subnuclear Physics, Erice, 1976, ed. A. Zichichi (Plenum, New York, 1978) p. 1;
Ultraviolet divergences in quantum theories of gravitation, *in* General relativity: an Einstein centenary survey, eds. S.W. Hawking and W. Israel (Cambridge University, Cambridge, 1979) p. 790.

[1.8] G. 't Hooft, Phys. Lett. 109B (1982) 474.

[1.9] T.H.R. Skyrme, Proc. Roy. Soc. A260 (1961) 227;
E. Witten, Nucl. Phys. B223 (1983) 433.

[1.10] D.J. Gross, A. Neveu, J. Scherk, and J.H. Schwarz, Phys. Lett. 31B (1970) 592.

[1.11] M.B. Green and J.H. Schwarz, Phys. Lett. 149B (1984) 117.

[1.12] M.B. Green and J.H.Schwarz, Phys. Lett. 109B (1982) 444.

[1.13] D.J. Gross, J.A. Harvey, E. Martinec, and R. Rohm, Phys. Rev. Lett. 54 (1985) 502, Nucl. Phys. B256 (1985) 253, Nucl. Phys. B267 (1986) 75.

[1.14] L.J. Dixon and J.A. Harvey, Nucl. Phys. B274 (1986) 93;
L. Alvarez-Gaumé, P. Ginsparg, G. Moore, and C. Vafa, Phys. Lett. 171B (1986) 155;
H. Itoyama and T.R. Taylor, Phys. Lett. 186B (1987) 129.

[1.15] E. Witten, Phys. Lett. 149B (1984) 351.

[1.16] P. Candelas, G.T. Horowitz, A. Strominger, and E. Witten, Nucl. Phys. B258 (1985)

46;
L. Dixon, J.A. Harvey, C. Vafa, and E. Witten, Nucl. Phys. B261 (1985) 678, 274 (1986) 285;
K.S. Narain, Phys. Lett. 169B (1986) 41;
H. Kawai, D.C. Lewellen, and S.-H.H. Tye, Nucl. Phys. B288 (1987) 1;
I. Antoniadis, C. Bachas, and C. Kounnas, Nucl. Phys. B289 (1987) 87;
R. Bluhm, L. Dolan, and P. Goddard, Nucl. Phys. B289 (1987) 364, Unitarity and modular invariance as constraints on four-dimensional superstrings, Rockefeller preprint RU87/B1/25 DAMTP 88/9 (Jan. 1988).

[1.17] S.J. Gates, Jr., M.T. Grisaru, M. Roček, and W. Siegel, Superspace, *or* One thousand and one lessons in supersymmetry (Benjamin/Cummings, Reading, 1983);
J. Wess and J. Bagger, Supersymmetry and supergravity (Princeton University, Princeton, 1983);
R.N. Mohapatra, Unification and supersymmetry: the frontiers of quark-lepton physics (Springer-Verlag, New York, 1986);
P. West, Introduction to supersymmetry and supergravity (World Scientific, Singapore, 1986).

Chapter 2

[2.1] S. Weinberg, Phys. Rev. 150 (1966) 1313;
Kogut and D.E. Soper, Phys. Rev. D1 (1970) 2901.

[2.2] K . Bardakçi and M.B. Halpern, Phys. Rev. 176 (1968) 1686.

[2.3] W. Siegel and B. Zwiebach, Nucl. Phys. B282 (1987) 125.

[2.4] S.J. Gates, Jr., M.T. Grisaru, M. Roček, and W. Siegel, Superspace, *or* One thousand and one lessons in supersymmetry (Benjamin/Cummings, Reading, 1983) p. 74.

[2.5] A. J. Bracken, Lett. Nuo. Cim. 2 (1971) 574;
A.J. Bracken and B. Jessup, J. Math. Phys. 23 (1982) 1925.

[2.6] W. Siegel, Nucl. Phys. B263 (1986) 93.

[2.7] W. Siegel, Phys. Lett. 142B (1984) 276.

[2.8] G. Parisi and N. Sourlas, Phys. Rev. Lett. 43 (1979) 744.

[2.9] R. Delbourgo and P.D. Jarvis, J. Phys. A15 (1982) 611;
J. Thierry-Mieg, Nucl. Phys. B261 (1985) 55;
J.A. Henderson and P.D. Jarvis, Class. and Quant. Grav. 3 (1986) L61.

Chapter 3

[3.1] P.A.M. Dirac, Proc. Roy. Soc. A246 (1958) 326;
L.D. Faddeev, Theo. Math. Phys. 1 (1969) 1.

[3.2] W. Siegel, Nucl. Phys. B238 (1984) 307.

[3.3] C. Becchi, A. Rouet, and R. Stora, Phys. Lett. 52B (1974) 344;
I.V. Tyutin, Gauge invariance in field theory and in statistical physics in the operator formulation, Lebedev preprint FIAN No. 39 (1975), in Russian, unpublished;
T. Kugo and I. Ojima, Phys. Lett. 73B (1978) 459;
L. Baulieu, Phys. Rep. 129 (1985) 1.

[3.4] M. Kato and K. Ogawa, Nucl. Phys. B212 (1983) 443;

S. Hwang, Phys. Rev. D28 (1983) 2614;
K. Fujikawa, Phys. Rev. D25 (1982) 2584.
[3.5] N. Nakanishi, Prog. Theor. Phys. 35 (1966) 1111;
B. Lautrup, K. Dan. Vidensk. Selsk. Mat. Fys. Medd. 34 (1967) No. 11, 1.
[3.6] G. Curci and R. Ferrari, Nuo. Cim. 32A (1976) 151, Phys. Lett. 63B (1976) 91;
I. Ojima, Prog. Theo. Phys. 64 (1980) 625;
L. Baulieu and J. Thierry-Mieg, Nucl. Phys. B197 (1982) 477.
[3.7] L. Baulieu, W. Siegel, and B. Zwiebach, Nucl. Phys. B287 (1987) 93.
[3.8] S. Ferrara, O. Piguet, and M. Schweda, Nucl. Phys. B119 (1977) 493.
[3.9] J. Thierry-Mieg, J. Math. Phys. 21 (1980) 2834.
[3.10] E.S. Fradkin and G.A. Vilkovisky, Phys. Lett. 55B (1975) 224;
I.A. Batalin and G.A. Vilkovisky, Phys. Lett. 69B (1977) 309;
E.S. Fradkin and T.E. Fradkina, Phys. Lett. 72B (1978) 343;
M. Henneaux, Phys. Rep. 126 (1985) 1.
[3.11] M. Quirós, F.J. de Urries, J. Hoyos, M.L. Mazón, and E. Rodriguez, J. Math. Phys. 22 (1981) 1767;
L. Bonora and M. Tonin, Phys. Lett. 98B (1981) 48.
[3.12] S. Hwang, Nucl. Phys. B231 (1984) 386;
F.R. Ore, Jr. and P. van Nieuwenhuizen, Nucl. Phys. B204 (1982) 317.
[3.13] W. Siegel and B. Zwiebach, Nucl. Phys. B288 (1987) 332.
[3.14] W. Siegel and B. Zwiebach, Nucl. Phys. B299 (1988) 206.
[3.15] W. Siegel, Nucl. Phys. B284 (1987) 632.
[3.16] W. Siegel, Universal supersymmetry by adding 4+4 dimensions to the light cone, Maryland preprint UMDEPP 88-231 (May 1988).

Chapter 4

[4.1] W. Siegel and B. Zwiebach, Nucl. Phys. B263 (1986) 105.
[4.2] T. Banks and M.E. Peskin, Nucl. Phys. B264 (1986) 513;
K. Itoh, T. Kugo, H. Kunitomo, and H. Ooguri, Prog. Theo. Phys. 75 (1986) 162.
[4.3] M. Fierz and W. Pauli, Proc. Roy. Soc. A173 (1939) 211;
S.J. Chang, Phys. Rev. 161 (1967) 1308;
L.P.S. Singh and C.R. Hagen, Phys. Rev. D9 (1974) 898;
C. Fronsdal, Phys. Rev. D18 (1978) 3624;
T. Curtright, Phys. Lett. 85B (1979) 219;
B. deWit and D.Z. Freedman, Phys. Rev. D21 (1980) 358;
T. Curtright and P.G.O. Freund, Nucl. Phys. B172 (1980) 413;
T. Curtright, Phys. Lett. 165B (1985) 304.
[4.4] W. Siegel, Phys. Lett. 149B (1984) 157, 151B (1985) 391.
[4.5] W. Siegel, Phys. Lett. 149B (1984) 162; 151B (1985) 396.
[4.6] E.S. Fradkin and V.I. Vilkovisky, Phys. Lett. 73B (1978) 209.
[4.7] W. Siegel and S.J. Gates, Jr., Nucl. Phys. B147 (1979) 77;
S.J. Gates, Jr., M.T. Grisaru, M. Roček, and W. Siegel, Superspace, *or* One thousand and one lessons in supersymmetry (Benjamin/Cummings, Reading, 1983) p. 242.
[4.8] E. Witten, Nucl. Phys. B268 (1986) 253.
[4.9] A. Neveu, H. Nicolai, and P.C. West, Phys. Lett. 167B (1986) 307.
[4.10] W. Siegel and B. Zwiebach, Phys. Lett. 184B (1987) 325.

[4.11] M. Scheunert, W. Nahm, and V. Rittenberg, J. Math. Phys. 18 (1977) 155.

[4.12] A. Galperin, E. Ivanov, S. Kalitzin, V. Ogievetsky, and E. Sokatchev, JETP Lett. 40 (1984) 912, Class. Quant. Grav. 1 (1984) 469, 2 (1985) 601;
W. Siegel, Class. Quant. Grav. 2 (1985) 439;
E. Nissimov, S. Pacheva, and S. Solomon, Nucl. Phys. B296 (1988) 462, 297 (1988) 349;
R.E. Kallosh and M.A. Rahmanov, Covariant quantization of the Green-Schwarz superstring, Lebedev preprint (March 1988).

[4.13] J.P. Yamron, Phys. Lett. 174B (1986) 69.

[4.14] G.D. Daté, M. Günaydin, M. Pernici, K. Pilch, and P. van Nieuwenhuizen, Phys. Lett. 171B (1986) 182.

[4.15] H. Terao and S. Uehara, Phys. Lett. 173B (1986) 134;
T. Banks, M.E. Peskin, C.R. Preitschopf, D. Friedan, and E. Martinec, Nucl. Phys. B274 (1986) 71;
A. LeClair and J. Distler, Nucl. Phys. B273 (1986) 552.

Chapter 5

[5.1] A. Barducci, R. Casalbuoni, and L. Lusanna, Nuo. Cim. 35A (1976) 377;
L. Brink, S. Deser, B. Zumino, P. DiVecchia, and P. Howe, Phys. Lett. 64B (1976) 435;
P.A. Collins and R.W. Tucker, Nucl. Phys. B121 (1977) 307.

[5.2] S. Deser and B. Zumino, Phys. Lett. 62B (1976) 335.

[5.3] E.C.G. Stueckelberg, Helv. Phys. Acta 15 (1942) 23;
R.P. Feynman, Phys. Rev. 80 (1950) 440.

[5.4] J. Polchinski, Commun. Math. Phys. 104 (1986) 37.

[5.5] J. Schwinger, Phys. Rev. 82 (1951) 664.

[5.6] R.J.Eden, P.V. Landshoff, D.I. Olive, and J.C. Polkinghorne, The analytic S-matrix (Cambridge University, Cambridge, 1966) p. 152.

[5.7] *ibid.*, p. 88;
J.D. Bjorken and S.D. Drell, Relativistic quantum fields (McGraw-Hill, New York, 1965) p. 225.

[5.8] J. Iliopoulos, G. Itzykson, and A. Martin, Rev. Mod. Phys. 47 (1975) 165.

[5.9] M.B. Halpern and W. Siegel, Phys. Rev. D16 (1977) 2486.

[5.10] F.A. Berezin and M.S. Marinov, Ann. Phys. (NY) 104 (1977) 336;
A. Barducci, R. Casalbuoni, and L. Lusanna, Nucl. Phys. B124 (1977) 93;
A.P. Balachandran, P. Salomonson, B.-S. Skagerstam, and J.-O. Winnberg, Phys. Rev. D15 (1977) 2308.

[5.11] H. Georgi, Lie algebras in particle physics: from isospin to unified theories (Benjamin/Cummings, Reading,1982) p. 192.

[5.12] S.J. Gates, Jr., M.T. Grisaru, M. Roček, and W. Siegel, Superspace, *or* One thousand and one lessons in supersymmetry (Benjamin/Cummings, Reading, 1983) pp. 70, 88.

[5.13] L. Brink, J.H. Schwarz, and J. Scherk, Nucl. Phys. B121 (1977) 77;
W. Siegel, Phys. Lett. 80B (1979) 220.

[5.14] M.T. Grisaru, W. Siegel, and M. Roček, Nucl. Phys. B159 (1979) 429;
M.T. Grisaru and W. Siegel, Nucl. Phys. B201 (1982) 292, B206 (1982) 496;
S.J. Gates, Jr., M.T. Grisaru, M. Roček, and W. Siegel, Superspace, pp. 25, 382, 383.

[5.15] W. Siegel, Nucl. Phys. B156 (1979) 135.
[5.16] J. Koller, Nucl. Phys. B222 (1983) 319;
P.S. Howe, G. Sierra, and P.K. Townsend, Nucl. Phys. B221 (1983) 331;
J.P. Yamron and W. Siegel, Nucl. Phys. B263 (1986) 70.
[5.17] J. Wess and B. Zumino, Nucl. Phys. B70 (1974) 39;
R. Haag, J.T. Łopuszański, and M. Sohnius, Nucl. Phys. B88 (1975) 257;
W. Nahm, Nucl. Phys. B135 (1978) 149.
[5.18] P. Ramond, Physica 15D (1985) 25.
[5.19] W. Siegel, Free field equations for everything, *in* Strings, Cosmology, Composite Structures, March 11-18, 1987, College Park, Maryland, eds. S.J. Gates, Jr. and R.N. Mohapatra (World Scientific, Singapore, 1987).
[5.20] A. Salam and J. Strathdee, Nucl. Phys. B76 (1974) 477;
S.J. Gates, Jr., M.T. Grisaru, M. Roček, and W. Siegel, Superspace, pp. 72-73;
R. Finkelstein and M. Villasante, J. Math. Phys. 27 (1986) 1595.
[5.21] W. Siegel, Phys. Lett. 203B (1988) 79.
[5.22] J. Strathdee, Inter. J. Mod. Phys. A2 (1987) 273.
[5.23] W. Siegel, Class. Quantum Grav. 2 (1985) L95.
[5.24] W. Siegel, Phys. Lett. 128B (1983) 397.
[5.25] W. Siegel and S.J. Gates, Jr., Nucl. Phys. B189 (1981) 295.
[5.26] S. Mandelstam, Nucl. Phys. B213 (1983) 149.
[5.27] L. Brink, O. Lindgren, and B.E.W. Nilsson, Nucl. Phys. B212 (1983) 401.
[5.28] R. Casalbuoni, Phys. Lett. 62B (1976) 49;
L. Brink and J.H. Schwarz, Phys. Lett. 100B (1981) 310.
[5.29] A.R. Miković and W. Siegel, On-shell equivalence of superstrings, Maryland preprint 88-218 (May 1988).
[5.30] K. Kamimura and M. Tatewaki, Phys. Lett. 205B (1988) 257.
[5.31] P. Penrose, J. Math. Phys. 8 (1967) 345, Int. J. Theor. Phys. 1 (1968) 61 ;
M.A.H. MacCallum and R. Penrose, Phys. Rep. 6C (1973) 241.
[5.32] A. Ferber, Nucl. Phys. B132 (1978) 55.
[5.33] T. Kugo and P. Townsend, Nucl. Phys. B221 (1983) 357;
A. Sudbery, J. Phys. A17 (1984) 939;
K.-W. Chung and A. Sudbery, Phys. Lett. 198B (1987) 161.
[5.34] Z. Hasiewicz and J. Lukierski, Phys. Lett. 145B (1984) 65;
I. Bengtsson and M. Cederwall, Particles, twistors, and the division algebras, Ecole Normale Supérieure preprint LPTENS-87/20 (May 1987).
[5.35] D.B. Fairlie and C.A. Manogue, Phys. Rev. D34 (1986) 1832, 36 (1987) 475;
J.M. Evans, Nucl. Phys. B298 (1988) 92.

Chapter 6

[6.1] P.A. Collins and R.W. Tucker, Phys. Lett. 64B (1976) 207;
L. Brink, P. Di Vecchia, and P. Howe, Phys. Lett. 65B (1976) 471;
S. Deser and B. Zumino, Phys. Lett. 65B (1976) 369.
[6.2] Y. Nambu, lectures at Copenhagen Symposium, 1970 (unpublished);
O. Hara, Prog. Theo. Phys. 46 (1971) 1549;
T. Goto, Prog. Theo. Phys. 46 (1971) 1560.
[6.3] S.J. Gates, Jr., R. Brooks, and F. Muhammad, Phys. Lett. 194B (1987) 35;

C. Imbimbo and A. Schwimmer, Phys. Lett. 193B (1987) 455;
J.M.F. Labastida and M. Pernici, Nucl. Phys. B297 (1988) 557;
L. Mezincescu and R.I. Nepomechie, Critical dimensions for chiral bosons, Miami preprint UMTG-140 (Aug. 1987);
Y. Frishman and J. Sonnenschein, Gauging of chiral bosonized actions, Weizmann preprint WIS-87/65/Sept.-PH (Sept. 1987);
R. Floreanini and R. Jackiw, Phys. Rev. Lett. 59 (1987) 1873;
S. Bellucci, R. Brooks, and J. Sonnenschein, Supersymmetric chiral bosons, SLAC preprint SLAC-PUB-4458 MIT-CTP-1548 UCD-87-35 (Oct. 1987);
S.J. Gates, Jr. and W. Siegel, Leftons, rightons, nonlinear σ-models, and superstrings, Maryland preprint UMDEPP 88-113 (Nov. 1987);
M. Henneaux and C. Teitelboim, Dynamics of chiral (self-dual) p-forms, Austin preprint (Dec. 1987);
M. Bernstein and J. Sonnenschein, A comment on the quantization of chiral bosons, SLAC preprint SLAC-PUB-4523 (Jan. 1988);
L. Faddeev and R. Jackiw, Hamiltonian reduction of unconstrained and constrained systems, MIT preprint CTP#1567 (Feb. 1988);
J. Sonnenschein, Chiral bosons, SLAC preprint SLAC-PUB-4570 (March 1988).

[6.4] C.M. Hull, Covariant quantization of chiral bosons and anomaly cancellation, London preprint Imperial/TP/87/88/9 (March 1988).

[6.5] P. Goddard, J. Goldstone, C. Rebbi, and C.B. Thorn, Nucl. Phys. B56 (1973) 109.

Chapter 7

[7.1] P. Ramond, Phys. Rev. D3 (1971) 86;
A. Neveu and J.H. Schwarz, Nucl. Phys. B31 (1971) 86, Phys. Rev. D4 (1971) 1109.

[7.2] D.B. Fairlie and D. Martin, Nuo. Cim. 18A (1973) 373, 21A (1974) 647;
L. Brink and J.-O. Winnberg, Nucl. Phys. B103 (1976) 445.

[7.3] F. Gliozzi, J. Scherk, and D.I. Olive, Phys. Lett. 65B (1976) 282, Nucl. Phys. B122 (1977) 253.

[7.4] Y. Iwasaki and K. Kikkawa, Phys. Rev. D8 (1973) 440.

[7.5] W. Siegel, Covariant approach to superstrings, *in* Symposium on anomalies, geometry and topology, eds. W.A. Bardeen and A.R. White (World Scientific, Singapore, 1985) p. 348.

[7.6] M.B. Green and J.H. Schwarz, Phys. Lett. 136B (1984) 367, Nucl. Phys. B243 (1984) 285.

Chapter 8

[8.1] M. Virasoro, Phys. Rev. D1 (1970) 2933;
I.M. Gelfand and D.B. Fuchs, Functs. Anal. Prilozhen 2 (1968) 92.

[8.2] E. Del Giudice, P. Di Vecchia, and S. Fubini, Ann. Phys. 70 (1972) 378.

[8.3] F. Gliozzi, Nuovo Cim. Lett. 2 (1969) 846.

[8.4] E. Witten, Some remarks about string field theory, *in* Marstrand Nobel Sympos. 1986, p. 70;
I.B. Frenkel, H. Garland, and G.J. Zuckerman, Proc. Natl. Acad. Sci. USA 83 (1986)

8442.
[8.5] P. Jordan, Z. Physik 93 (1935) 464;
M. Born and N.S. Nagendra Nath, Proc. Ind. Acad. Sci. 3 (1936) 318;
A. Sokolow, Phys. Z. der Sowj. 12 (1937) 148;
S. Tomonaga, Prog. Theo. Phys. 5 (1950) 544;
T.H.R. Skyrme, Proc. Roy. Soc. A262 (1961) 237;
D. Mattis and E. Lieb, J. Math. Phys. 6 (1965) 304;
B. Klaiber, *in* Lectures in theoretical physics, eds. A.O. Barut and W.E. Brittin (Gordon and Breach, New York, 1968) v. X-A, p. 141;
R.F. Streater and I.F. Wilde, Nucl. Phys. B24 (1970) 561;
J. Lowenstein and J. Swieca, Ann. Phys. (NY) 68 (1971) 172;
K. Bardakçi and M.B. Halpern, Phys. Rev. D3 (1971) 2493;
G. Dell'Antonio, Y. Frishman, and D. Zwanziger, Phys. Rev. D6 (1972) 988;
A. Casher, J. Kogut, and L. Susskind, Phys. Rev. Lett. 31 (1973) 792, Phys. Rev. D10 (1974) 732;
A. Luther and I. Peschel, Phys. Rev. B9 (1974) 2911;
A. Luther and V. Emery, Phys. Rev. Lett. 33 (1974) 598;
S. Coleman, Phys. Rev. D11 (1975) 2088;
B. Schroer, Phys. Rep. 23 (1976) 314;
S. Mandelstam, Phys. Rev. D11 (1975) 3026;
J. Kogut and L. Susskind, Phys. Rev. D11 (1975) 3594.
[8.6] M.B. Halpern, Phys. Rev. D12 (1975) 1684;
I.B. Frenkel and V.G. Kač, Inv. Math. 62 (1980) 23;
I.B. Frenkel, J. Func. Anal. 44 (1981) 259;
G. Segal, Comm. Math. Phys. 80 (1981) 301;
P. Goddard and D. Olive, Vertex operators in mathematics and physics, eds. J. Lepowsky, S. Mandelstam, and I.M. Singer (Springer-Verlag, New York, 1985) p. 51.
[8.7] R. Marnelius, Nucl. Phys. B211 (1983) 14.
[8.8] W. Siegel, Phys. Lett. 134B (1984) 318.
[8.9] A.M. Polyakov, Phys. Lett. 103B (1981) 207.
[8.10] A.A. Tseytlin, Phys. Lett. 168B (1986) 63;
S.R. Das and M.A. Rubin, Phys. Lett. 169B (1986) 182, Prog. Theor. Phys. Suppl. 86 (1986) 143, Phys. Lett. 181B (1986) 81;
T. Banks, D. Nemeschansky, and A. Sen, Nucl. Phys. B277 (1986) 67;
G.T. Horowitz and A. Strominger, Phys. Rev. Lett. 57 (1986) 519;
G. Münster, Geometric string field theory, DESY preprint DESY 86-045 (Apr. 1986).
[8.11] H. Aoyama, Nucl. Phys. B299 (1988) 379.
[8.12] D.Z. Freedman, J.I. Latorre, and K. Pilch, Global aspects of the harmonic gauge in bosonic string theory, MIT preprint CTP#1559 (Jan. 1988).

Chapter 9

[9.1] K. Bardakçi and H. Ruegg, Phys. Rev. 181 (1969) 1884;
C.J. Goebel and B. Sakita, Phys. Rev. Lett. 22 (1969) 257;
Chan H.-M. and T.S. Tsun, Phys. Lett. 28B (1969) 485;
Z. Koba and H.B. Nielsen, Nucl. Phys. B10 (1969) 633, 12 (1969) 517.
[9.2] G. Veneziano, Nuo. Cim. 57A (1968) 190.

[9.3] C. Lovelace, Phys. Lett. 135B (1984) 75;
D. Friedan, Z. Qiu, and S. Shenker, proc. APS Div. of Particles and Fields Conf., Santa Fe, eds. T. Goldman and M. Nieto (World Scientific, Singapore, 1984);
E. Fradkin and A. Tseytlin, Phys. Lett. 158B (1985) 316;
A. Sen, Phys. Rev. D32 (1985) 2102, Phys. Rev. Lett. 55 (1985) 1846;
C.G. Callan, E.J. Martinec, M.J. Perry, and D. Friedan, Nucl. Phys. B262 (1985) 593.

[9.4] S. Mandelstam, Nucl. Phys. B64 (1973) 205;
J.L. Torres-Hernández, Phys. Rev. D11 (1975) 3565;
H. Arfaei, Nucl. Phys. B85 (1975) 535, 112 (1976) 256.

[9.5] E. Cremmer and J.-L. Gervais, Nucl. Phys. B90 (1975) 410.

[9.6] W. Siegel, Nucl. Phys. B109 (1976) 244.

[9.7] M.B. Green and J.H. Schwarz, Phys. Lett. 151B (1985) 21.

[9.8] P.H. Frampton, P. Moxhay, and Y.J. Ng, Phys. Rev. Lett. 55 (1985) 2107, Nucl. Phys. B276 (1986) 599;
L. Clavelli, Phys. Rev. D33 (1986) 1098, 34 (1986) 3262, Prog. Theor. Phys. Suppl. 86 (1986) 135.

[9.9] O. Alvarez, Nucl. Phys. B216 (1983) 125;
E. Martinec, Phys. Rev. D28 (1983) 2604;
G. Moore and P. Nelson, Nucl. Phys. B266 (1986) 58;
A. Cohen, G. Moore, P. Nelson, and J. Polchinski, Nucl. Phys. B267 (1986) 143;
S. Mandelstam, The interacting-string picture and functional integration, *in* Unified string theories, eds. M. Green and D. Gross (World Scientific, Singapore, 1986) pp. 46;
E. D'Hoker and D.H. Phong, Nucl. Phys. B269 (1986) 205;
M.A. Namazie and S. Rajeev, Nucl. Phys. B277 (1986) 332;
D. Friedan, E. Martinec, and S. Shenker, Nucl. Phys. B271 (1986) 93;
E. D'Hoker and D.H. Phong, Commun. Math. Phys. 104 (1986) 537, Phys. Rev. Lett. 56 (1986) 912;
E. Gava, R. Jengo, T. Jayaraman, and R. Ramachandran, Phys. Lett. 168B (1986) 207;
G. Moore, P. Nelson, and J. Polchinski, Phys. Lett. 169B (1986) 47;
R. Catenacci, M. Cornalba, M. Martellini, C. Reina, Phys. Lett. 172B (1986) 328;
E. Martinec, Phys. Lett. 171B (1986) 189;
E. D'Hoker and D.H. Phong, Nucl. Phys. B278 (1986) 225;
S.B. Giddings and S.A. Wolpert, Comm. Math. Phys. 109 (1987) 177;
E. D'Hoker and S.B. Giddings, Nucl. Phys. B291 (1987) 90.

[9.10] E. D'Hoker and D.H. Phong, The geometry of string perturbation theory, Princeton preprint PUPT-1039 (Feb. 1988), to appear in Rev. Mod. Phys. 60 (1988).

[9.11] J. Shapiro, Phys. Rev. D5 (1972) 1945;
E. Witten, Global anomalies in string theory, *in* Symposium on anomalies, geometry and topology, eds. W.A. Bardeen and A.R. White (World Scientific, Singapore, 1985) p. 348.

Chapter 10

[10.1] M. Kaku and K. Kikkawa, Phys. Rev. D10 (1974) 1110, 1823, M. Kaku, Phys. Rev. D10 (1974) 3943;

E. Cremmer and J.-L. Gervais, Nucl. Phys. B76 (1974) 209.
[10.2] M.B. Green and J.H. Schwarz, Nucl. Phys. B243 (1984) 475.
[10.3] Chan H.M. and J. Paton, Nucl. Phys. B10 (1969) 519;
N. Marcus and A. Sagnotti, Phys. Lett. 119B (1982) 97.
[10.4] N. Berkovits, Nucl. Phys. B276 (1986) 650; Supersheet functional integration and the interacting Neveu-Schwarz string, Berkeley preprint UCB-PTH-87/31 LBL-23727 (July 1987);
S.-J. Sin, Geometry of super lightcone diagrams and Lorentz invariance of lightcone string field theory, (II) closed Neveu-Schwarz string, Berkeley preprint LBL-25120 UCB-PTH-88/09 (April 1988).
[10.5] S. Mandelstam, Nucl. Phys. B83 (1974) 413.
[10.6] T. Kugo, Prog. Theor. Phys. 78 (1987) 690;
S. Uehara, Phys. Lett. 196B (1987) 47;
S.-J. Sin, Lorentz invariance of light cone string field theories, Berkeley preprint LBL-23715 (June 1987).
[10.7] N. Linden, Nucl. Phys. B286 (1987) 429.
[10.8] D.J. Gross and V. Periwal, Nucl. Phys. B287 (1987) 1.
[10.9] J. Greensite and F.R. Klinkhamer, Nucl. Phys. B281 (1987) 269, 291 (1987) 557; Superstring amplitudes and contact interactions, Berkeley preprint LBL-23830 NBI-HE-87-58 (Aug. 1987).
[10.10] M. Saadi, Intermediate closed strings as open strings going backwards in time, MIT preprint CTP#1575 (April 1988).

Chapter 11

[11.1] C. Marshall and P. Ramond, Nucl. Phys. B85 (1975) 375;
A. Neveu, H. Nicolai, and P.C. West, Nucl. Phys. B264 (1986) 573;
A. Neveu, J.H. Schwarz, and P.C. West, Phys. Lett. 164B (1985) 51,
M. Kaku, Phys. Lett. 162B (1985) 97;
S. Raby, R. Slansky, and G. West, Toward a covariant string field theory, *in* Lewes string theory workshop, eds. L. Clavelli and A. Halprin (World Scientific, Singapore, 1986).
[11.2] T. Banks and M.E. Peskin, Gauge invariant functional field theory for bosonic strings, *in* Symposium on anomalies, geometry, topology, eds. W.A. Bardeen and A.R. White, Mar. 28-30, 1985 (World Scientific, Singapore, 1985);
M. Kaku, Nucl. Phys. B267 (1986) 125;
D. Friedan, Nucl. Phys. B271 (1986) 540;
A. Neveu and P.C. West, Nucl. Phys. B268 (1986) 125;
K. Bardakçi, Nucl. Phys. B271 (1986) 561;
J.-L. Gervais, Nucl. Phys. B276 (1986) 339.
[11.3] E. Witten, Nucl. Phys. B276 (1986) 291;
T. Kugo and H. Terao, New gauge symmetries in Witten's Ramond string field theory, Kyoto preprint KUNS 911 HE(TH)88/01 (Feb. 1988).

Chapter 12

[12.1] A. Strominger, Phys. Rev. Lett. 58 (1987) 629.

[12.2] C.-Y. Ye, Covariantized closed string fields from light-cone, MIT preprint CTP#1566 (Feb. 1988).

[12.3] H. Hata, K. Itoh, T. Kugo, H. Kunitomo, and K. Ogawa, Phys. Lett. 172B (1986) 186, 195; Nucl. Phys. B283 (1987) 433; Phys. Rev. D34 (1986) 2360, 35 (1987) 1318, 1356; Prog. Theor. Phys. 77 (1987) 443, 78 (1987) 453.

[12.4] M.A. Awada, Phys. Lett. 172B (1986) 32;
A. Neveu and P. West, Phys. Lett. 168B (1986) 192.

[12.5] A. Hosoya and H. Itoyama, The vertex as a Bogoliubov transformed vacuum state in string field theory, Fermilab preprint FERMILAB-PUB-87/111-T OUHET-12 (Jan. 1988);
T. Kugo, H. Kunitomo, and K. Suehiro, Prog. Theor. Phys. 78 (1987) 923.

[12.6] D.J. Gross and A. Jevicki, Nucl. Phys. B283 (1987) 1, 287 (1987) 225;
Z. Hlousek and A. Jevicki, Nucl. Phys. B288 (1987) 131.

[12.7] E. Cremmer, C.B. Thorn, and A. Schwimmer, Phys. Lett. 179B (1986) 57;
C.B. Thorn, The oscillator representation of Witten's three open string vertex function, *in* Proc. XXIII Int. Conf. on High Energy Physics, July 16-23, 1986, Berkeley, CA, ed. S.C. Loken (World Scientific, Singapore, 1987) v.1, p. 374.

[12.8] S. Samuel, Phys. Lett. 181B (1986) 255;
N. Ohta, Phys. Rev. D34 (1986) 3785;
K. Itoh, K. Ogawa, and K. Suehiro, Nucl. Phys. B289 (1987) 127;
A. Eastaugh and J.G. McCarthy, Nucl. Phys. B294 (1987) 845.

[12.9] S.B. Giddings, Nucl. Phys. B278 (1986) 242.

[12.10] S.B. Giddings and E. Martinec, Nucl. Phys. B278 (1986) 91;
S.B. Giddings, E. Martinec, and E. Witten, Phys. Lett. 176B (1986) 362.

[12.11] J.A. Shapiro and C.B. Thorn, Phys. Rev. D36 (1987) 432, Phys. Lett. 194B (1987) 43;
D.Z. Freedman, S.B. Giddings, J.A. Shapiro, and C.B. Thorn, Nucl. Phys. B298 (1988) 253.

[12.12] B. Zwiebach, Constraints on covariant theories for closed string fields, MIT preprint CTP#1583 (April 1988); A note on covariant Feynman rules for closed strings, MIT preprint CTP#1598 (May 1988).

[12.13] A. LeClair, M.E. Peskin, and C.R. Preitschopf, String field theory on the conformal plane, I. Kinematical principles, II. Generalized gluing, SLAC preprints SLAC-PUB-4306, 4307 (Jan. 1988), C.R. Preitschopf, The gluing theorem in the operator formulation of string field theory, Maryland preprint UMDEPP-88-087 (Oct. 1987);
A. Eskin, Conformal transformations and string field redefinitions, MIT preprint CTP#1560 (Feb. 1988).

[12.14] M. Bochicchio, Phys. Lett. 193B (1987) 31, 188B (1987) 330;
C.B. Thorn, Nucl. Phys. B287 (1987) 61.

[12.15] H. Hata, K. Itoh, T. Kugo, H. Kunitomo, and K. Ogawa, Phys. Lett. 175B (1986) 138.

[12.16] G.T. Horowitz, J. Lykken, R. Rohm, and A. Strominger, Phys. Rev. Lett. 57 (1986) 283.

[12.17] D.J. Gross and A. Jevicki, Nucl. Phys. B293 (1987) 29;
S. Samuel, Nucl. Phys. B296 (1988) 187.

[12.18] M.B. Green and N. Seiberg, Nucl. Phys. B299 (1988) 559;
C. Wendt, Scattering amplitudes and contact interactions in Witten's superstring field theory, SLAC preprint SLAC-PUB-4442 (Nov. 1987).

INDEX